P9-CCR-536

HISTORICAL GEOLOGY

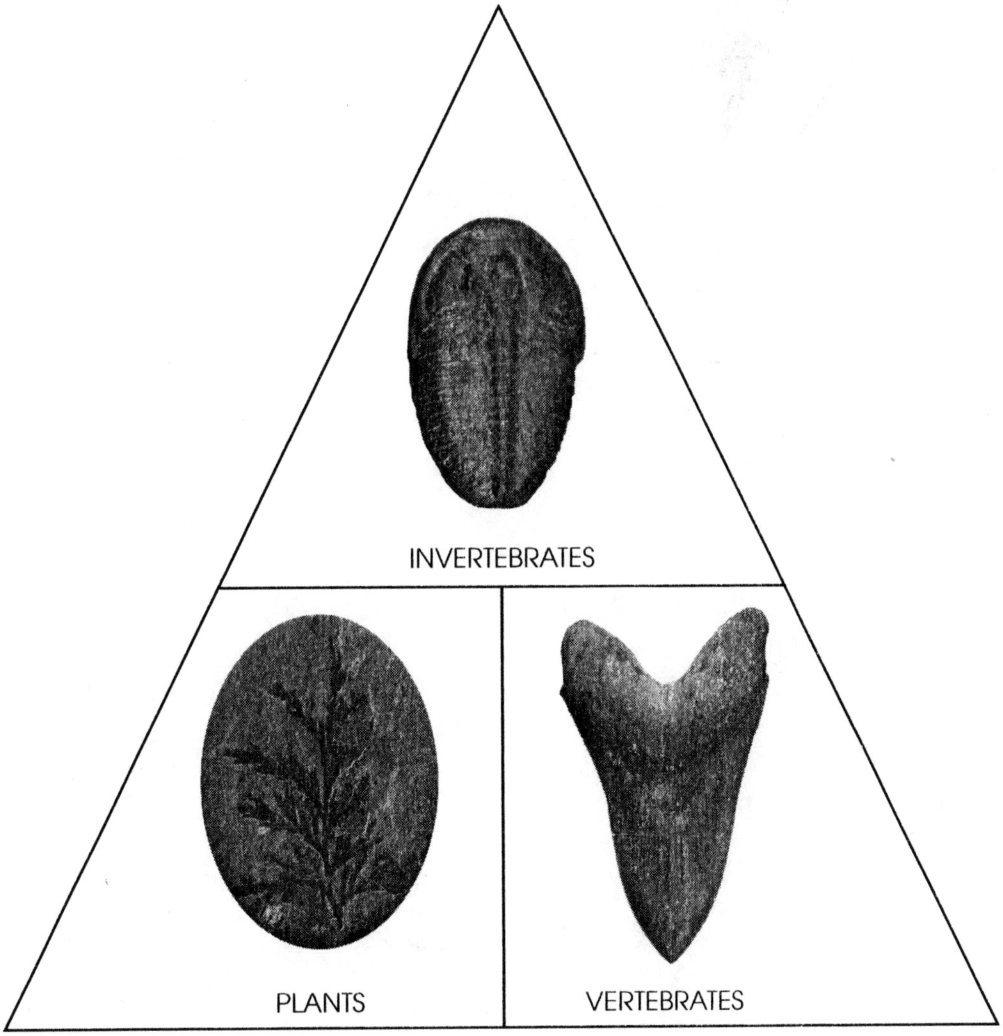

HISTORICAL GEOLOGY

Interpretations and Applications

Sixth Edition

Jon M. Poort
DeGolyer and MacNaughton
Dallas, Texas

Roseann J. Carlson
Department of Geophysical Sciences
Tidewater Community College
Virginia Beach, Virginia

PEARSON
Prentice Hall

Upper Saddle River, NJ 07458

Library of Congress Cataloging-in-Publication Data

Poort, Jon M.
Historical geology: interpretations and applications / Jon M. Poort, Roseann J. Carlson. - - 6th ed.
p. cm.
ISBN 0-13-144786-6
1. Historical geology. I. Carlson, Roseann J. II. Title.

QE28.3.P66 2005
551.7- -dc22

2004005637

Editor-in-Chief: John Challice
Executive Editor: Patrick Lynch
Assistant Editor: Melanie Cutler
Vice President and Director of Production and Manufacturing, ESM: David W. Riccardi
Executive Managing Editor: Kathleen Schiaparelli
Production Editor: nSight/Mark Corsey
Director of Creative Services: Paul Belfanti
Art Director: Jayne Conte
Cover Designer: Bruce Kenselaar
Art Editor: Abby Bass
Manufacturing Manager: Trudy Pisciotti
Manufacturing Buyer: Lynda Castillo
Cover Images: (*top left*) The fossilized skull of an Albertosaurus dinosaur, courtesy of Dorling Kindersley; (*top right*) Earth with a single landmass, the supercontinent Pangea, courtesy of Dorling Kindersley/David Donkin; (*bottom left*) Fossil in mudstone, featuring "Y" forked leaf, courtesy of Dorling Kindersley/Natural History Museum, London.; (*bottom right*) Pyritized ammonite in cross section, courtesy of Dorling Kindersley/Natural History Museum, London.

© 2005, 1998 Pearson Education, Inc.
Pearson Prentice Hall
Pearson Education, Inc
Upper Saddle River, NJ 07458

All rights reserved. No part of this book may be reproduced in any form or by any means, without permission in writing from the publisher.

Earlier editions © 1992, 1985, 1980 by Macmillan Publishing Company, and © 1976, 1973 by Burgess Publishing Company

Pearson Prentice Hall® is a trademark of Pearson Education, Inc.

Printed in the United States of America

V0N4 10 9

ISBN 0-13-144786-6

Pearson Education Ltd., *London*
Pearson Education Australia Pty. Ltd., *Sydney*
Pearson Education Singapore, Pte. Ltd.
Pearson Education North Asia Ltd., *Hong Kong*
Pearson Education Canada, Inc., *Toronto*
Pearson Educación de Mexico, S.A. de C.V.
Pearson Education—Japan, *Tokyo*
Pearson Education Malaysia, Pte. Ltd.
Pearson Education, Inc., *Upper Saddle River, New Jersey*

CONTENTS

PREFACE vii

ACKNOWLEDGMENTS viii

CHAPTER ONE **ROCK CYCLE AND SEDIMENTARY ROCKS** 1

Scope and Methods of Historical Geology 1

Rock Cycle Review 2

Sedimentary Rocks 2

Development, Transportation, and Lithification of Clastic Sedimentary Rocks 5

Classification and Framework Geometry of Sedimentary Rocks 8

Clastic Sedimentary Rocks: Sandstones 8

Clastic Sedimentary Rocks: Shales 25

Clastic Sedimentary Rocks: Volcanic Materials 26

Classification and Framework Geometry of Carbonate and Chemical Rocks 27

Carbonate Sedimentary Rocks 27

Sedimentary Environments of Deposition 35

Exercises 38

CHAPTER TWO **FUNDAMENTAL CONCEPTS** 53

Fundamental Laws of Geology 53

Review of Unconformities 56

Review of Geologic Structures 59

Application of Radiometric Geochronology 60

Radiometric Dating 64

Construction of a Geologic Cross Section 67

Reconstruction of a sequence of geologic events: The geologic history of an area 69

Exercises 70

CHAPTER THREE **PHYSICAL STRATIGRAPHY 92**
Stratigraphy and Correlation 92
Stratigraphy 92
Exercises 95
Lithostratigraphic Analysis 97
Exercises 101
Correlation 107
Exercises 112

CHAPTER FOUR **PALEONTOLOGY 142**
Definition of a Fossil 142
Nature of the Fossil Record 143
Major Concepts of Ecology and Paleoecology 147
Biostratigraphy and Biozones 150
Use of Fossil Assemblages in Age Determinations 151
Exercises 152
Classification of Animals and Plants 156
Key to Identifying Major Invertebrate Fossil Phyla 157
Classification and Descriptions of Selected Kingdoms, Phyla, and Classes 159
Kingdoms: Monera, Protista, Animalia, Plantae 159
Exercises 198

CHAPTER FIVE **GEOLOGIC MAP INTERPRETATION 212**
Geologic Map Interpretation 214
Exercises 217
Geologic Maps and Cross Sections 226
Exercises 226

CHAPTER SIX **PLATE TECTONICS 244**
Exercises 246

GLOSSARY 258

PREFACE

This new and revised edition of *Historical Geology: Interpretations and Applications* has been written as a guide to the laboratory study of historical geology and can be used in conjunction with most of the current textbooks in the field. This manual helps students understand the fundamental concepts of historical geology by providing outcrop and realistic situations to which they can apply geologic concepts leading to an interpretation of the available data. The direct application of abstract principles to concrete situations and practical problems reinforces the learning process while instilling a strong sense of the purpose for geologic study. This method is of particular value to students for whom a course in historical geology will be their last academic encounter with the physical sciences.

COMPREHENSIVENESS

The manual is divided into six chapters. The first chapter covers the great group of sedimentary rocks that are vital to the interpretation of most geologic structures and sequence of events, and that act as the preservation medium for the majority of fossils. Chapters 2 and 3 provide discussions, examples, and exercises in such basic principles of historical geology as the concepts of geologic time, the use of sedimentary rock layering and stratification during the ordering of geologic events, and the interpretation of environmental and sea level changes. Chapter 4 presents a systemic overview of the three groups of life forms that are commonly preserved as fossils: the invertebrates, the vertebrates, and plants. Chapter 5 covers the development and reading of geologic maps. Chapter 6 deals with plate tectonics, an important topic for both physical and historical geology. Within these chapters, exercise sets are available that require students to use the principles they have learned to solve problems involving both physical and historical geological concepts and methodologies.

USE OF THE SCIENTIFIC METHOD

Historical geology is best learned through fieldwork supervised by a professional. When field studies are not feasible, diagrams, maps, photographs, and slides of fossils, landforms, and sedimentary structures can be used effectively to model a field experience. In this respect, the manual permits students to visualize how geologic data are collected, tabulated, synthesized, interpreted, and applied to solutions of various geologic problems.

NEW ORGANIZATION

As suggested by several reviewers and users of the manual, a comprehensive change has been made to this edition. A greatly expanded section concerning the development, classification, and interpretation of common depositional attributes of sedimentary rocks has been added as a new Chapter 1. Selected problems from the previous Chapter 7 have been dispersed within the revised first four chapters. Extensive new photographs and new problems have been added, and many problems have been reorganized and moved to be near their explanatory text material.

SCHEDULING FLEXIBILITY

In most cases, this manual will be used during a one-semester or one-quarter term. It provides students with numerous exercises and problems. Some of these, such as the exercise questions that conclude each chapter, will require only a few minutes to complete. Others can be worked in the laboratory or at home, when more time is available. Variations in the length and content of exercises give instructors considerable latitude in choosing topics they wish to stress.

This manual is written and illustrated with outcrop or rock feature photographs with the aim of helping students readily understand the principles of historical geology. These illustrations should help build student awareness that the geologic problems presented in the manual are based on real data, often in contexts that could typically affect the nongeologist in his or her professional life. This manual is not, therefore, *just* an academic exercise; it is designed to demonstrate that geology is a profession that relates directly to the world outside of college and that basic principles of geology can be used to understand professional situations as well as postcollege travel experiences. A glossary is also included.

INSTRUCTOR AIDS AND INTERNET SITES

One of this manual's principal teaching strengths is that it does not depend on extensive supplemental teaching aids; its coverage of the basic principles of historical geology can be thought of as self-contained. A complimentary Instructor's Guide accompanies the manual and is available from the publisher on request. To obtain a copy contact your local Prentice Hall sales representative. The guide provides lists of a few additional items that may be necessary for a historical geology laboratory. It also offers suggested answers to the manual's exercises and problems.

Instructors should also encourage students to use the Internet as a geologic resource. Many Web sites change through time, new sites are rapidly being added, and topical links are continually being developed. Instructors can provide key words, associated topics, and names of major museums and research organizations for students to expand in-depth analysis of topics provided in the classroom or in a laboratory.

Jon M. Poort
Roseann J. Carlson

ACKNOWLEDGMENTS

We would like to express our sincere appreciation to the following, who reviewed the previous edition and made many helpful suggestions for improvement of the manual: Sharon Choens, San Jacinto College, Central; Chris Dewey, Mississippi State University; M. Jane Knaus, Southwest Texas State University; Arthur C. Lee, University of Texas, Brownsville; Kenneth Rasmussen, Northern Virginia Community College; Sarah Ulerick, Lane Community College; and Paul Vespucci, Northern Virginia Community College. We would like to acknowledge Lucy L. Mauger and Terry Chase, who have contributed many of the line drawings and illustrations appearing in this edition.

Many significant changes have been made to this edition of the manual as we developed new material and accepted many proposed changes envisioned by the group of reviewers. Professor Poort collected many new photographs from the Colorado Plateau area to support details presented in the revised text and to be used as illustrations within problem sets. As part of this effort, Professor Poort would especially like to thank Mr. and Mrs. Carl Ulrich of Ulrich's Fossil Gallery west of Kemmerer, Wyoming, for their insight into the geology and fossils in this region of Wyoming. He would also like to express his thanks to the staff working in the nearby Fossil Butte National Monument. Professor Poort would also like to express his appreciation to Creties Jenkins, at DeGolyer and MacNaughton, for providing Cretaceous outcrop localities in central Utah, and to offer a special thanks to his wife for her understanding and the many roads traveled through the course of the manual's revisions.

Professor Carlson would like to thank her family for continued support and Dr. James Coble for helpful suggestions in the effort to continually improve this manual.

CHAPTER 1

Rock Cycle and Sedimentary Rocks

SCOPE AND METHODS OF HISTORICAL GEOLOGY

Historical geology is a natural extension of a student's introductory course in geology, physical geology. There are several major differences between physical and historical geology, and the goals of this workbook are to review and increase students' geologic knowledge by expanding on the use of rock strata, geologic events, and fossils to investigate and better understand the earth's geologic history through time. The formulation of meaningful geological interpretations requires that both the student and the professional geologist possess an inquisitive mind and a working knowledge of geologic processes, topographic map interpretations, and the rock and mineral identification methods. Historical geology can be defined as a dynamic investigation of the earth's crust and of all its inhabitants through time. Typically, the historical geology curriculum includes discussions of (a) the historical development and preservation of the rock units that comprise the earth's crust; (b) the preservation and documentation of evolving plant and animal evidence in the rock record; and (c) the development of the concept of geologic time. As geological data are collected, they are either systematically interpreted to derive a logical, sequential explanation of an area's geologic history or are interpreted to understand a geological process as demonstrated in a laboratory experiment. This is putting scientific method into practice. Reconstruction of geologic history is very important where economic, environmental, and political decisions are required in industrial and governmental activity.

In an initial geology or earth science course, the student becomes familiar with the important dynamic processes that shape the earth's crust and the basic rock-forming and economic minerals that compose it. In historical geology, earth processes such as erosion and mountain building are placed within a framework of geologic time; students will investigate the difference between *measured* geologic time and *relative* geologic time. These two methods of deriving geologic time involve analysis and understanding of both the sequence of life forms on the earth's crust and the concept of radiometric dating. Radiometric dating techniques enable earth scientists to construct a *measurable* geologic time framework in which most major geologic events can be accurately placed. For example, through radiometric studies of moon rocks and meteorites, scientists have discovered that the earth–moon system was formed approximately 4.5–5 billion years ago. Radiometric dating methods are associated primarily with minerals found in igneous and metamorphic rocks.

Because radiometric dates have to be determined through detailed and costly laboratory analyses, they are not practical for most geological studies. Thus, geologists very often have to rely on the equally important *relative* dating techniques, which involve using a combination of animals and plants preserved in rocks of the earth's crust (fossils) and on interpretation of geologic events as can be interpreted from a sequence of preserved rocks. Most living animal and plant groups today are believed to have developed within only the last approximately 600 million years of the earth's history; therefore, relative dating using fossils can be effective only within the most likely portion of the geologic record, that is, relative to the measured age of the earth–moon system.

Scientists are continually refining radiometric dating techniques and discovering new fossils, both of which contribute to the more complete understanding of the concept of geologic time. From all these data, geologists have compiled a detailed sequential developmental history of the earth's crust. Additionally, scientists are now using these data to develop major concepts and models for past climatic and tectonic changes and to determine how these events may have caused evolutionary changes in the fossil record through geologic time.

The time framework used in historical geology is called the *geologic time scale* (see inside back cover). Note that seven-eighths of all the geologic time shown is Precambrian. This time interval includes all the earth's history preceding the observed preservation of most multicelled plant and animal groups in the rock record.

ROCK CYCLE REVIEW

One of the most important concepts in physical geology is the *rock cycle*, defined as the cyclic change of existing rock materials from one form into another form. In the course of this cycle, rock materials are created, altered, and destroyed by various processes operating on the surface and deep within the earth's interior (see Fig. 1.1). One of the central themes in historical geology is the constant recycling of crustal materials associated with the continuous resculpting of the earth's surface over time. Thus, in historical geology, the earth materials currently existing across the crust of the earth are examined to develop an ordered historical record for sequences of geologic events and cycles preserved in the rock and fossil record.

Within historical geology, much of the focus is centered on the layered rocks seen across the exposed areas of the earth's crust: sedimentary rocks. It is also important to note that nearly always the preservation of geologic life forms and the wide array of depositional environments with which they are directly associated are generally best preserved within sedimentary rocks. The pressures and temperatures present within the normal igneous and metamorphic development realm are not conducive to growth and preservation of most life forms. Therefore, since most sedimentary rocks are formed and preserved at the earth's normal lower surface temperatures and pressures, this chapter will be focused on the most important aspects of sedimentary rocks and sedimentary depositional environments. Some of the critical elements of igneous and metamorphic rock development and interpretation relative to historical geology will be reviewed in Chapter 2 when the concept of the ordering of geologic events is discussed.

SEDIMENTARY ROCKS

Overview

Sediments and sedimentary rocks are the most common rock components spread across the earth's continental surface, along the coastlines, and over the ocean's sea floor. On land they form the majority of the ridges and valleys, the farmlands, the beaches, and vacation spots that we visit or see daily. Sedimentary materials are being formed and accumulated every day at the bases of mountains, along river channels and deltas, and on the beaches and shelf regions of continental margins. This supply of sediment is formed across the earth's surface as the result of mechanical and chemical weathering, abrasion, and solution processes acting upon preexisting igneous rocks, metamorphic rocks, and sedimentary rocks. In addition to human activities such as mining, farming, and construction activity, a dynamic, tectonically active earth's crust is also constantly adding

ROCK CYCLE REVIEW

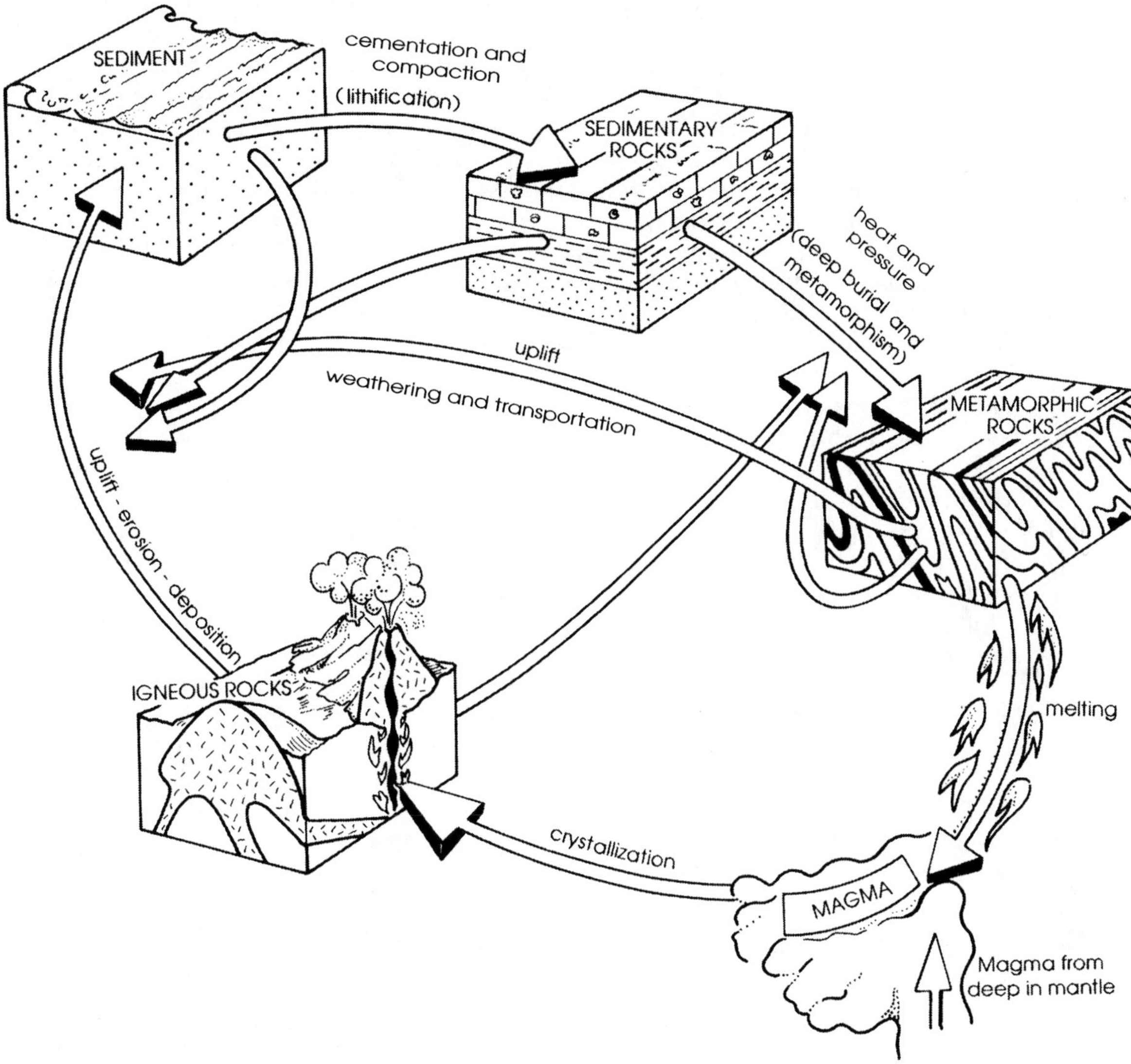

FIGURE 1.1 The rock cycle diagram. Drawn by Lucy Mauger.

debris through volcanic activity, undersea vents, faulting, and mass wasting. On the earth's surface, the process of sedimentary rock building and destruction represents a large segment of the dynamic rock cycle. The emphasis of this chapter will be placed on the origin, textures, and classifications of sedimentary rocks and how this vast and complex volume of earth materials is interpreted and used in historical geology.

The process of developing sedimentary rocks is complex. It brings together not only the variables of chemical and physical processes of weathering on the earth's surface but also the complex alterations occurring during the transportation, accumulation, and lithification of these unconsolidated materials. The physical and mechanical weathering processes develop large volumes of both granular (*clastic*) materials and also the dissolved components in ionic form within fluid systems, all operating within the earth's gravitational system. Most commonly the fluid transporting medium is water; however, wind and ice can also move large volumes of sedimentary materials. Once these unconsolidated materials of any origin come to rest at temporary or permanent sites of deposition, another chapter in their history begins to take place. These transported clastic sediments may initially undergo a progressive process of compaction. The compaction may be mechanical, where the grains are jostled and nestled together as tightly as possible. Alternatively, in an aqueous system, the granular materials may be continually buried and tightly squeezed by subsequent sediment accumulation. During accumulation of more sediment load, the burial weight squeezes or expels the water from between the grains, allowing the sediment grains to be more consolidated or compacted. Later, dissolved materials such as silica and calcite may be precipitated in any remaining pores between the grains, or perhaps very fine, soft clay minerals may be squeezed into the pores, forming a binding agent between the grains. Gradually these cementation processes help turn the loose unconsolidated sediments into rock: the process of *lithification*. For sedimentary rocks, all of these weathering, transportation, depositional, and lithification processes commonly take place in the low-temperature and low-pressure environments present at or near the earth's surface.

Within these same environments of sediment accumulation and transformation, another aspect of sedimentary rocks that is of great importance to historical geology can be developed: the accumulation of organic and inorganic life forms that are preserved as fossils. Early invertebrate animals and algal plant material can be traced back more than a billion years. Only a minimal chronological history exists for the erosion, sedimentary rock accumulation, and tectonic events that occurred during the nearly 400 million years that elapsed between the earliest indications of life on earth and the explosion of life forms that occurred in the Cambrian. Animal and plant fossil life forms that are in part similar to those present on the earth's surface today were not extensively preserved in the geologic record until the explosion of life forms occurred in the Cambrian-aged sedimentary rocks accumulated in the geologic record approximately 600 million years ago.

The use of sedimentary rocks in historical geology is very important because they often contain links between the past and the present. Before sedimentary rocks can be utilized in the interpretation of the past conditions of weathering and accumulation of earth materials, a geologist needs to investigate and interpret the mechanical and chemical processes of analogous modern systems. This concept is associated with one of the oldest "laws" of geology: the law of uniformitarianism—*the present is the key to the past*. Thus, the bedding and textures being formed within modern sediment accumulations provide models for interpreting and translating those data into an understanding of the preserved rock strata of the geologic past. Additionally, the interactions of animals and plants observed within modern environments and their association with specific sediment types must be understood prior to translating correlations and environmental interpretations of similar suites of fossils from the ancient sedimentary rock record.

As previously mentioned, another important concept associated with historical geology is that of time as related to a sequence of geologic events. Sequences of sedimentary rocks are often involved with igneous or metamorphic materials that contain combinations of minerals that can be radiometrically dated by means of decay curves. The use of radiometric data is associated with *measured* geologic time. Within a sedimentary rock framework, the change of fossil groups or distinctive morphological shapes within a fossil genre over time in the preserved record is associated with *relative* geologic time. When either igneous rocks cut through or metamorphism alters sedimentary rocks containing fossils, an associative relationship is developed that geologists can use to develop an ordering of geologic events within a framework of time. The concept of relative time is vitally important to the reconstruction of past geologic events and the ordering of these events into a sequence, which becomes an interpreted geologic history of the earth: historical geology.

The study of sedimentary rocks can involve many scientific disciplines. Considerable knowledge of mathematics, biology, and physics is required to fully understand the mechanics and processes associated with weathering, transportation, lithification, the preservation of life forms, and the postdepositional alterations and changes that may occur. This chapter provides an overview of many of the important definitions, insights, and interpretations of sedimentary rocks used by geologists in historical geology. Chapter 2 focuses on the ordering of geologic events; Chapter 3 focuses on the spatial and sequential distribution of sedimentary rocks using the various methods of correlation and stratigraphy. Chapter 4 reviews the very important aspects of fauna and flora, the fossils, as preserved in sedimentary rocks. These four chapters present data that are critical to the understanding of the concept of geologic time and its application to historical geology. Not only do the concepts discussed herein provide a basis for geologists to interpret the earth's history but they also provide insights into the exploration of the earth's crust for natural resources as well as into environmental solutions to situations resulting from human activities.

DEVELOPMENT, TRANSPORTATION, AND LITHIFICATION OF CLASTIC SEDIMENTARY ROCKS

In this section, the various processes of creating clastic sedimentary rocks are investigated. The largest primary group of sedimentary rocks are those called *clastic rocks*. Clastic rocks are composed of the bits and pieces of preexisting rocks of any origin plus any added fragments of organic debris. Because all of the products of mechanical and chemical weathering become available across the crust of the earth, one of the first processes to be involved is the transportation of these materials from the site of their origin until, eventually, they come to rest in various sites of accumulation. Once these materials have accumulated, several processes occur that lead to the lithification of the material, resulting in the development of a clastic sedimentary rock. This building of a sedimentary rock is discussed in the following paragraphs.

Weathering of Parent Rock Material

Weathering and Transport The products of mechanical and chemical weathering of existing materials across the earth's crust are of vital importance to the development of sedimentary rocks. Existing rocks across the earth's exposed continental surfaces are dominantly combinations of igneous, metamorphic, and preexisting sedimentary rocks. The major physical and chemical weathering processes involved in breaking down rock materials into smaller particle sizes and/or into materials in solution or suspension were previously studied in physical geology. In physical geology, it was shown that the weathered and abraded debris, the *regolith*, either remained near the site of weathering (Figs. 1.2 and 1.3) or was transported by fluids, wind, ice, or volcanic activity, under the earth's gravitational conditions, downward to lower base-level sites of accumulation. These transported rock materials could be solids or could be ions dissolved in solution. The transportation of these solid or dissolved earth materials takes place in the lower-temperature and lower-pressure environments present on the earth's crust. New information on the development of solid and chemical materials issuing forth from seafloor vents indicates that some materials are now being derived from localized sites of higher environmental pressures and temperatures.

Accumulation and Compaction The weathered material being transported will eventually come to rest. This base-level site of accumulation may be at the base of a mountain, in a lake or swamp, in open valleys, or below tidal and current levels offshore or in deltas. These unconsolidated materials normally accumulate in a sequence of filling episodes over a span of time. This newly accumulated stack of sediments normally contains a high percentage of pore space between particles that is most often filled with *interstitial* water

FIGURE 1.2 Sandstone talus from two Cretaceous channel sands, Utah.

(water filling the pores between particles) from the transportation medium. As more sediment is moved into these sites of deposition, more overbearing weight begins to squeeze or compress the materials, resulting in less space for the interstitial waters, and the sediment undergoes an expulsion, or dewatering, process. This compaction leads to a reduction of the bulk volume of sediment and a net reduction in layer thickness and pore space within the sediment. This process of compression and compaction is usually associated with continual loading due to the increased weight of additional overlying material over time; however, in some instances, tectonic activity can help achieve the same result. During tectonic activity such as folding or faulting, the accumulated sediments can experience greatly increased formation pressures as they are compressed or contorted by structural forces. The net result could be the rendering of similar compaction forces and expulsion of formation waters. Because of either the slower normal loading and burial compaction or compaction driven by tectonic forces, the mass of loosely packed sedimentary particles will undergo a decrease in porosity as the grains are more tightly packed and water is lost.

Transformation of a Sediment Deposit into a Sedimentary Rock

Once the transported sediment has reached its site of accumulation and has undergone varying degrees of compaction, the sediments will slowly be transformed or lithified into layers of stone, a process called *diagenesis*. Diagenesis involves all of the physical, chemical, and biological transformations that take place within sediment after its initial deposition

FIGURE 1.3 Talus slopes, central Colorado.

and compaction. During compaction, these "soft" sediments may undergo differential squeezing, resulting in the contortion of thin sediment layers. Additionally, residual clays may be compressed into many of the pore spaces between grains. Animal boring and burrowing (*bioturbation*) and plant activity may actively contribute to mixing of materials and to disruption of the bedding of layers that are undergoing the overall lithification process.

One of the important steps in lithification is *cementation*. This diagenetic process alters the soft clastic sediments into consolidated, hard lithified rocks. Cementation results when precipitation of minerals in solution takes place within the interstitial pore spaces between the individual grains of the sediment. The interstitial water between the grains may become saturated and begin to precipitate and build a crystalline mass along the edges of the grains and gradually fill the space between the grains, thus bonding them in place. This greatly decreases the remaining porosity and lithifies the sediments into a stone or rock. The most common cement is the mineral calcite, which is abundant in the crust and is both easily dissolved and easily precipitated to form crystal material within the pressures and temperatures present at the earth's surface. Other fairly common cements are limonite (iron), silica (quartz), and clay minerals. Cementation may take place simultaneously with sediment accumulation at the site of deposition or later during the compaction and lithification process. In some cases, the cementing agents may be introduced at a later time as secondary cements, with overgrowth of the original grains. In many types of sandstone, geologists sometimes find that the original cements have been partially removed, thus creating secondary porosities at some later date.

CLASSIFICATION AND FRAMEWORK GEOMETRY OF SEDIMENTARY ROCKS

Overview

As previously discussed, sedimentary rocks are a framework of grains derived from preexisting rocks that have undergone lithification and cementation. These grains and chemical materials have been transported to sites of accumulation by any of several fluids within the earth's gravitational field. The accumulated clastic materials at the site of deposition will usually initially have a high porosity that will most often be later reduced with subsequent compaction and cementation. Even though compacted, most of the internal structures and fossils will be preserved.

For centuries geologists have studied sedimentary rocks, and the fossils often included within them, in detail. Many scholarly classification systems have been designed and used by geologists, only to be amended or restated as the result of new discoveries of rocks or fossils in the field or due to the expansion of new scientific instrumentation and technical analysis methodologies. As more detailed chemical and physical analysis methodologies become available, geologists develop a more detailed understanding of the sedimentary rock components and the overall lithification processes. Consequently, geologists have added more variations or combinations of rock components, which in turn has led to new classifications of sedimentary rocks that often have become both more inclusive and more complex over time.

Sedimentary rocks make up only about 5% of the total rock volume comprising the total earth's crust; however, they make up approximately 75% of the rock materials exposed across the earth's surface area. Of the sedimentary rocks present, various scientists have estimated, approximately 40–60% of the sedimentary rocks are shale, 20–25% sandstone, and 20–25% limestone. These percentages vary from one scientific study to another, depending on their initial estimate of the total volume of sedimentary rock material across the earth's crust. The two major building blocks for any classification of sedimentary rocks are the 75% of the sedimentary rocks that are clastic (i.e., sandstones and shales), with the remaining 25% consisting of the chemical rocks (limestones and dolostones) and the precipitates (gypsum and halite).

CLASTIC SEDIMENTARY ROCKS: SANDSTONES

Framework Grain Geometry and Attributes of Clastic Rocks

The framework, or arrangement, and size of the particles in clastic sedimentary rocks are very important to geologists. These data are instrumental in the interpretation and understanding of both the rock's depositional history and its classification. Key analysis factors that must be investigated for sedimentary rock interpretations and classifications include (a) analysis of what preexisting rocks may have been the source material and where the particles were derived (*provenance*), (b) the mode and relative length of transport, (c) any alterations of the sediment during and after burial, and (d) the style, extent, and maturity of the depositional environment. To better understand and describe the internal geometry of clastic sedimentary rocks, geologists use unconsolidated sediments collected on the earth's modern surface as analogues for the interpretation of older sediments. Analyses of the internal arrangement geometry of the clastic particles, the size and shape (*morphology*) of the particles, and what mineral or grain compositions have survived are all attributes to be documented prior to development of the sedimentary rock classification process. For example, a geologist taking a piece of the Lower Triassic Moenkopi Formation from a layer exposed in a southeastern Utah outcrop may enter the following description in his or her field notebook: "sandstone, thin-bedded, brown to red-brown, fine-grained, with red-brown siltstone and thin beds of gypsum. The formation has thin regular bedding, with sands having abundant ripple marks and scattered raindrop imprints." For the overlying basal Shinarump member of the Upper Triassic Chinle Formation, the description could be "sandstone, cross-bedded, gray to yellowish orange, fine- to coarse-grained, irregularly interbedded with lenses of buff and dark grey conglomerate and green to grey siltstone. Pebbles are predominantly chert and quartzite." In these descriptions of clastic rocks of various ages, key framework attributes, such as the particle sorting, particle shape, kinds and composition of rock fragments, the bedding, and color, are all part of the rock descriptive, interpretation, and classification process. Fossils, another important attribute, can be a sparsely scattered minor component in a sedimentary rock or can be the major rock constituent, like that of a coquina. In most cases the fragile shells of calcareous fossils do not remain in sandstones, due to the grinding/abrasive actions of sand grains (usually harder quartz grains) during transport. In the following paragraphs, the primary framework parameters of clastic rocks will be reviewed.

Sorting Sorting of sedimentary particles is a process that segregates or selects particles that have similar sets of characteristics. The particles may have similarity of particle size, shape, or specific gravity. The selection of the particles is a function of the energy within the transportation environment and the length of time that a depositional environment or process remains stable. A stream, for example, at a specific velocity may transport particles of similar size or similar mass as a function of the medium's velocity and density (i.e., air versus water). The degree of sorting within a clastic rock is one of the environmental parameters that can help interpret the overall energy within a depositional system. While some depositional environments may have signature particle size distributions, sorting is seldom unique to specific environments alone. Well-sorted sands often occur in such environments as stream channels, sand dunes, and beaches (Fig. 1.4). Poorly-sorted sediments are associated with glacial tills, alluvial fans, and high-energy stream and flood deposits.

FIGURE 1.4 Eolian cross-bedding in Jurassic Entrada Formation, Utah.

Porosity and Permeability *Porosity* is associated with the amount of space or voids between the grains of either consolidated or unconsolidated rocks, soils, or other material. Porosity is usually expressed as the ratio of the void pore volume to the bulk volume of material (as a percentage), no matter whether the pore voids are isolated or connected. Permeability is associated with the porosity. *Permeability* is a measure of the relative interconnectivity of the pore spaces within a porous rock, sediment, soil, or other material. Therefore, permeability relates to the transmission of fluids through the material under unequal pressure conditions. In sandstones, the permeability can be affected by the particle sorting and grain size, the presence of cements, or whether clay minerals may be inhibiting the interconnection of the pore spaces, causing restriction to the flow of the fluids. Most sandstone has some degree of porosity and permeability. However, examples such as vesicular lavas or volcanic rocks such as pumice may have excellent porosity formed by the gas bubbles; but once ejected material is cooled, the bubble walls solidify and can reduce the permeability to zero. Porosity and permeability are very important in the storage and the production of oil and gas and in groundwater in the ground as a natural resource. These same factors are also equally important in designing and developing barriers or channels for the environmental disposal of fluids and materials.

Packing Packing is the framework arrangement of clastic grains. Under ideal conditions, when all the particles are the same size and are arranged with a cubic geometry, then the packing, the porosity, and the permeability would be at a maximum. However, in nature, the porosity, permeability, and packing are usually greatly reduced due to the presence of random particle sizes.

Particle Rounding Particle rounding is the degree to which the original angular edges or corners of a sedimentary particle have been rounded off or smoothed to curving faces. This rounding of angular corners usually occurs by abrasion during transportation.

Particle Shape Particle shape refers to the overall geometric shape or form of the particle in relation to its length, width, and height. During the transportation process, the abrasion of rolling, bouncing, and sliding particles will try to attain a "better sphericity" as related to its initial shape, mass, and surface area. Clastic particles, of course, come in every conceivable combination of shapes, from very angular to nearly spherical. Normally, clastic particles can be categorized into four groups: tabular, bladed, elongate, and spherical (*equant*) shapes (Fig. 1.5). In sediments under transport, the tabular, bladed, or elongate particle shapes often may align their long axis parallel to the transporting fluid's currents. Occasionally, these particles will shingle, or overlap, forming what is called an *imbrication* texture.

Rock Fragments, Micas, and Heavy Minerals The weathering of preexisting rock materials on the earth's surface produces rock fragments from multiple sources and compositions that will be mixed together during transportation. Generally, particles derived from coarse-grained igneous and metamorphic rocks will not exist for long distances because abrasion and crushing may be enhanced due to their specific mineral cleavages and hardness. Mica fragments are often found in many types of sandstone. Even though originally the micas may have been in a large crystal, or "book" due to micas' nearly perfect cleavage in one direction, each sheet is quickly reduced in size to small plates, even down to clay-sized particles. Often the fine-grained rock fragments, such as mica flakes, basalt, and materials with high hardness and low or no cleavage, such as quartz, will survive the longest. As the particle size decreases, the identification of original source rock becomes very difficult. Clastic sedimentary rocks also often contain a small percentage of accessory minerals that have high specific gravities and hardnesses. Occasionally a

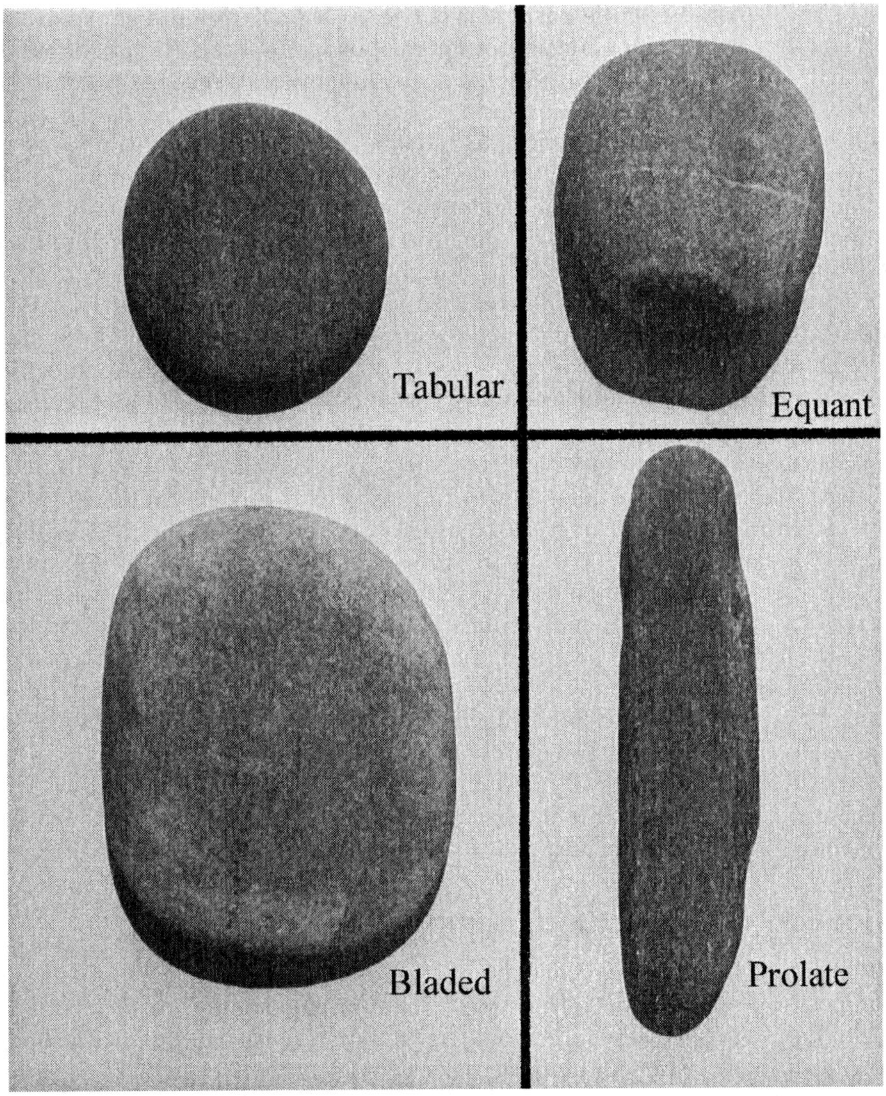

FIGURE 1.5 Classification of pebble shapes.

"unique" suite of heavy minerals may be preserved in sandstone, which can provide insight into the type of sediment dispersal from a source area and may indicate relative transportation distances.

Color Color in sedimentary rocks is a double-edged sword. At first glance, it can be very useful in tracing layers of sedimentary rocks from one outcrop to another. Colors can be identified with a "unique" set of environmental conditions present during the accumulation or lithification process, e.g., a black shale. However, alternatively, the color of an unweathered sediment or formation may change from place to place as weathering takes place. This is because the color can be dependent upon the composition of the grains or their cements. The first factor, their *composition*, is probably the most important in determining the color of rocks composed of fine-to-coarse detrital debris. A relatively pure quartz sandstone would be a light color, whereas an arkose containing abundant orthoclase would most likely be pink. Sediment containing a high organic content would often tend to be dark brown or black. Generally, as the particle size becomes finer, the darker the overall color. The second factor causing color consists of the various *cements*. The sediment or rock color may be due directly to inclusion of cements or to clay minerals that may coat the surfaces of the matrix minerals. The major cementing agents in sedimentary rocks are calcium carbonate, silica, and iron. Calcium carbonate and silica cements do not usually impart a distinctive color to the sediments. Oxidation-derived coloring is especially important when the coating mineral is iron. The color will be dependent on the oxidation state of the iron, not on the amount of iron. For example, ferrous iron will impart a green, blue, or grey color to a rock, but upon oxidation to a ferric state, this color will change to reddish hues. Hydrated iron oxide (the mineral limonite) will produce a tan or brown color. Further oxidation and dehydration will produce a typical dark red to purple color as the iron oxide changes to the mineral hematite. Overall, color can be a useful tool; but during geologic interpretations, because of many depositional and secondary variables, caution should be exercised.

Maturity Maturity of sediments is often envisioned as the influence that geologic processes have had upon a particle's path from provenance to the final stages of particle deposition and lithification. The degree of maturity depends on whether the resulting clasts are composed of chemically stable minerals, are spheroidal in shape, and are well rounded and well sorted. However, for the maturity of the rock, there are other important considerations. If the original sediment acquisition site was a low-relief, warm, high-moisture area where deep in situ weathering was taking place, then the sedimentary material being produced might not have unstable minerals because of a lack of mechanical weathering and dominance of active chemical weathering. The provenance site could be in an active tectonic area where large amounts of highly angular, highly unsorted, immature-sediment, mixed-composition clastic materials are being generated and accumulating. Therefore one must consider the duration of time the particle has been in the system and environmental conditions in the area where the particle was derived, if possible.

Classification of Clastic Sedimentary Rocks

Clastic sedimentary rocks involve collections of discrete particles of earth materials of any size or composition that have been transported, accumulated, and transformed into a rock. The particle can range from very large boulder-sized materials to those microscopic in size. In the classification of clastic sedimentary rocks, geologists have settled on a size range scale initially produced by Wentworth ("A scale of grade and class terms for clastic sediments," *J. Geol.*, 1922, v. 30, pp. 377–392) that is used to define and name groups of particle sizes. These particle size groups were based on laboratory analysis using modern unconsolidated sediment material. Dry sediment materials were processed in the laboratory using sets of sieves with different screen hole diameters. Sand samples, such as beach and river sands, were used to determine natural size distributions. The distributions were developed by plotting particle size distribution histograms [a *histogram* is a bar graph where each bar, representing the percentage of each particle size (y-axis), is plotted against each particle size present within the sample (x-axis)]. From these data, a size scale was developed. Wentworth defined, for example, that particles that fell into a size range of between 1/16 millimeter and 2 mm in diameter would be known as sand-sized material (See Table 1.1). The different size categories cover huge boulders, common sand and silt sizes that can either be seen by normal visual observations, down to particulate material that needs microscopic or X-ray diffraction analysis. Under gravitational conditions during transport in a moving fluid system, such as a stream, clastic particles can be transported simultaneously by traction (pebbles and gravels moving along the bottom by rolling, sliding, dragging, or bouncing), inertia suspension (sand-sized particles), viscous suspension (silts), and colloidal suspension (clays). If the transporting medium is wind, the lower density of the wind current, compared to water,

TABLE 1.1 Classification of Terrigenous Clastic Sedimentary Rocks

Wentworth size class	*Particle size range*		*Common constituents in composition*	*Rock name*
Boulder	>256 mm			
Cobble	64 mm up to 256 mm			
Pebble	4 mm up to 64 mm	Gravel	Rounded fragments of any rock type Igneous, quartzite, chert dominant	Conglomerate
Granule	2 mm up to 4 mm		Angular fragments of any rock type Igneous, quartzite, chert dominant	Breccia
Very coarse sand	between 1 mm and 2 mm			
Coarse sand	1/2 to 1 mm		Quartz with minor accessory minerals	Quartz sandstone
Medium sand	1/4 to 1/2 mm	Sand	Quartz with at least 25% feldspar	Arkose
Fine sand	1/8 to 1/4 mm		Quartz with at least 25% rock fragments	Greywacke
Very fine sand	1/16 to 1/8 mm			
Silt	between 0.0039 and 0.0625 mm	Mud	Clay minerals, quartz, and rock fragments	Siltstone
Clay	<0.0039 mm		Clay minerals, quartz, and rock fragments	Shale

will selectively limit the traction load to smaller sand sizes, with the silt and clay particle sizes being carried farther away from the sand accumulations. Ice can transport tremendous loads of particles with an extensive range of particle sizes.

Specific names such as coarse or fine sand are simply particle size subdivisions of the overall sand size category. Remember, there is no correlation within the definition of *sand* that directly relates to any particle composition, just a certain size range. However, since quartz is the most common mineral "sand"-sized particle found in sandstones, the name *sandstone* has become synonymous with sandstones dominantly composed of only quartz particles. Examples of alternatives to a quartz-based "sand" include the white sands of New Mexico (which are sand-sized gypsum particles) and the black "sand" beaches of Hawaii (which are sand-sized fragments of volcanic basalts) or, perhaps, many southern Florida beach "sands" (which are sand-sized fragments of organic corals and marine animal shells).

Bedding and Textures of Clastic Rocks

An important part of studying clastic sedimentary rocks is an understanding of bedding associated with rock units. Bedding can be viewed in the laboratory under a microscope, in a hand specimen, or in outcrops of a sequence of sedimentary rock layers. Bedding depends on many factors, such as variable current strengths, sediment supply, arrangement of particles, and biologic activity within the depositional environments.

As is discussed later, the term *bedding* is sometimes associated with either the top or the basal bedding boundary of the sedimentary unit, while at other times it is used to describe aspects of the internal structure within a bed, such as the term *cross-bedding* or, in another case, special bedding, termed *graded bedding*. While only a few key sedimentary textures and structures are reviewed here, remember that a great variety of combinations usually exist within sedimentary deposits. Some of these sedimentary structures and textures found in clastic rocks can also be found in the carbonate rocks, such as limestones and dolostones.

Common Sedimentary Bedding and Stratification: Clastic Rocks

Bedding, or layering, of the sediment is a feature of the rock unit that is important to the understanding of the origin of the sedimentary deposit. The basic unit of an outcrop is the single bed or rock unit. Each rock unit is preserved evidence of a site for a clastic detrital material accumulation. A rock layer can be of almost any thickness, because thickness is dependent on the continuity of the sediment supply, the rate of deposition, the compaction, and the accumulation time. Generally, a single lithological unit or bed was accumulated under one set of depositional or physical conditions. The distribution and areal extent of the bed are related to the environment of deposition. A stream channel would have a restricted, narrower distribution as compared to a floodplain or deltaic accumulation area. A distinct layer of rock should have a similar mineral composition throughout the bedding thickness and similar internal textures and bedding structures. Bedding thicknesses are often described by geologists via a variety of terms, such as *thick-bedded, thin-bedded, wavy-bedded, laminated bedding*, and *discontinuous bedding*. Thickness variations can be described in detail, such as very thin laminar bedding, or in other cases in very general terms, such as massive bedding for thick sequences of rocks. Some geologists have tried to define thicknesses in terms of centimeters of thickness accumulation, but because of different depositional rates and textures within a bed, the thickness often varies laterally and the classification becomes fraught with exceptions. Another factor that can cause bedding thickness variations is postdepositional erosion. Usually a given sedimentary unit or bed will have a definable top and base that denote the start and finish of the set of given sedimentary conditions or environments represented by the accumulation of the specific bed (Figs. 1.6 and 1.7).

Common Sedimentary Bedding: Internal Bedding

Conglomerates Conglomerates are a type of bedding important to historical geologic studies. Conglomerates involve the larger particle sizes as seen in the classification chart, those in the gravel, pebble, and boulder size ranges. Conglomerates can be of nearly any composition because they are derived from preexisting rocks that are suddenly put into "motion." Generally, when a conglomerate is found in a sequence of stratigraphic units, geologists would note that a combination of major changes of energy within the environment of deposition has taken place and that the source area for the transport materials has probably undergone an episode of structural activity. If, in an outcrop, a sandstone or a shale were suddenly overlain by a conglomerate containing pebbles and cobbles, the normal fluvial or stream transportation system would

FIGURE 1.6 Discontinuous bedding, Jurassic Kayenta Formation, Utah.

FIGURE 1.7 Bedding of Cretaceous fluvial sandstones, shales, and coals, Utah.

have had to have a sudden major energy increase to flood conditions. Alternatively, in the area, the source area is uplifted, a new episode of erosion can be started with larger particle sizes becoming available. As can be seen in Figure 1.8(a), even within a conglomerate, there will be some lower-energy zones, as distinguished by pockets of sandstone within the overall high-energy conglomerate accumulation. Generally, a conglomerate has a chaotic mixture of particle sizes and very little internal bedding. A close-up of the same conglomerate is illustrated in Figure 1.8(b). Conglomerates can be developed during floods, volcanic events, tectonic activity, slumping, and by glacial activity. All of these activities can move large quantities suddenly into an adjacent, calmer depositional environment. In nearly every instance, at the basal boundary of the conglomerate varying amounts of erosion will occur, often forming an unconformity (see Chapter 2, p. 56 and Figs. 2.2 to 2.8).

An example of a Pennsylvanian-aged conglomerate involved in a region's geologic history would be one associated with the formation of the Arbuckle Mountains in south central Oklahoma. In Figure 2.5 (p. 58) the conglomerate (the Collings Ranch conglomerate) is unconformably resting on the tilted Ordovician units. The conglomerate is comprised primarily of eroded fragments of Ordovician-to-Mississippian-aged carbonate rocks. The uppermost, or youngest, portion of this conglomerate is called the Vanoss conglomerate. The composition of the Vanoss conglomerate changes from the carbonate-rich Collings Ranch conglomerate to a unit comprised of a majority of conglomeritic Precambrian granitic materials. Therefore, geologists have surmised that as the Arbuckle Mountains underwent uplift and erosion in the Pennsylvanian Period, the Paleozoic carbonates were the first to be eroded, filling adjacent basins. Then, as the uplift proceeded, the granitic Precambrian rocks in the core of the uplift were progressively eroded, forming the last part of the conglomerate accumulated in the basins. Thus, even in the chaotic bedding of a conglomerate, geologic history is preserved for later interpretation.

FIGURE 1.8(a) Cretaceous North Horn Conglomerate, Utah.

FIGURE 1.8(b) Close-up, North Horn Conglomerate, Utah.

Cross-bedding One of the classic structures associated with the movement of sand-sized particles on the earth's surface are the large fields of sand dunes in the arid areas of the world, such as those in the Saharan Desert of North Africa. The building and movement of sand in sand dunes is easily understood. The wind-blown sand is bounced or rolled up the long, moderately sloping windward face of the dune, only to be deposited on the low-energy, steep-sloped leeward or downwind face of the dune. As the sand dune moves, successive steep leeward faces are built over the one before. Eventually, the internal structures of the dune are high-angle layers of sand relative to the flat bed under the dune. These inclined internal dune beds are known as *cross-beds* (Fig. 1.4 and 1.10(e)). Similar bedding phenomena and internal structures form in moving waters of a stream or lake where water currents move particles along a surface, building small-scale dune-like features called ripple marks (Fig. 1.10c, d). These internal cross-bedding features can be seen at a very large scale, such as those very large, spectacular ancient sand dunes in the Jurassic sediments of Zion National Park or Arches National Park in Utah.

Graded Bedding Graded bedding is the particular type of bedding most often associated with *turbidity currents*, that is, pulses of higher-density water where the sediment is being held in suspension by turbulent flow conditions. These turbidity flows move down slopes (often continental slopes) under water at relatively high speeds into deeper, more tranquil water. As the sediment-laden current slows along the ocean floor, the energy decreases and particles begin to settle out of the turbidity current. The largest or most dense particles settle first, followed in order upward to the finest particles left in suspension. The resulting deposit is graded in particle size from coarse at the base upward to very fine particle sizes at the top. Clay particles continue to settle, forming an overlying shale unit. The basal boundary of a turbidite is usually very sharp because the coarse particles of the graded bed rest on the uppermost clays of the underlying turbidite flow. Thus, a single turbidite is comprised of a sediment couplet; a basal sand layer that grades from coarser to finer particle size in an upward direction and an upper shale unit. Often these turbidite layers will be sequentially repeated, forming thick deposits. In Figure 1.9, note two turbidite sands; the lower (at the hammer) has sands grading upward to a series of laminated shale-sand layers and then finally to an upper shale layer that has eroded back in the outcrop. The second turbidite sequence starts with a much more massive

FIGURE 1.9 Graded bedding in Cretaceous turbidite sandstones, Utah (note how the bedding becomes shaley toward the top of the sand unit, followed by a shale).

sand, to a laminated zone and an upper shale. In cases where many turbidite sands are accumulating, the upper shale unit is often missing as other successive turbidites interrupt the completion of the previous turbidites' development. Often, graded bedded sandstones appear as lenses of sand in thick sequences of deeper-water shales [Figs. 1.9 and 1.10(f)].

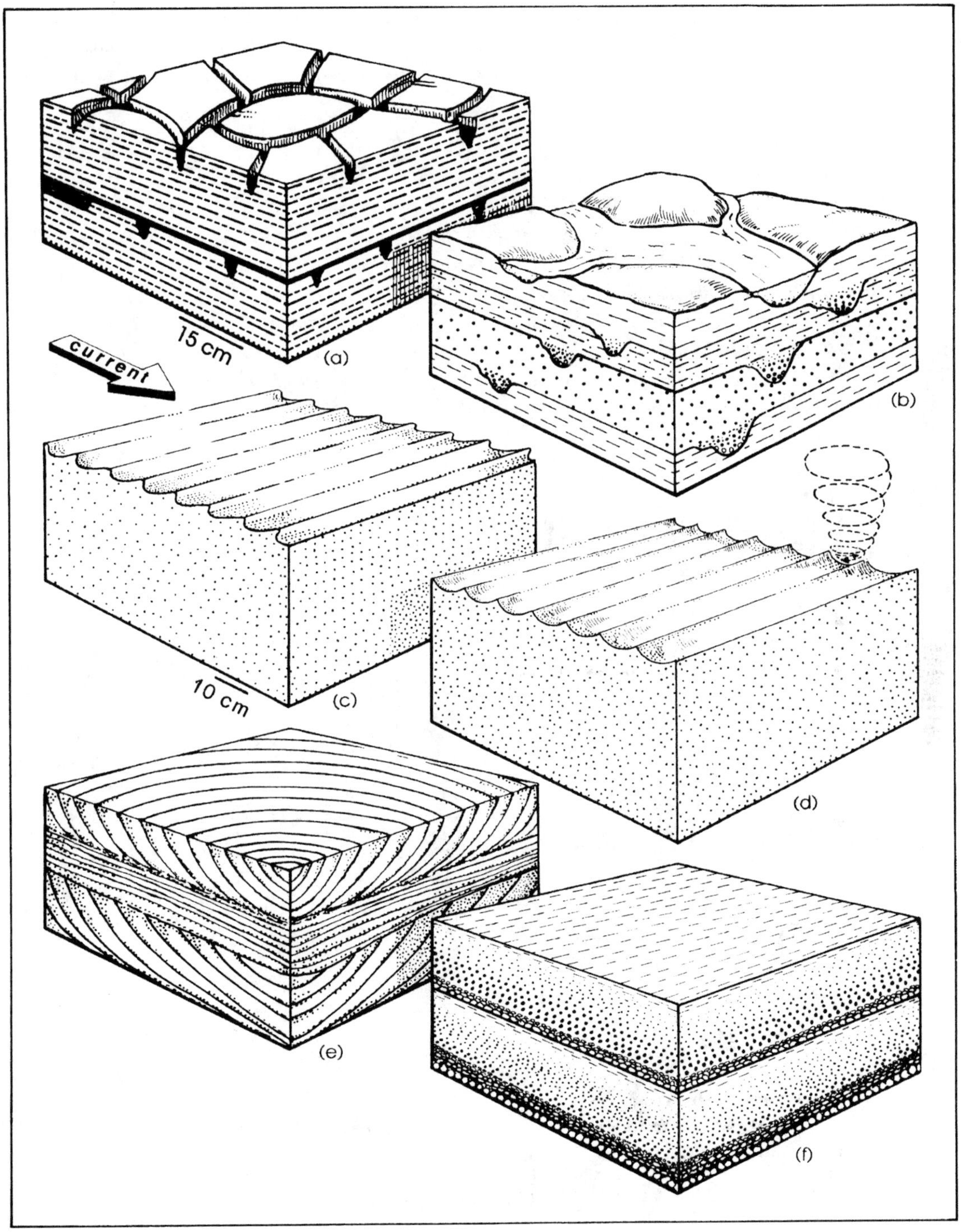

FIGURE 1.10 Sedimentary structures. Drawn by Lucy Mauger.

Bedding Plane Surface (Base, or "Sole," of Bed) Sole marks are usually found associated with the basal bedding surface of sandstones and limestones that overlie shales. These sole marks, or *casts*, are formed when a fresh supply of sand for the next bedding unit initially covers the top of the underlying mud surface. To form these sole marks, some kind of depression needs to be formed on the underlying mud surface that will then be filled by the overlying sand materials. Later, after lithification of the sandstone, these sole marks will be in the form of a *positive*, or raised, feature when the bottom of the sandstone is viewed. These marks can help in the interpretation as to which is the top or base of the sandstone unit. Sole marks can fall into two primary categories, due to different causes.

Load Casts These sole marks are formed by a loading of sand onto the preceding sedimentary unit. As the sands of the overlying unit accumulate, any soft portion of the underlying unit will allow accumulating sand to sink into the underlying muds. Load casts are of varying size and shape but are more symmetrical and generally have few direct indications of current direction associated with them (Fig. 1.11). Many descriptive names have been associated with them.

Scour Marks and Cut-and-Fill Structures These features rely on the currents in the overlying sand regime to cause scouring and/or cutting due to current fluctuations or by some material or object being carried by the current. The hydraulic action of the current forms scour marks, which are later filled with overlying sand. One of the most common types of scour marks is called *flute casts* (Fig. 1.12). Scour marks can

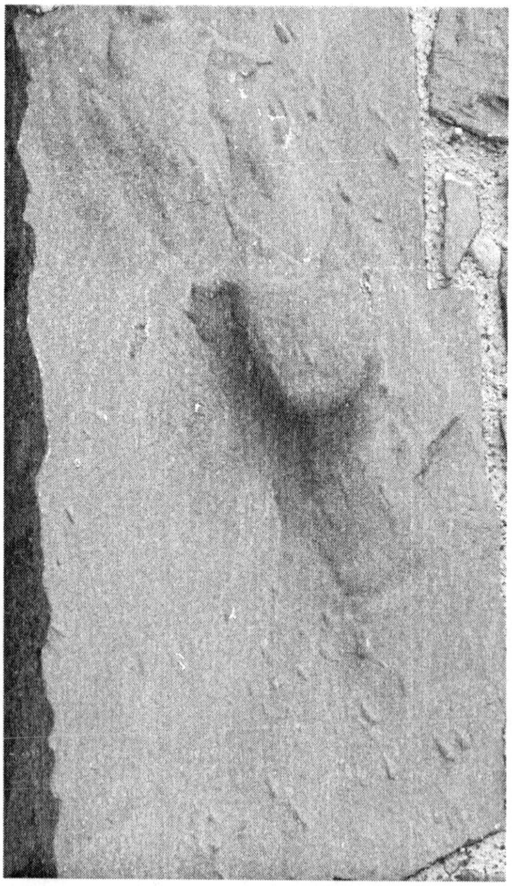

FIGURE 1.11 Load cast, Pennsylvanian Atoka Formation, Oklahoma (slab 18" × 24").

FIGURE 1.12 Flute cast, Pennsylvanian Atoka Formation, Oklahoma (slab 13" × 17").

FIGURE 1.13 Cut- and-fill structure (stream channel), Cretaceous Mesa Verde Formation, Utah (Note: Follow the thick basal sand and note that it cuts across underlying layers to the right. Above the sand, note the arcuate bedded sand accumulating right to left. As the channel is filled and abandoned, a uniform floodplain shale overlies and buries the channel).

vary greatly in size, shape, and pattern. Another variety of scour mark is the *tool mark* (see Fig. 1.17). The "tools" can be pieces of shell, sticks, or other debris carried by the current that drag, slide, or bounce along the base of the bed. These telltale marks may be lines crossing other marks or odd-shaped pits. On a much larger scale, a stream could be flowing across a floodplain and cutting a channel into the underlying finer silt and sand sediments. Later, the sand-filled channel could be covered by successive floodplain silts and shales [Figs. 1.13 and 1.10(b)].

Bedding Plane Surface (Top of Bed) The upper surface of a bed is where many sedimentary features can be found, as well as where extensive organic activity can be present, where drying of the surface might occur, or be a place where shallow water currents might develop distinctive features like ripple marks. The following four groups have been selected for review.

Current Ripples and Current Marks One of the most common surface bedding features are current ripple marks. These current-related ripple marks are part of a surface texture of a bed and can be formed by flowing water or oscillation of water currents by wave activity [Fig. 1.10(c,d)]. Current ripple marks are wavy surfaces that look like a miniature field of sand dunes with smaller wave lengths, 1–6 inches being most common. Current ripple marks can be used to indicate the direction of the current flow, similar to a sand dune; the short, steep slope is down current (Figs. 1.14 and 1.15).

Mud Cracks As surface waters recede along streams, beaches, etc., thin layers of mud may develop on coarser underlying silts and sands. When this surface of mud accumulation is exposed to air and drying begins to take place, mud cracks can be formed. As the thin mud layers dry, the water within the clay is expelled and the clay surface shrinks. The shrinkage causes cracks to form, developing odd-shaped or polygonal mud

FIGURE 1.14 Current ripple marks, Triassic Shinarump Formation, Utah (note the two current directions on the successive rock layers).

FIGURE 1.15 Ripple marks, Cretaceous Mesa Verde Formation, Utah (note the animal trail over the ripple crest near the marker and the pockmarks of burrowings).

FIGURE 1.16 Modern mud cracks, bank of the Llano River, Mason, Texas.

fragments [Figs. 1.10(a) and 1.16]. As the clay fragments continue to dry, the cracks enlarge and the mud clasts begins to curl concave upward, separating the mud fragment from the sandy, nonshrinking layer under the mud. When this mud-cracked surface is buried by a younger sequence of sands and clays, the sands of the overlying next bedding unit progressively fill the cracks between the clay clasts. Preserved mud cracks are not common. But once preserved and found, they will appear on bedding surfaces as polygonal "rings" of sand filled with shale (imagine slowly covering Fig. 1.16 with a layer of sand to the top of the clay).

Tool Marks and Raindrop Imprints Occasionally, raindrop imprints may be formed and preserved on the clay-rich upper bedding surface of a sand unit. These features are usually preserved as indentations or dimples on the bedding surface. Tool marks are formed by objects such as sticks and fossil fragments that are carried by a current and dragged or bounced along the bottom. Often, they are linear marks made by currents that bounce, roll, or drag the objects along the underlying mud surface, leaving a variety of trails, or *signatures*, that are then filled by sands of the overlying bed (Fig. 1.17).

Biogenic Activity: Tracks, Borings, Bioturbation, and Footprints While sandstones are usually very poor environments for the preservation of fossils or fossil activity, the muddy, fine-sand, low-energy upper bedding surface is often a site for both animal and plant organic activity. At this surface, animals can burrow into the underlying fine

FIGURE 1.17 Tool marks, Cretaceous Mesa Verde Formation, Utah.

sand–muddy substrate looking for nutrients, or they can walk along the surface looking for food or prey. Thus, a series of burrowings, called *bioturbation*, can be developed at the upper surface and can also obscure other bedding features. Animal tracks on the surface can be found in a wide variety of forms, from tracks or trails of small organisms (trilobites), to large dinosaur footprints and tail drags, to sinuous trails of wading and shore bird footprints. Often it is hard to differentiate the much more common current-related scour marks made by sticks from tail drags of animals unless other, associated environmental evidence can be found (Figs. 1.15, 1.17, 1.18, and 1.19).

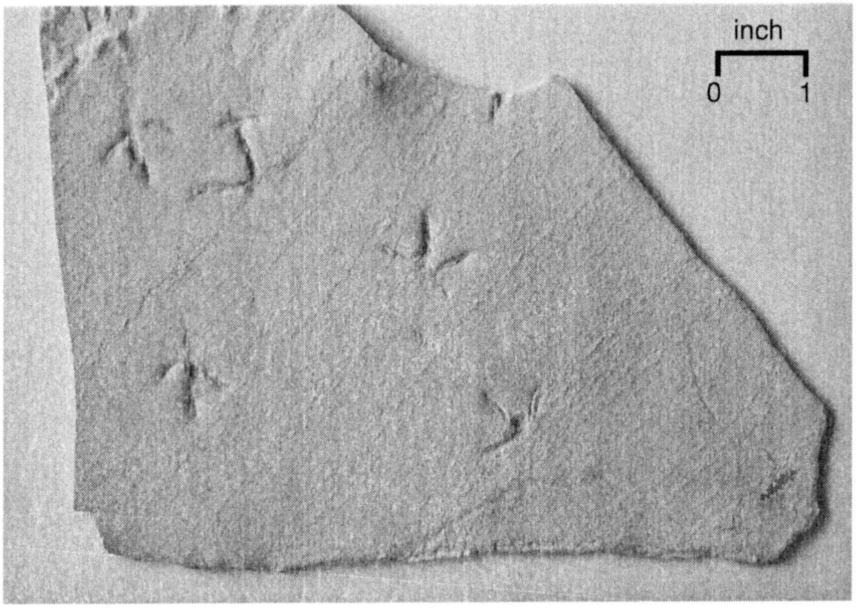

FIGURE 1.18 Fossil bird tracks, Eocene Green River Formation, Utah.

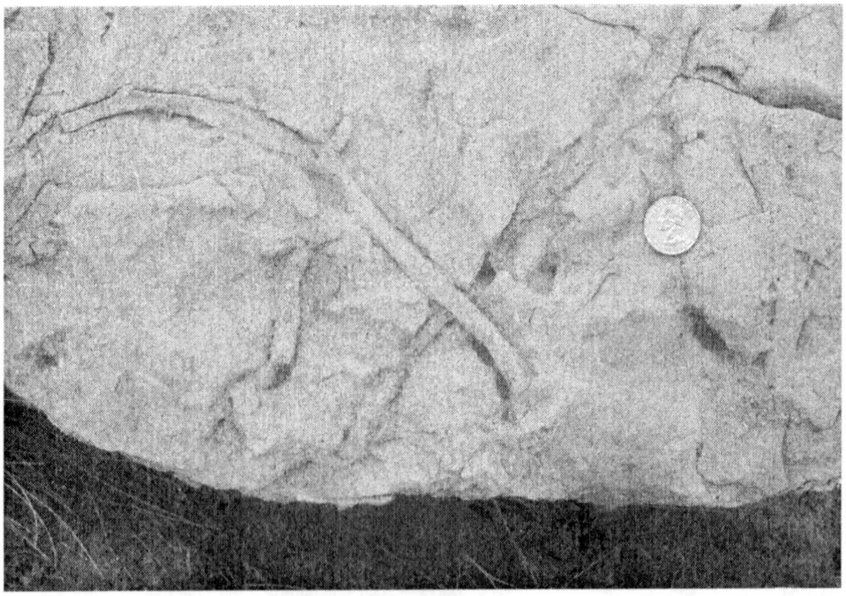

FIGURE 1.19 Sand-filled burrows, Cretaceous Mesa Verde Group, Utah.

CLASTIC SEDIMENTARY ROCKS: SHALES

Composition, Bedding, and Textures

Even though shales are very common and widespread in the geologic record, they are often hard to map unless they are in a nearly vertical road cut or stream cut, because they erode easily. Consequently, shales are generally the tree- and grass-covered "valley formers," as compared to the more resistant ledge- or cliff-forming sandstones and limestones. The earth materials that comprise shales are hard to define and classify on strict size terms as are those for sand and pebbles. In size terms only, the silt-sized particles would be less than 1/16 millimeter in diameter, and clay particles should be less than 1/256 millimeter in diameter. Lithified silt would be cemented and form a siltstone. Overall, it is the finer-sized materials comprising the clays that make them hard to classify. Rocks of this particle size are very difficult to analyze, even in thin section, and they often require analysis by X-ray diffraction or thermal analysis equipment. Generally, the resulting rock falls into two generic "mapping" categories: a *mudstone*, if the bedding is massive, or blocky or a *shale*, if the rock has a lamination character or *fissility*. Compositionally, silica is the dominant mineral constituent of both shale and clay.

Shales can be deposited as very thick sequences of rock. Some geologists have estimated that 40–70% of all the clastic sedimentary rocks accumulated since the beginning of the Cambrian are shales. Newly deposited clays accumulations have very high water content as the small particles settle into the aqueous environment of deposition. As compaction gradually takes place, the water is expelled, leaving the platy mica-like clay particles to orient themselves parallel to the bedding. This mineral or particle orientation is called *fissility*, that is, the tendency for the rock to split along surfaces nearly parallel to the primary bedding. Very good fissility is often associated with shales that are rich in organic material. The layering, or lamination, of a shale is often described as massive when there is little or no splitting, grading into terms such as platy and thin-bedded and down to the very easily split shales, the fissile shales.

In some cases, shales can maintain a cyclic lamination, such as in varves. Varves have been shown to develop where annual climatic cycles may induce "blooms" of microorganisms or change the silt content of glacial waters as seasonal temperatures change. Bedding destruction of shales can take place as minerals grow along the bedding planes or, as previously mentioned, where bioturbation has taken place.

Important Varieties and Economic Uses

Shales have never been comprehensively classified as have the clastic sedimentary sandstones and limestones. Often a field geologist will use the general color or chemical content to describe a shale unit in an outcrop. The most prominent color for shales as exposed in outcrops is a medium-to-dark grey. In the field, the geologist commonly will apply the term *shale* to describe siltstones, mudstones, and claystones, without detailed analysis. However, there are times that the geologist must note and perhaps map one of the following several special shales varieties.

Red Shales Red shales are most often a dark red, although the color can be light red to pink. These red shales and red mudstones are often described by geologists as belonging to a *red bed* facies. Most of these red shales are massively bedded deposits, nearly always unfossiliferous and silty, and associated with terrestrial or continental depositional environments where arid climatic conditions are dominant.

Black Shales, or Carbonaceous Shales Black shales often contain an unusually high percentage of organic material. They are frequently highly fissile and can usually be split into large, thin sheets. Most of the time they do not have visible fossils, but they may have some phosphate-rich shelled organisms that live only in restricted environments. The black shales often contain small grains of disseminated pyrite. At times, pyrite crystals and pyrite nodules will be formed. Also, a few famous localities exist where crinoids, gastropods, cephalopods, and other fossils have been replaced with pyrite.

Siliceous Shales Siliceous shales are not extensively preserved in the geologic record. However, when found, they tend to be hard and nonfissile. Shales are naturally high in silica because of the mineral compounds that comprise the majority of the rock materials; however, siliceous shales have added silica. The dominant sources for increased silica content in the shales are precipitated *amorphous* (no crystalline structure) silica, added volcanic ash or very fine-grained volcanic glass, and, in some cases, organic silica derived from radiolarians or sponge spicules (see Chapter 4).

Calcareous Shales Calcareous cements can be added to shales by direct precipitation of calcite into the shale. Alternatively, calcite can be dispersed from dissolution of fossils into the shale. As the calcite content increases, the shale becomes less fissile and will react when tested by hydrochloric acid. The calcite content may increase to the point where shaly carbonate streaks or nodule zones are developed as marls.

Loess Loess is a special clay or silt-rich deposit most often associated with accumulations of tan-to-brown wind-blown soils. Loess accumulations frequently form thin layers from 2 to 10 feet thick, but locally some accumulations are over 100 feet thick. Layers of loess are distinct from an engineering standpoint. When loess is sliced into for a road cut, the standing slope will approach vertical without slumping (good examples are found in the Vicksburg, Mississippi area). Loess is associated primarily with winds selectively picking up silt-sized particles from glacial outwash beds, fluvial floodplain soils, or arid desert surfaces and redepositing them downwind. The finer, clay-sized "dust" is carried further downwind by the air currents and is often dispersed into widespread depositional environments. Currently, no loess deposits have been identified in the ancient geologic record. All of the known accumulations of loess deposits in the world are found in the Pleistocene Period.

Shales are important for economic reasons. Shale is the key ingredient in products such as brick, all kinds of building tiles, ceramic ware, pottery, and, as kaolinite, a filler for paper. Additionally, when shales are mixed with limestone, fired, and ground to a powder, they produce Portland cement. The metamorphosed equivalent of shale is slate, which is used for roofing and decorative stone. In limited areas of the world, such as zones within the Green River shales of Eocene age and the Cretaceous Mesa Verde Group in the western United States, there are thick deposits of bituminous, or oil-bearing, shales. These oil shales must be mined and then distilled, yielding a product that later can be refined to produce heating oils and motor fuels.

CLASTIC SEDIMENTARY ROCKS: VOLCANIC MATERIALS

Volcanic-based sedimentary materials are present on the earth's surface in some places in very large quantities. The problem, however, is that it is often difficult to determine whether the preserved volcanic material is a sedimentary deposit or a pyroclastic deposit. Once volcanic activity starts, pyroclastic materials of all sizes and shapes are ejected into the atmosphere. The ejected material can range from airborne debris to lava flows.

Coarser pyroclastic debris like blocks, bombs, lapilli, and pumice flows will most likely initially stay close to the source area until weathering and aqueous transportation systems are established. The finer material, the volcanic ash, may be distributed in large quantities over very large areas. This finer airborne volcanic ash will fall into all of the many land-, water-, and ice-based depositional environments across the earth's surface. Occasionally, a distinctive ash fall can be preserved across a widespread area and become a *marker* bed for geologic studies. The pyroclastic event from a volcano is very short-lived in geologic time, so geologists can often use an identified volcanic ash bed essentially as a time line or marker while correlating rocks and establishing a sequence of geologic events. Once the volcanic material becomes part of the normal weathering cycle away from the volcanic activity, it is subject to weathering, reworking into sand and shale deposits, or perhaps dissolved and precipitated as cement or chemically altered silica product. As a clastic sedimentary rock, if volcanic ash comprises greater than 80% of the deposit, then the sedimentary rock would be called a *tuff*. Tuffaceous sandstones would be formed if 50–80% of the sand-sized materials are of volcanic origin.

CLASSIFICATION AND FRAMEWORK GEOMETRY OF CARBONATE AND CHEMICAL ROCKS

Overview of Carbonate and Chemical Sedimentary Rocks

The second primary group of sedimentary rocks is a diverse group mostly composed of weathering products that have been in solution for part or all of their transportation history plus some organic-based deposits. This group of sedimentary rocks is subdivided into two primary subgroups. The most common and widespread subgroup is the calcium- and magnesium-based carbonate rocks. These carbonates constitute the large family of limestones and dolostones. The second subgroup comprises the accumulations of nonclastic sedimentary materials that include the organic and chemical sedimentary rocks. This second, nonclastic, group of sedimentary rocks can be divided into three significant types: (1) the chemical precipitates (gypsum and halite), (2) the noncarbonate group consisting of the cherts and iron deposits, and (3) the organic plant debris accumulations comprising all the various kinds of coal deposits. The following discussion classifies and reviews these two diverse subgroups of rocks.

The first discussion is directed toward the large subgroup of carbonate-based sedimentary rocks. A classification of these rocks is presented, followed by a brief discussion of where and what kind of carbonate environments exist today. This is followed by some of the key attributes of these rocks in the field and environmental interpretations. Following is an overview discussion of the importance of cherts and coal deposits within the geological record and their depositional and environmental significance.

CARBONATE SEDIMENTARY ROCKS

Classification of Carbonate Sedimentary Rocks

The classification of these carbonate and chemical rocks are shown in Table 1.2. Carbonate sedimentary rock has been very abundant on the earth's surface. At times in the geologic past, especially as preserved in the Ordovician, carbonates are believed to have covered most of the continental surface areas. From Precambrian to recent times, limestones and dolostones are calculated to have made up approximately 25% of the total sediments deposited and preserved on the earth's surface.

TABLE 1.2 Classification of Organic and Chemical Sedimentary Rocks (Marine and/or Freshwater)

Texture	*Composition*	*Rock name*
Medium- to coarse-grained		Crystalline limestone
Microcrystalline conchoidal fracture		Micrite
Aggregates of oolites		Oolitic limestone
Shells and/or fossil fragments loosely cemented	Calcite ($CaCo_3$)	Coquina
Abundant fossils in calcareous matrix		Fossiliferous limestone
Shells of microscopic organisms, plus precipitated calcite		Chalk
Banded calcite		Travertine
Textures similar to those in limestone	Dolomite ($CaMg\ (CO_3)_2$)	Dolostone (dolomite)
Cryptocrystalline, dense	Silica (SiO_2)	Chert, opal, agate, flint
Fine to coarse crystalline	Gypsum ($CaSO_4 \cdot 2H_2O$)	Rock gypsum
Fine to coarse crystalline	Halite (NaCl)	Rock salt

Limestones are dominantly composed of the mineral calcite. Occasionally, a second mineral, aragonite, will be combined with the calcite in small quantities. Dolostone is primarily composed of the mineral dolomite. All three of these minerals are known as carbonates. Briefly, here are the comparative differences for these three carbonate-forming minerals:

Calcite (calcium carbonate: $CaCO_3$)—hexagonal, occurs in a variety of colors (often white to yellow to tan), hardness = 3, very common, actively reacts to hydrochloric acid, forms many invertebrate shells and is the dominant mineral of limestones.

Aragonite (calcium carbonate: $CaCO_3$)—orthorhombic, is a less stable form of calcite, mainly secreted by selected organisms to form skeletal frameworks.

Dolomite (calcium magnesium carbonate: $CaMgCO_3$)—hexagonal, slightly harder and denser than calcite, less reaction to acid, often mixed with limestone, often secondary in origin because magnesium is added under special environmental conditions, weathered texture or surface in outcrops often differs from limestone.

The main overall group of carbonate rocks may often be associated with common accessory carbonate-based minerals. When these accessory minerals are present in sufficient quantities they can be mined as important economic deposits. Examples of the most common economically mined accessory minerals are siderite (iron carbonate), rhodochrosite (manganese carbonate), smithsonite (zinc carbonate), witherite (barium carbonate), cerussite (lead carbonate), and two extensively mined copper carbonate minerals, malachite and azurite. Dissemination of these accessory carbonate minerals within localized accumulations of limestones or dolostones can add color to these mineral-bearing strata along veins, on the weathered surfaces of outcrops, or within the deposit itself (see museum mineral collections from mining areas like Bisbee and Morenci in Arizona, mines near Tsumeb, South Africa, and the Ural Mountains of Russia).

Rocks that have over 75–80% carbonate minerals are considered limestones. The remaining compositional material is often varying amounts of clay minerals, fine sands, and fossil debris. Calcite is also used by many invertebrate organisms in the making of shells and internal body structures; thus, fossils are often associated with many limestones, which tells us that the original organisms lived within the same environmental conditions (Figs. 1.20 and 1.21). Limestones, in general, can have various origins. This can easily be seen in Table 1.2, where several specific types of limestone have been named for special "unique" accumulations of clastic carbonate materials. Examples of these are the oolite-rich oolitic limestones, the calcareous microscopic organisms that comprise the chalks, and the fossil hash of the coquinas. In other cases, limestones are a direct crystalline calcite precipitate yielding common crystalline limestones called sparry or micrite limestones, depending on the developed crystal size. Lastly are the banded precipitates we associate with groundwater activity, like the travertine stalactite deposits in caves.

FIGURE 1.20 Fossiliferous limestone, Oklahoma.

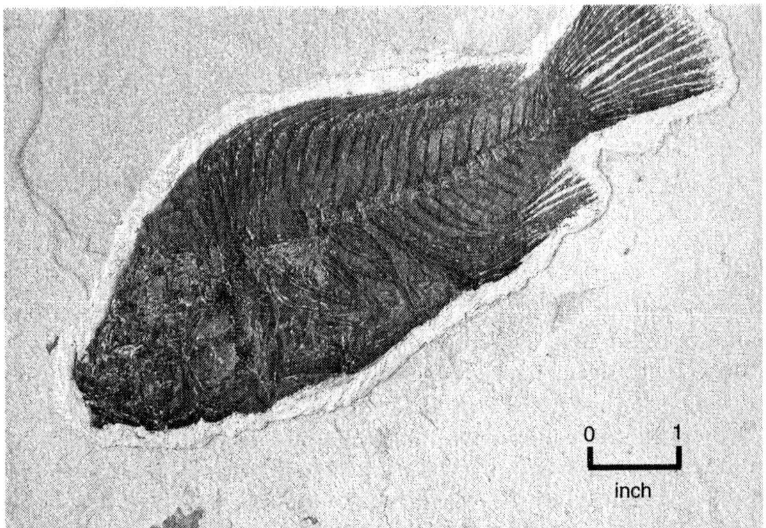

FIGURE 1.21 Fossil fish, Eocene Green River Formation, Wyoming.

Deposits of clastic carbonates often contain many of the same bedding and textures as clastic rocks, such as internal and bed boundary attributes and bedding features. Because of the chemical nature of the calcite-based carbonates, limestones and dolostones are often subject to postdepositional mineral replacement and diagenetic alterations. The fluids that can cause the changes may be associated with hydrothermal activity or perhaps groundwater, because both can easily transport chemical compounds in solution that may react with the original carbonates. The resulting alterations by chemical reactions to these secondary solutions may produce significant changes or complete loss of many of the original crystal, bedding, and fossil characteristics.

Modern Carbonate Environments

The volume of limestones in the geologic past from the late Precambrian through the balance of the Paleozoic is impressive, with thick widespread accumulations present across the majority of the world's continental areas. The overall depositional environments across the continents underwent significant changes during the Mesozoic, resulting in these massive limestone accumulations becoming measurably thinner and interbedded with extensive sequences of clastic rocks. Compared to the preserved volume of carbonates in the geologic past, especially during the Paleozoic, few carbonate environments exist today. After decades of study, geologists are convinced that their analysis of the limited varieties of modern carbonate and chemical systems available today can be successfully used as analogues for older, preserved carbonate systems. From these studies, for the majority of the carbonate developments and reef formation, the environmental restraints for the development of a carbonate depositional system could need clean, warm oceanic waters with active wave action and abundant organic activity, both plants and animals.

Across the world today, the variety of modern carbonate occurrences fall into four major groups that are used by geologists for interpretative and analogous analysis purposes. These four groups of carbonate environments are briefly examined next.

Shallow-Water Carbonates The vast majority of carbonates in the geologic record belong to this group of shallow-water carbonates. Some of the best modern examples occurring today are located in the Bahama Islands, the Florida Bay and Keys area, and the Campeche Banks of the Yucatan region of eastern Mexico and the Great Barrier Reef system along the northeastern coast of Australia. In modern shallow-water carbonate environments, the primary clastic constituents produced are calcareous sands composed of skeletal biologic debris, small spherical oolites and organic pellets developed in the tidal flats, and carbonate muds derived from algal materials in the very shallow waters of lagoons and mud flat areas.

> *Preserved, possibly large-scale, coastal margin, atoll-like reef and platform carbonate systems*: These are the largest of the modern carbonate-accumulation reefs and atolls. Both plants and animals are involved in the construction of a reef's wave-resistant framework of carbonate material. The modern reef framework is constructed by corals and invertebrate organisms that build a carbonate structure up above the surrounding shallow coastal oceanic bottom into surface waters where there is abundant wave activity and sunlight. These coastal reef buildups generally have three constructive areas: (1) The central highly fossiliferous *reef core* usually does not have preserved bedding. This core area can be very porous and is often dolomitic. (2) The rather narrow apron of wave-broken organic debris, the reef breccia, forms the *fore-reef* area on the deeper-water oceanfront side of the reef. One characteristic of this fore-reef zone is its steep bedding of up to 45° into the deeper water zones, perhaps due in part to slumping and compaction in an area where the reef debris quickly goes from surface waters to the deeper shelf areas. (3) The *back-reef*, or shelf lagoon, extends from the

active reef core back toward the tidal flats along the coastal margin. These back-reef lagoons are areas where carbonate muds are formed and accumulated; they seldom contain the larger reef-type invertebrate reef builders. The bedding in these platform carbonate sediment accumulations is usually flat-lying. Modern examples would be the Great Barrier Reef of Australia and some of the platform reefs of the Yucatan Peninsula of Mexico. Many ancient barrier reefs have been identified in the sedimentary record, but one well-documented barrier reef carbonate accumulation famous in the United States is the Permian Reef in southwest Texas.

Patch reefs, such as the bioherm and biostrome deposits: A smaller type of carbonate buildup comprises nonreef carbonates that have a much smaller size and are more restricted in their dimensions. In outcrops, these smaller nonreef buildups are typically surrounded by rock materials of a different composition. These structures are associated with intense organic activity in a localized area. These organic limestone buildups have been associated with a large accumulations of crinoids, bryozoans, algae, stromatoporoids, mollusks, or corals that form an organic colony on the seafloor. The organic activity and its associated debris may build up enough to act as a current barrier within an area. These buildups are preserved as lenticular carbonate bodies in a sequence of sands or shales. If the top and bottom of the carbonate buildup are essentially the same as major bedding planes for the surrounding sands or shales, the buildup is called a *biostrome*. If, however, the top of the buildup continues upward through several sand or shale layers, the dome or mound-shaped buildup is called a *bioherm* (Fig. 1.22).

FIGURE 1.22 Cambrian bioherm, Llano River, Mason, Texas.

Smaller algal stromatolitic accumulations: The smallest organic carbonate accumulation variety is the pinnacle reefs or moundlike buildups of limited areal extent called *stromatolites*. The small cyanobacteria algal mounds found in Hamelin Pool of the Shark Bay World Heritage Area, Western Australia, have been suggested to be modern analogues for the 2–3 billion-year-old shallow-water blue-green algal stromatolite structures found in the carbonates and iron ores formed in the Precambrian. The vertical sections of the Precambrian stromatolites, as seen in outcrops, have the same cabbage-like domal or columnar-internal features as those mounds in Shark Bay. The algae colonies trap the carbonate muds; after lithification as a limestone, the stromatolites appear as apparent fossils.

A special modern shallow-water carbonate environment exists in the Middle East. Here, the shallow-water carbonate environment is associated with the hot, shallow evaporitic tidal flats along the coastlines of the Persian Gulf countries. These very low-relief shoreline tidal flats, called *sabkha* flats, are covered daily during high tide by a shallow sheet of seawater. Later, as the tidal waters slowly retreat back across these wide, flat areas during low tide, evaporation takes place. During the waters' retreat, algal growth, mud cracks, and other irregularities can trap some the returning waters still on the flats. The high temperatures cause quite rapid evaporation rates that result in concentration and eventual precipitation of the salts and carbonates in solution.

A third major group of shallow-water carbonate systems of significant interest to historical geologic studies includes the following subgroups.

Deep-Sea Carbonates Deep-sea carbonates are not associated with the deepest regions of the ocean floor below approximately 4000-meter water depth. At these deep, high-pressure, low-temperature environments, the carbonate ions would remain in solution. However, in some instances, along coastal regions, skeletal material is transported into deeper waters along coastal margins by turbidity currents. Of interest to historical geology, ocean floor exploration, and oil exploration are the accumulations of planktonic foraminifera shells and pelagic oozes or muds including pteropod and globigerina remains. While not common, these carbonates often make good marker layers used in regional stratigraphic correlations and interpretations of offshore depositional environments.

Evaporitic Carbonates Evaporitic carbonates are associated with continental areas where arid climates are present. In these arid conditions, water may temporarily accumulate after rains in flat, open areas, and relatively rapid evaporation can take place. Not only do the ephemeral surface waters evaporate, but as the ground heats up it causes the moisture in the soil to move toward the surface by capillary action. As these groundwaters are evaporated, a residue of carbonate-rich materials is built up. While these are minor contributors to the vast volume of carbonates, they are distinctive. These impure lime-rich evaporitic deposits are known as *caliche*. Geologists use the presence of these limestones as climatic indicators when interpreting paleoenvironments.

Freshwater Carbonates This group of carbonates is associated with freshwater. The common sources for these carbonate deposits are in freshwater lakes, hard-water springs, and fumaroles, where both supersaturations and evaporation cause deposits like tufa to be formed. Additionally, thick, banded layers of dense freshwater carbonates, known as *travertine,* can form in caves. In caves and underground passageways, travertine can take many very ornate forms, such as stalagmites, stalactites, and many strawlike features.

Textures and Structures of Carbonate Rocks

Sparry Calcite and Micrite Sparry calcite and micrite muds are on opposite ends of the crystal size scale. Sparry calcite is coarsely crystalline calcite where crystal boundaries can be seen. These larger crystal sizes usually are not present within the rocks except as vein fillings and often as calcite filling of voids, such as the internal cavities of fossils. The micrite muds, on the other hand, are carbonate muds composed of microcrystalline calcite ooze. The micron-sized crystals are derived from precipitation in seawater in the shallow flats and then are trapped on the flats in beds of algal growth. Mixed in with the micrite ooze may also be very fine aragonite crystals produced both by some invertebrate animals and within plants, with the crystals being released upon their death. The lithified rock composed of these carbonate muds is termed a *micrite*. Micrite limestones found in outcrops are usually very uniform, dense, with buff to grey to black colors, and have a characteristic conchoidial fracture when chipped.

Secondary Porosity Limestones usually do not have significant porosity. However, since limestones are subject to solution by acidic waters, secondary solution of limestones after deposition (often after fracturing related to folding or faulting) can take place, creating porosity and even caves. Additional porosity within a carbonate rock unit can occur during the process of dolomitization. Chemically active hydrothermal fluids can cause recrystallization and replacement of crystals. The secondary porosity is very important when it is associated with increased permeability, especially in the production of hydrocarbons, for example.

Oolites and Pellets These two terms are related to small particles formed in shallow waters with wave-induced current activity. The *oolites* refer to spherical balls of calcite with concentric layering. An individual piece of material is an oolith, and a rock made of these pieces is called an *oolitic limestone*. These pieces form in waters where carbonate is being precipitated and are rolled back and forth on the bottom of the shallow water. *Pellets* may be similar in size, usually about half a millimeter in diameter; however, they have no internal structure and may be clots of micrite mud or perhaps fecal material from organic activity. Thin sections of limestones often clearly show these features along with the fossil hash, sparry calcite, and micritic muds.

Limestones can be found across the earth's surface. Carbonates often form as a series of thick layers in outcrops that can be quarried and used as an important source of stone for both construction and facing of buildings. Another related carbonate building stone is the metamorphosed limestone, marble. Perhaps even more important is the crushing of carbonates for the production of concrete powder and aggregate.

Fossils The majority of carbonates have been accumulated within marine environments in the geologic past, particularly in the Paleozoic. This long Paleozoic interval of geologic time was also the heyday of the invertebrate animals. The vast array of invertebrate fossil types and ages is explored in greater depth in Chapter 4. However, fossil (or fragments of fossils) associated with these shallow, relatively clean-water environments present during limestone accumulations can be a key portion of the rock's fabric. The fossils in limestones serve as environmental indicators, and many fossils associated with carbonates are excellent index fossils, which are very useful in developing relative time for sequences of rocks and tectonic events.

Chemical Precipitates and Organic Deposits

A complex suite of minerals contributes to the formations of these separate groups of chemical precipitates and evaporitic sedimentary rocks. These deposits are primarily developed through evaporation of aqueous systems. The evaporation concentrates the ion-bearing solutions into brines; later, after precipitation, evaporitic deposits are formed. These evaporitic deposits fall into three primary groups: the previously discussed carbonates (calcite, magnesite, and dolomite), the chlorides (primarily halite or salt), and the sulfates (primarily anhydrite and gypsum). The purely chemical precipitates, such as gypsum, anhydrite, and halite, are generally formed under special environmental conditions where ionic concentration, evaporation, and precipitation of these chemical compounds occur. These sedimentary rocks are often accumulated in economic quantities. Gypsum occurs not only in thick beds interbedded with limestones and shales, but also as layers underlying salt deposits, one of the first minerals to be formed in an evaporitic environment. It also commonly occurs as secondary recrystallized deposits filling cross-cutting veins, where it is called *satin spar*.

Another group of noncarbonate chemical precipitates of importance contains silica cherts, banded irons, the phosphorites, and a specialized iron silicate called *glauconite*.

Organic material preserved in the geologic record comprises only a very small percentage of the total accumulated sediment volume, but it is important in both economic and historical geology terms. This group of rocks contains materials that originated as growing organisms, especially plants; these are called *carbonaceous* materials. These materials contain varying amounts of carbon. Carbonaceous materials (carbon-bearing) should not be confused with the carbonates (calcite-bearing). Carbonaceous organic materials preserved in sediments can range from small residues of individual sticks or leaves to the development of massive coal deposits. The following subsections briefly review the significance of cherts and organic materials in historical geology.

Chert There is no definitive classification of chert. Chert occurs in outcrops primarily as either nodules or thin layers in thick, massive zones. The nodules are usually parallel to bedding and often associated with limestones and shales. The nodules can be of nearly any shape but are often dense, rounded masses of cryptocrystalline (grains too small to be seen with a microscope) silica that is resistant to weathering and exhibits conchoidial fracture. The colors range from white to black to various shades of grey, tan, red, and yellow. Locally, chert goes by the name *jasper, chalcedony, Tripoli*, or, if bedded, *novaculite*. The origin of chert is ambiguous because it occurs with some tuff-bearing rocks, deep-sea sediments associated with sponge spicule accumulations, and amorphous to fossiliferous silica precipitates. Chert fragments can be found in all sizes of clastic rock particles.

Coals Coals start as organic materials such as woods, plants, and peat. The composition is primarily carbon, hydrogen, and oxygen in varying proportions, with some minor constituents like nitrogen and sulfur. Coals are quite rare in the geologic record, with the first significant coals found in the Devonian, probably associated with the development of the woody plants at that time. The great development of coals was in the Carboniferous (or the Coal Measures), primarily during the Pennsylvanian. Coals are combustible, opaque noncrystalline solids. Coals have been classified according to a rank based on the degree of change or metamorphism of the initial constituents.

The lowest grade is lignite, followed, in order, by subbituminous, bituminous, and the highest coal grade, anthracite. The lowest-grade coals are brown, never black, in color. They display the basic structures of the original woody material and burn with a smoky fire. Most lignites, or brown coal, are Cretaceous or younger in age. Bituminous coals are generally higher in carbon content and contain less water. The bituminous coals burn easily and are the most common coal being mined. Bituminous coal beds are mined in the subsurface as well as in large, open-pit strip mines. Bituminous coal has banding or layering and both bright layers and dull layers when a fresh break is viewed.

Anthracite coal is the highest grade of coal. It is harder to ignite but burns with the most heat and little smoke. It has the highest carbon content and has a bright luster and a conchoidial fracture.

SEDIMENTARY ENVIRONMENTS OF DEPOSITION

Overview

Now that the primary facets of clastic, carbonate, and chemical sedimentary rocks have been reviewed, the next vital step in understanding historical geology is to use these data to develop insight into the interpretation of the many depositional environments across the earth's surface. Features of bedding, particle size, sedimentary features, composition, and incorporated fossils are all clues to the understanding of a suite of sedimentary rocks. Fossils found within beds of rock can be interpreted in three principal ways. First, fossils can help define the type of sedimentary environment of deposition. Many fossils live only where specific environmental conditions prevail. For example, corals are most often associated with shallow, nearly sediment-free, circulating ocean waters along the continental shelf or island margins, where plant and animal nutrients are abundant. Therefore, geologists would not expect to find corals associated with river or delta front sediment accumulations. Second, index fossils can be traced in different sedimentary rock systems in different geographic areas to develop a perspective of how complex rock units can be related in space and time. And third, changes in the fossils' distribution, morphology, and associations can signify changing environmental and depositional conditions through time.

Sedimentary rocks from the geologic past can be found in rock outcrops worldwide. These layered sedimentary rocks are the pages of a geologic "history book." The "pages," or sedimentary layers, in outcrops are the preserved information available for interpretation of life forms, climates, and environments of the geologic past. If we want to interpret and map the locations of seas, mountains, deserts, lakes, rivers, swamps, and coastlines in the geologic past, for example, in the Devonian (approximately 400 million years ago), we must catalog and map the preserved Devonian-aged strata in a given research area. After mapping, the detailed analysis of those sedimentary aspects and features of the mapped Devonian must be interpreted. Paleogeographic and interpretive maps can be constructed using analoges derived from modern geologic environments as tabulated in Fig. 1.23.

Sedimentary environments are commonly subdivided into the three general categories: (1) nonmarine (terrestrial), (2) transitional, and (3) marine. These environments, their sediments, and their typical sedimentary features are described in Table 1.3. The geologist can use these summary data for the various environments for the mapping of the development of lithofacies and environments of deposition mapping.

TABLE 1.3 Sedimentary Environments

Environment	*Common sediment*	*Sedimentary features*
NONMARINE (terrestrial)		
Fluvial (stream)	Gravel, sand, mud	Mudcracked silty flood-plain deposits Round pebbles Lenses of sandstone Cross-bedding
Lacustrine (lake)	Silt, clay, lime, mud	Laminated beds Freshwater fauna
Paludal (swamp)	Silt, clay, organic debris	Decayed vegetation, peat
Desert		Well-sorted material
Dune	Sand	Frosted, rounded grains Large-scale cross-bedding
Playa lake	Mud, evaporite salts	Laminated beds Mud cracks
Alluvial fan and debris flow	Boulders, gravel, sand	Poorly sorted arkose Beds of limited extent Lenticular or wedge-shaped units Cross-bedding
Glacial		Unsorted debris
Ice deposits	Morainal debris of all sizes	Striated boulders
Outwash plain	Sand, silt, clay	Braided stream deposits
Glacial lake	Silt, mud	Varved (annual) layers
TRANSITIONAL		
Delta	Sand, silt, mud, wood fragments	Delta plain: sand channels, coal, silt Delta front: silt, wood fragments, marine fossils Prodelta: mud, marine fossils
Beach and barrier island	Quartz sand Carbonate/shelly sand	Well-sorted cross-bedding Ripple marks Shell material
Coastal lagoons, bays, estuaries	Silt, mud, sand	Thin beds Brackish-water founa
MARINE		
Supratidal (typically) broad flats (sabhkas) occasionally flooded by storm surge	Evaporites, dolomite, silt, mud	Laminated beds Mud cracks Stromatolites (carbonate areas)
Littoral (intertidal)	Cobbles, sand, mud	Laminated beds Mud cracks Trace fossils Stromatolites (carbonate areas) Well-sorted sands
Sublittoral shelf	Sand, silt, clay	Wide geographic extent Thin-to-massive bedding Cross-bedding Diverse fauna

TABLE 1.3 *Continued*

Environment	*Common sediment*	*Sedimentary features*
Carbonate platform	Calcareous sediments (lack of land-derived clastic debris)	Ooids Shell fragment sands Lime mud Intraformational conglomerates
Organic reef	Shelly limestone	Porous limestone Fossiliferous Reef talus breccias
Bathyal depths (slope, rise) (3000–12,000 feet)	Mud Sand and gravel in submarine fans	Graded bedding Thick, coarse-grained beds Sole marks Deep-water fossils
Abyssal depths (abyssal plains) (more than 12,000 feet)	Mud, ooze (biogenic)	Relatively thin and even bedding Planktonic fossils

Source: Modified from: *Trip Through Time*, 2/e, by A. Cooper et al., © 1990. Reprinted by permission of Prentice-Hall, Inc., Upper Saddle River, NJ.

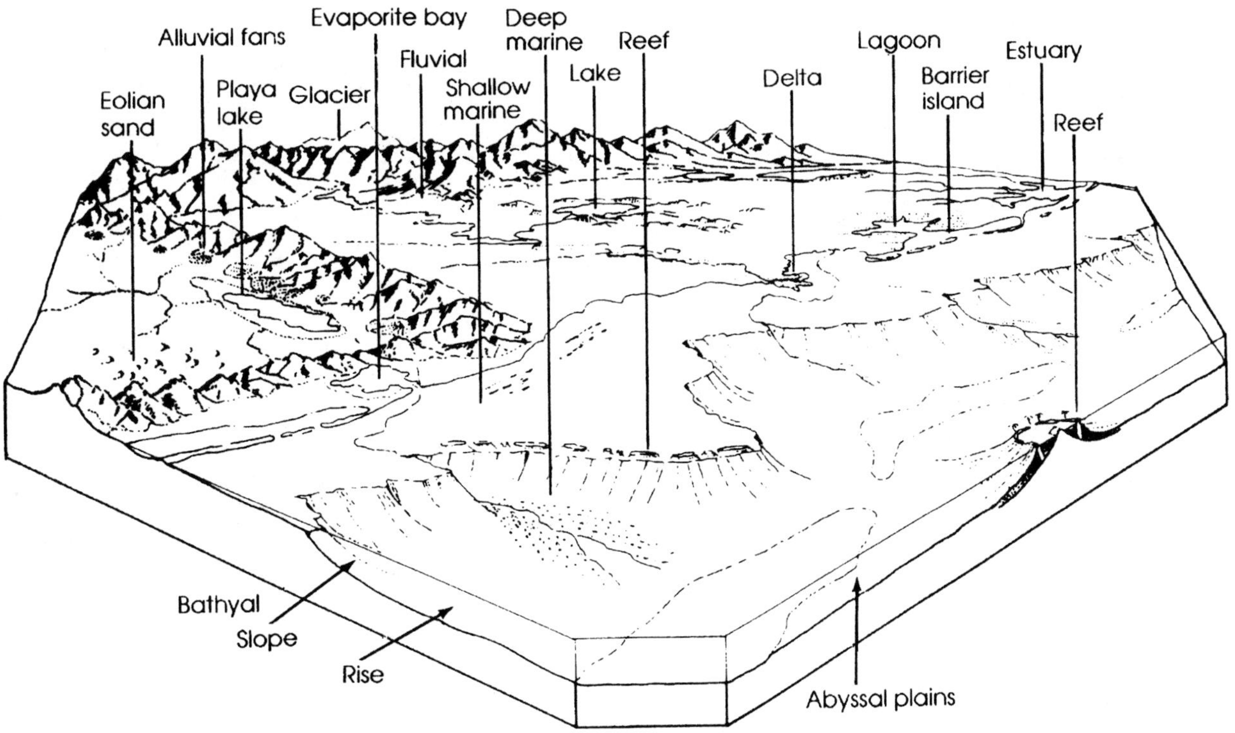

FIGURE 1.23 Modern sedimentary environments. Source: *Exercises in Physical Geology* by Kenneth Hamblin, © 1980. Reprinted by permission of Prentice-Hall, Inc., Upper Saddle River, NJ.

EXERCISES

Exercise 1-1 CONTINENTAL SEDIMENTARY ENVIRONMENTS, UTAH

This problem asks you to interpret continental sedimentary environments from field data collected in Utah. Use Table 1.3 for reference.

You are a secondary school science teacher for most of the year but have a job as a park guide at Dinosaur National Monument in northeastern Utah during the summer. Toward the end of the summer, an inquisitive tourist (a rare variety) remarks, "You have just explained that Morrison Formation, in which the dinosaurs are found, is made up of continental deposits. Does this mean that dinosaurs are always found in continental deposits like sand dunes?" You reply, "Since the dinosaurs were large terrestrial reptiles, they would most likely be found in continental deposits." The tourist is not satisfied and says, "Give me some additional evidence, other than the dinosaurs, that the Morrison Formation is continental." Fortunately, another tourist interrupts, but the question has posed a problem for you because the same question may well come up again. To avoid being caught off guard a second time, you tour the park and describe the rock types of the Morrison Formation in various places. Listed next are your descriptions.

Several indicators of terrestrial depositional environments are found in your samples. In the space after each, indicate what they are, and evaluate the strength and weakness of the evidence.

Location 1. Shale and siltstone, finely laminated, sandy layers quite common, some fragments of shale with raindrop imprints on the surface.

Location 2. Shale and siltstone, finely laminated, some pebbles, some pollen grains found with a microscope.

Location 3. Shale, blocky, red, some nodules of gypsum, a few lenses of very finely cross-bedded sandstone with asymmetrical ripple marks.

Location 4. Siltstone, some shale and sand, a few thin beds of conglomerate with fragments of dinosaur bones, clams, and tortoise shell.

Location 5. Sandstone, well-sorted, fine-grained, evidence of large-scale cross-bedding, frosted sand grains, a few thin layers of shale.

Location 6. Claystone, dark grey, platy, a few fragments of large leaves.

Location 7. Sandstone, well-sorted, fine-grained, a few pieces of petrified wood present.

Location 8. Siltstone and shale, grey, a few snails, bivalves, and ammonite fragments.

Location 9. Coarse sandstone, well-defined cross-bedding, crocodile bone fragments.

Exercise 1-2 PALEOGEOGRAPHIC MAP

This exercise combines sedimentary environment analysis with the creation of lithofacies maps. Compare the descriptions of rocks in Table 1.4 with the sediment and sedimentary features columns in Table 1.3. Determine the depositional environment for each of the rock descriptions, and write the depositional environment's name next to each number on the map in Figure 1.24. Then draw a boundary line to separate the depositional environments from each other, making a map of the area's paleogeography.

TABLE 1.4 Sedimentary Rock Descriptions

Localities	*Rock descriptions*
1, 2, 3, 17, 18, 19, 20	Poorly sorted arkose sandstone, wedge-shaped deposits, cross-bedded
4, 5, 6, 7, 8, 9, 14, 15, 16, 21	Well-sorted sand, frosted grains, large-scale cross-bedding
10, 11, 12, 13	Laminated muds, evaporites, mud cracks
22, 23, 24, 25	Sands, silts, clays, braided stream deposits
26, 27, 28, 29, 30, 31, 32	Unsorted conglomerates, morainal deposits, striated boulders, striated bedrock

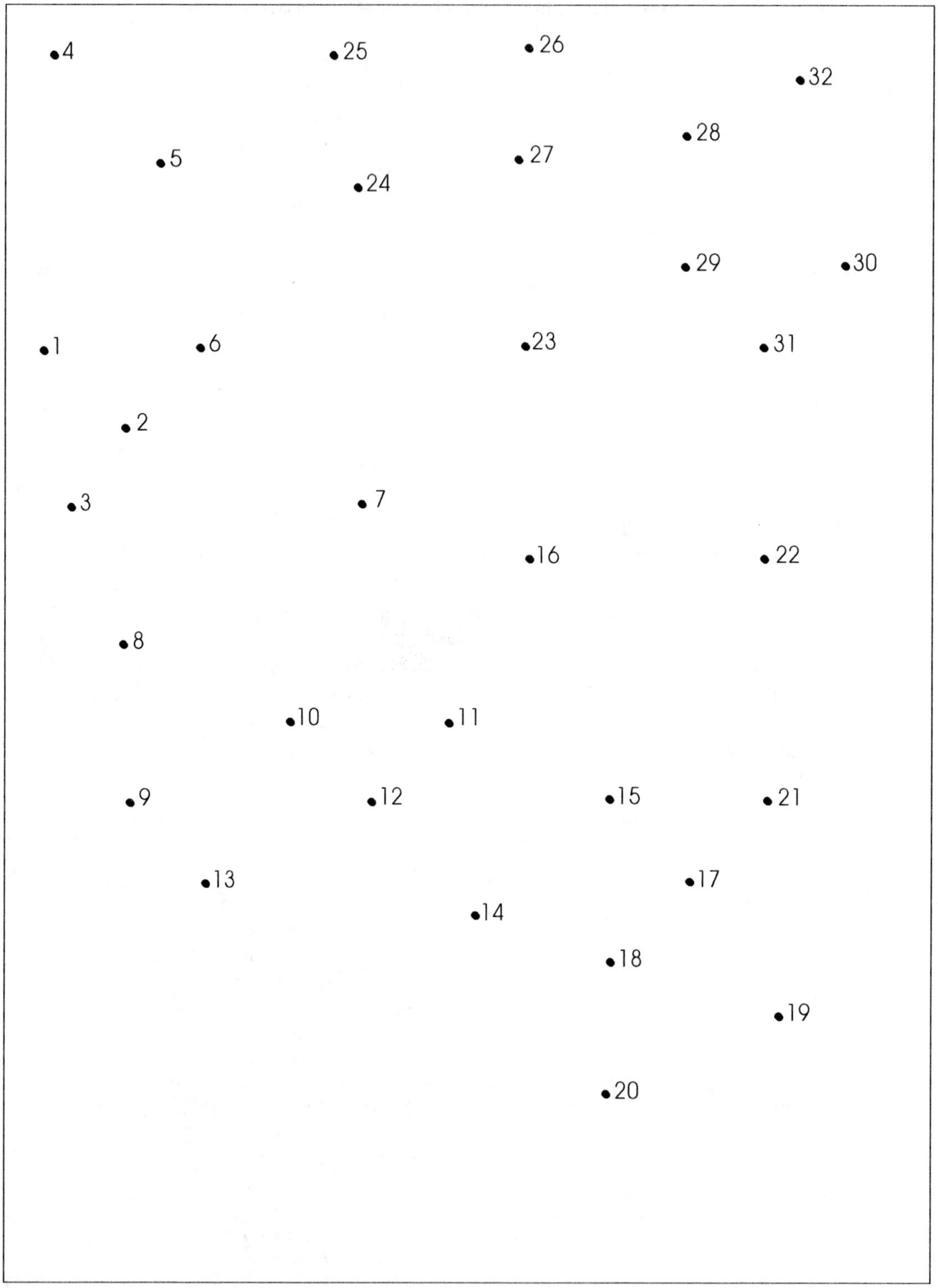

FIGURE 1.24 Paleogeographic map of sedimentary environments.

Exercise 1-3 PALEOGRAPHIC MAP OF CHINLE FORMATION, UTAH

This problem involves a selection of data accumulated to define the location for part of an old continental channel system within a basin. The drilling in the area has been designed to target the distribution of the sandstone as it occurs in the subsurface. The database, Table 1.5, has been derived from both shallow drilling and a few outcrops. In addition to the database in Table 1.5, a location map, Figure 1.25, is provided for the data points. Figure 1.26 shows an outcrop cross section of a channel sandstone within the Chinle Formation.

The Chinle Formation is interpreted as a widespread fluvial and floodplain sedimentary unit comprised of shale, channel sandstones, and some local conglomerates.

Typically, the unconformably underlying Moenkopi is considered to be a marine shale and sandstone unit with some lenses of limestone. From these data points, prepare a paleogeographic map of the Chinle Formation in this region.

TABLE 1.5 Chinle Formation Sample Database

Sample number	*Lithologic description*	*Sand thickness (feet)*	*Sample number*	*Lithologic description*	*Sand thickness (feet)*
1	Shale	0	31	Sandstone	11
2	Sandstone	11	32	Sandstone, some	1
3	Sandstone, some shale	4	33	Shale	0
4	Sandstone some shale	1	34	Sandstone,	7
5	Sandstone	9	35	Sandstone, coarse	18
6	Sandstone	11	36	Sandstone	9
7	Sandstone,	6	37	Sandstone	7
8	Shale	0	38	Sandstone	9
9	Shale	0	39	Sandstone,	14
10	Sandstone some shale	1	40	Shale	0
11	Sandstone	14	41	Shale	0
12	Shale	0	42	Shale	0
13	Sandstone,	7	43	Sandstone	4
14	Shale	0	44	Sandstone course	16
15	Sandstone	12	45	Sandstone	14
16	Sandstone	9	46	Sandstone	7
17	Sandstone	7	47	Shale	0
18	Sandstone	8	48	Shale,	0
19	Shale	0	49	Shale	0
20	Sandstone,	6	50	Sandstone coarse	18
21	Shale	0	51	Shale	0
22	Shale	0	52	Shale	0
23	Shale	0	53	Sandstone	4
24	Shale	0	54	Sandstone	11
25	Shale	0	55	Sandstone,coarse	16
26	Sandstone some shale	1	56	Sandstone	8
27	Sandstone	9	57	Sandstone some shale	1
28	Sandstone	13	58	Shale	0
29	Sandstone, couse	16	59	Sandstone	20
30	Sandstone couse	16	60	Shale	0

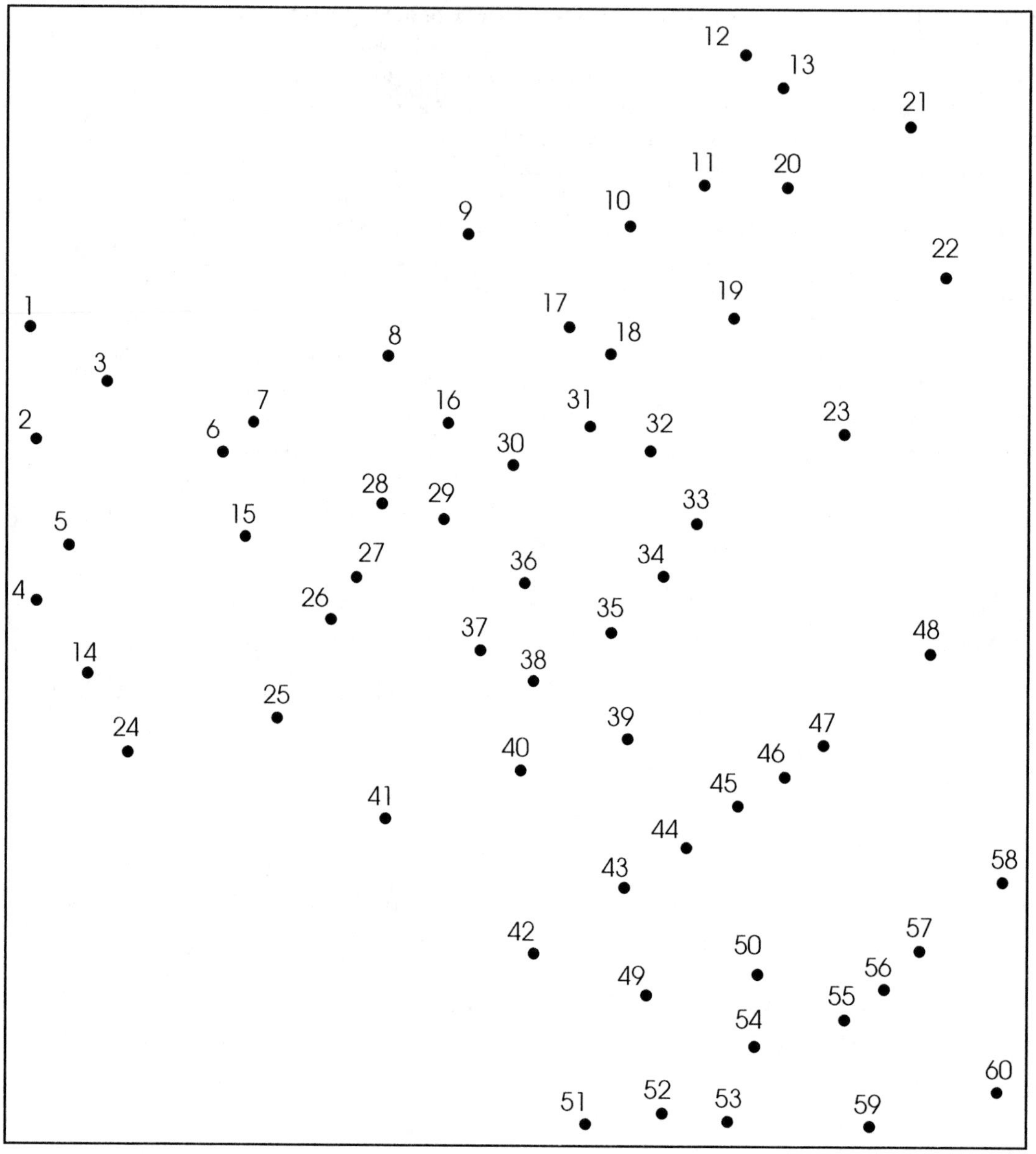

FIGURE 1.25 Location map for Chinle Formation sample database.

a. Transfer the data from Table 1.5 onto the location map, making note of the sand thickness and the lithologic composition of the sample.

b. Contour the data (using a contour interval of 5 feet) to develop an isopach map of the Chinle Formation thickness. Remember that you are trying to define a channel sandstone while contouring.

FIGURE 1.26 Upper Triassic Chinle Formation sandstone unconformably cutting down into the underlying Lower Triassic Moenkopi Formation.

c. The resulting map will indicate the shape and thickness of the channel sand and also result in the construction of a paleogeographic map at the time of deposition of the Chinle Formation in this specific area.

Questions

1. Describe the depositional nature of the Chinle sandstone as developed by the isopach map.

2. What kind of surface exists between the Chinle and Moenkopi Formations? Cite evidence for your conclusion.

3. Draw an arrow on your map showing the direction of the stream flow as derived from the contouring.

Exercise 1-4 CARLSBAD CAVERNS, NEW MEXICO

This exercise gives you the opportunity to interpret a carbonate sedimentary environment. You are working in the business office of Sage Brush Electric Company in Alpine, Texas. To supplement your income, you are an active land speculator in southeastern New Mexico and in the adjacent west Texas region. During a lunch break, you overhear a lineman for the company relating the possible discovery of a large system of caverns that may be even larger than those in Carlsbad Caverns National Park. He also mentions an approximate location. As soon as possible, you get the land ownership plots for the area immediately surrounding the lineman's discovery. In college you took geology, and you know that the west Texas region is famous for a large structure called the Permian Reef. In your old notes you find a cross-sectional sketch of the Permian Reef. Immediately you begin to compile pertinent facts on the Permian strata from the numerous dry oil tests in the region.

The three objectives of the task before you are to (1) identify the distribution of the often-cavernous Capitan Limestone in the subsurface, (2) define the orientation of the main reef limestone mass, and (3) locate any areas of high porosity within the reef mass that could form caverns. Once these objectives have been achieved, you can begin your speculative land purchases with some assurance of potential success.

DATA

1. A cross section of the Permian Reef (see Fig. 1.27).
2. Localities where data were collected (see Fig. 1.28).
3. Lithologic descriptions of the Permian strata at each locality.

SAMPLES

1. Anhydrite, white, sparsely fossiliferous, contains some salt.
2. Shale, grey, few fossils, some anhydrite.
3. Anhydrite, white, sparsely fossiliferous, some dolostone.
4. Limestone, grey, very fossiliferous, very porous.
5. Limestone, grey, very fossiliferous, very porous.
6. Shale, grey, a few fossils.
7. Limestone, grey, very fossiliferous, some dolostone and anhydrite lenses.
8. Limestone, grey, fossiliferous.
9. Limestone and shale, black to grey-black, clastic, very fossiliferous.
10. Limestone, black, dense, no fossils.
11. Limestone, grey-black, very fossiliferous, porous.
12. Limestone, dark grey, very fossiliferous, dense, some breccia zones.
13. Limestone and shale, black to grey, a few fossils, breccia.
14. Shale, black, platy, siliceous, nonfossiliferous.
15. Shale, black, platy, siliceous, nonfossiliferous.

a. On the location map, Figure 1.28, construct a lithofacies map of this region. Label the fore-reef, reef, and back-reef areas. (See the Permian Reef cross section, Fig. 1.27)

b. In which direction was the shoreline (land) during this interval of the Permian?

c. Contrast the water conditions that probably existed in the northwest corner of the map area with those in the southeastern corner.

d. Where would you make your speculative land purchase? (Show the location on the lithofacies map.) What were your reasons for choosing this area?

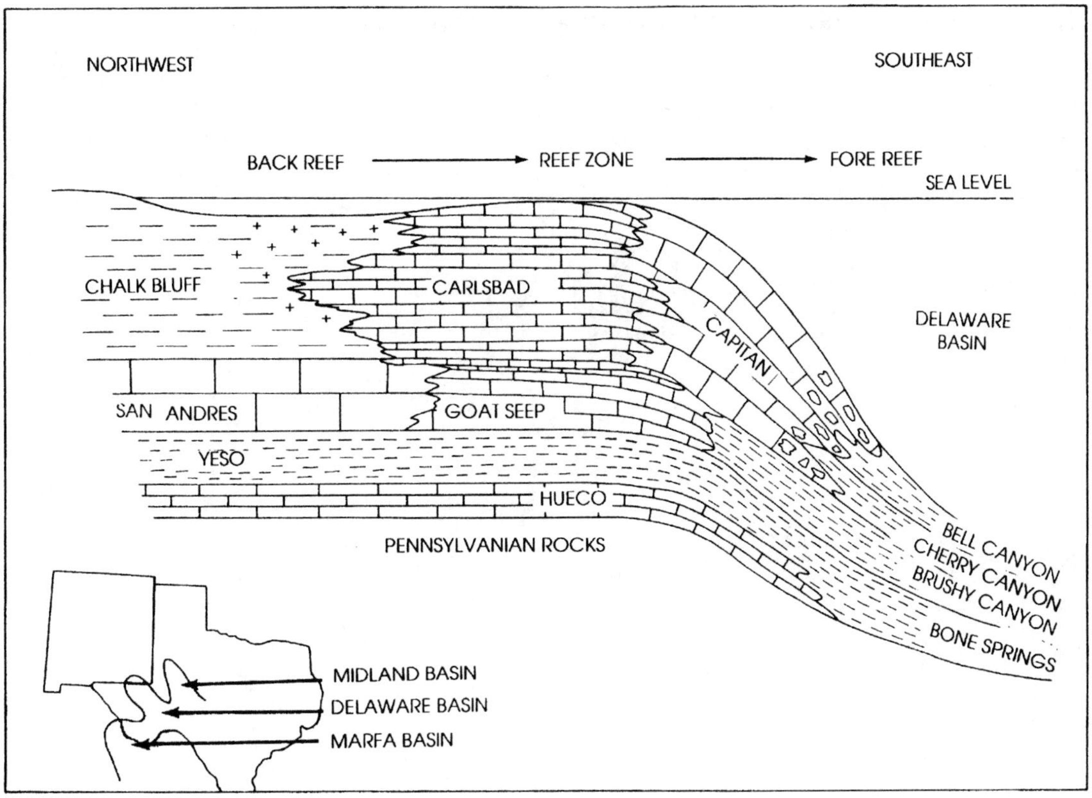

FIGURE 1.27 Cross-sectional view, Permian Reef.

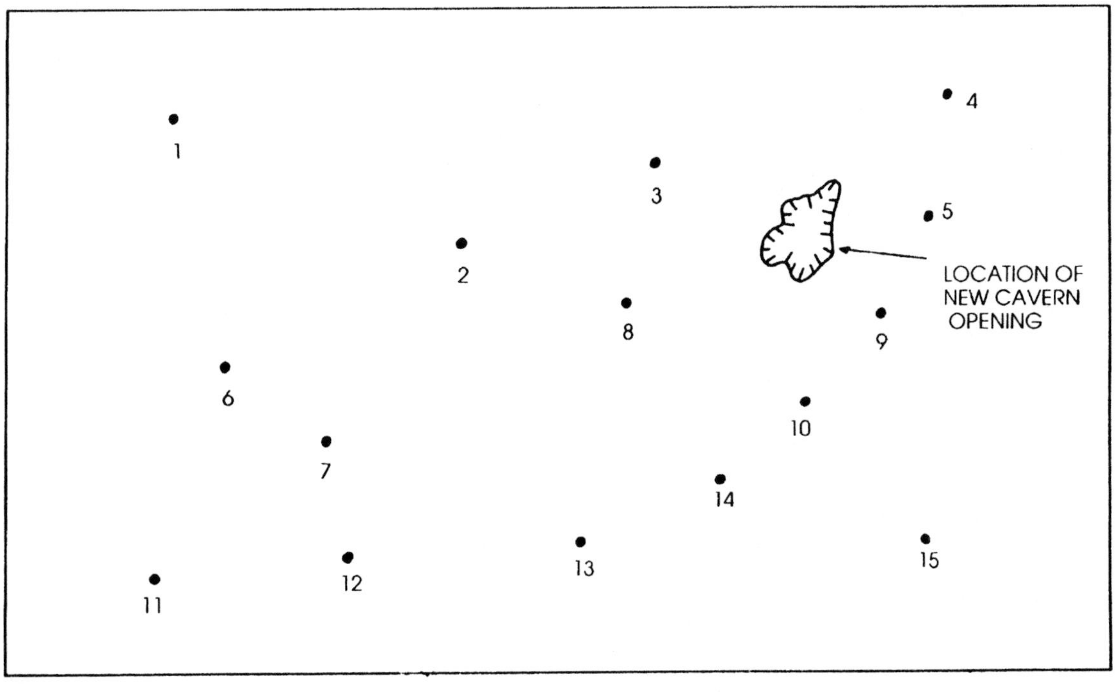

FIGURE 1.28 Map view of sample data locations.

Exercise 1-5 ENVIRONMENTAL ANALYSIS, MISSISSIPPI RIVER DELTA

As president of Southern Oysters, Inc., you are responsible for operating a small fleet of fishing boats in the East Bay area of the Mississippi River Delta (see Fig. 1.29). You are concerned over reports that, following a recent hurricane, the productivity of oyster beds is declining. You therefore consult the captain of your fleet and decide to search for new oyster beds in adjacent parts of the delta. You instruct your captain to look for sandy bottom areas, because oysters can tolerate no more than 10% clay in the bottom sediment.

Your captain returns with samples collected from a prospective harvest locality in Blind Bay at the mouth of Pass a Loutre near Thomasin Lumps (see Fig. 1.30 for locations). Here are the numbered locations and descriptions of the captain's samples.

1. Sand
2. Muddy sand
3. Muddy sand
4. Mud
5. Silty, dark grey sand, gastropods, small crabs
6. Muddy sand
7. Muddy sand
8. Mud
9. Mud
10. Silty sand, sea pansies, echinoids, bivalves
11. Clean, well-sorted sand; shrimp, oysters
12. Mud
13. Mud
14. Well-sorted, fine-grained sand; a few oyster shells
15. Sand, several bivalves, echinoids, small starfish
16. Silty grey sand
17. Muddy sand
18. Muddy sand, gastropods, bivalves, shrimp
19. Silty sand
20. Mud, black
21. Mud
22. Muddy sand
23. Silty sand
24. Mud
25. Mud
26. Muddy sand
27. Mud
28. Mud
29. Muddy sand
30. Silty sand

Questions

1. Using Figure 1.30 as a base map, draw a sediment distribution map showing clean sand, silty sand, muddy sand, and mud. (See Fig. 3.7.)

2. Outline with a red pencil the approximate boundary of the potential oyster bed. Describe its bottom sediments.

3. Starfish are a natural predator of oysters. Suggest how a starfish could kill and eat such a bivalve.

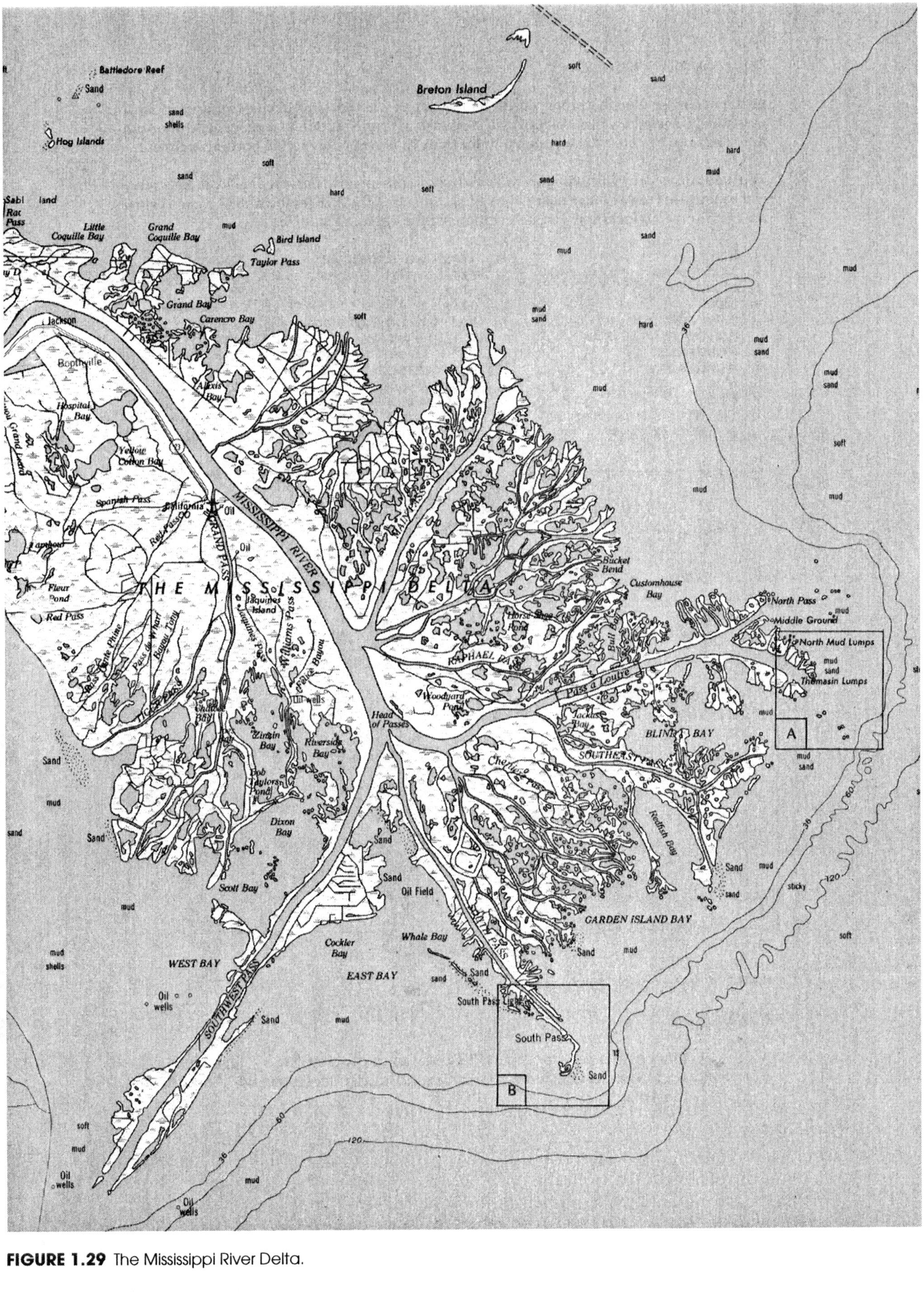

FIGURE 1.29 The Mississippi River Delta.

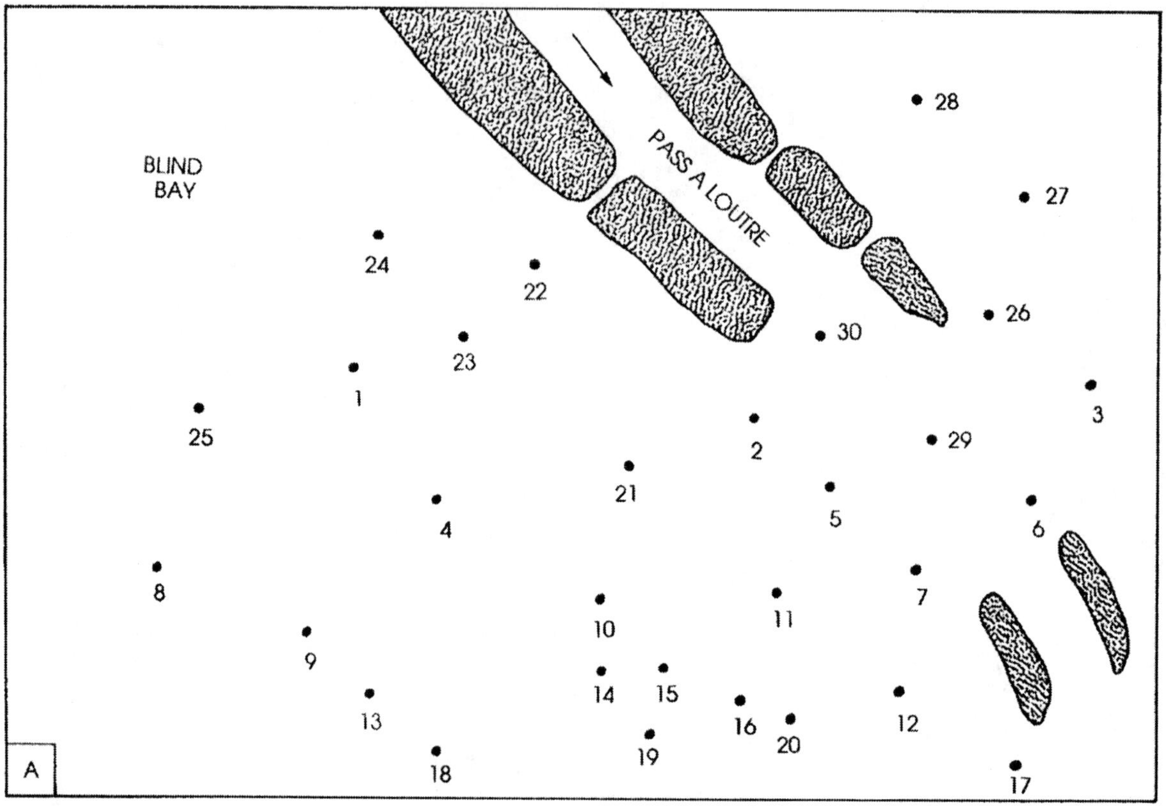

FIGURE 1.30 Sample location map.

4. What gives the muddy samples the dark grey to black coloration?

5. Why is sand concentrated at the mouth of Pass a Loutre, Southeast Pass, and South Pass?

6. Why is the sand concentrated on the southern side of the passes? (Note the spit development at South Pass, area **B** on the U.S. Geological Survey map, Fig. 1.29.)

7. Pass a Loutre is one of the principal distributaries on the Mississippi Delta. How does such a distributary network differ from a normal dendritic stream drainage pattern?

Exercise 1-6 MODERN COASTAL SEDIMENTS, NORTH CAROLINA

Figure 1.31 represents a coastal area with a river emptying into an estuary (Pamlico Sound), which in turn is separated from the open ocean by a series of offshore bars (barrier islands). The location of this map is easternmost North Carolina (Cape Hatteras area), but

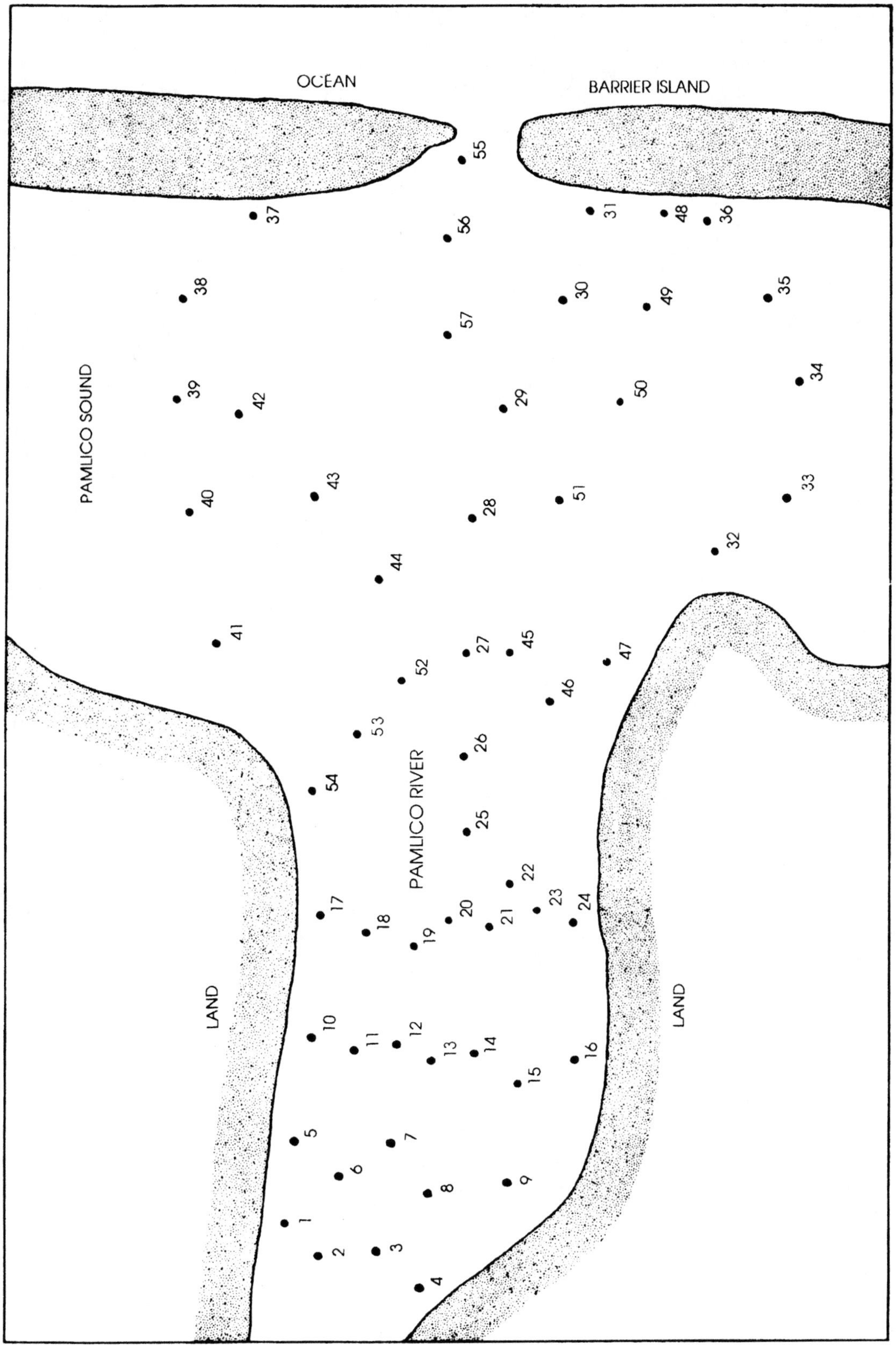

FIGURE 1.31 Sample location map.

it could represent many areas along the Atlantic and Gulf coasts of the southeastern United States that have well-developed spits and barrier islands.

To understand past geologic history, the geologist must have a working knowledge of the present conditions of deposition and depositional environments. The following problem will enable you to gain insight into the relationships of the depositional process, sources of sediments, and distribution of sediments in a modern environment.

DATA

Several cruises in the bay area were made to collect 57 sediment samples, whose locations were mapped (see Fig. 1.31) by using landmarks and sonic triangulation. A Ross snapper sampler with sediment capacity of 1/2 liter was used to collect the bottom samples. The bottom sediments were taken to the laboratory and separated into sand, silt, and clay components (fractions) by using sieves. The data listed in Table 1.6 represent the results derived from the sediment analysis.

EXERCISES

1. Construct maps showing the distribution of each of the sediment components. See Table 1.6.

- **a.** Draw a lithofacies map showing the distribution of sand. (Use a piece of tracing paper to overlay the location map. Then transfer the values from the preceding table onto the tracing paper.) Contour the data using a contour interval of 10%.
- **b.** Draw a lithofacies map showing the distribution of silt. Use a contour interval of 5%.
- **c.** Draw a lithofacies map showing the distribution of clay. Use a contour interval of 5%.

2. Compile a general lithofacies map showing the distribution of the sand-silt-clay sediment fractions, using boundaries defined by the following percentages:

- **a.** Sand: greater than 60% sand = color in yellow.
- **b.** Silt: greater than 15% silt = color in green.
- **c.** Clay: greater than 5% clay = color in brown.

3. Where is the general source for the sediments in the Pamlico River? Does this agree with your first impression? Why or why not?

4. If gold were associated with the coarsest sediment fraction, where would you recommend dredging? (Give two areas in order of preference.)

5. Oysters cannot tolerate clay concentration greater than 8–10% of total sediment. If a company wished to experiment with starfish in a closed pen, where would you recommend that it set up its experiment so as not to interfere with the oyster business? (Starfish are voracious predators of oysters.) Show the location on the lithofacies map in Question 2 with a red pencil.

TABLE 1.6 Samples

The samples are numbered according to their location. The following list of sieve analyses for each sample gives the percentage of sediment in each of the three size fractions. (See sample 1 in list for example.)

1. 95 (sand)	16. 85	31. 99	46. 85
3 (silt)	10	1	7
2 (clay)	5	0	8
2. 80	17. 84	32. 98	47. 93
17	6	2	4
3	10	0	3
3. 78	18. 80	33. 90	48. 100
17	9	7	0
5	11	3	0
4. 92	19. 69	34. 89	49. 97
7	11	6	2
1	20	5	1
5. 98	20. 46	35. 90	50. 95
2	38	8	4
0	16	2	1
6. 96	21. 50	36. 99	51. 87
3	40	1	10
1	10	0	3
7. 64	22. 67	37. 98	52. 85
27	13	1	11
9	20	1	4
8. 84	23. 82	38. 98	53. 85
10	11	2	8
6	7	0	7
9. 90	24. 90	39. 87	54. 94
6	9	7	4
4	1	6	2
10. 98	25. 67	40. 91	55. 100
1	12	5	0
1	21	4	0
11. 92	26. 64	41. 94	56. 98
6	15	3	1
2	21	3	1
12. 79	27. 78	42. 85	57. 93
11	12	8	5
10	10	7	2
13. 60	28. 86	43. 83	
31	10	10	
9	4	7	
14. 57	29. 88	44. 86	
34	9	10	
9	3	4	
15. 74	30. 97	45. 83	
21	2	11	
5	1	6	

6. Would you expect nearby high mountains to the west to be the sediment source for the estuary of the Pamlico River? Explain your reasoning.

7. Following are water depths in feet at the various sample locations. On an overlay, draw a bathymetric map of the river and the bay. Use a contour interval of 5 feet.

1. 10	**20.** 21	**39.** 13
2. 11	**21.** 24	**40.** 12
3. 14	**22.** 17	**41.** 8
4. 10	**23.** 14	**42.** 13
5. 6	**24.** 7	**43.** 14
6. 7	**25.** 25	**44.** 21
7. 13	**26.** 21	**45.** 25
8. 9	**27.** 28	**46.** 17
9. 11	**28.** 23	**47.** 10
10. 8	**29.** 20	**48.** 4
11. 10	**30.** 17	**49.** 17
12. 17	**31.** 2	**50.** 18
13. 21	**32.** 8	**51.** 20
14. 17	**33.** 9	**52.** 23
15. 12	**34.** 12	**53.** 16
16. 10	**35.** 9	**54.** 9
17. 8	**36.** 5	**55.** 28
18. 7	**37.** 4	**56.** 6
19. 14	**38.** 9	**57.** 15

8. What seems to be the relationship between the depth of water and each of the three sediment facies (sand, silt, and clay) in the bay? Explain in your own terms what you consider to be the primary reasons for these relationships.

9. Describe the topography of the river and sound.

CHAPTER

2

Fundamental Concepts

FUNDAMENTAL LAWS OF GEOLOGY

Many years ago, a farmer out plowing croplands near Mexico City witnessed the birth of a volcano in a corner of his cornfield. To the farmer, the lava and ash covering his field no doubt represented dramatic proof of nature's inexplicable ways and perhaps a financial loss. To the geologist, however, this volcanic eruption represented a logical sequence of geologic events. As the molten lava moved through the earth's crust toward the cornfield, the existing rocks in the area where the eruption occurred were metamorphosed by the heat. The surface area was then locally covered by extrusive lava and volcanic ash, which will, in time, weather and develop a soil zone, probably very similar to the one in which the farmer was planting corn. Therefore the farmer unwittingly witnessed the principles of cross-cutting relationships, lateral continuity, components, and superposition. These principles, or "laws," are fundamental to the geologic interpretation of a sequence of events and are reviewed in the following sections.

Uniformitarianism

The law of uniformitarianism as stated in the *Glossary of Geology* (© 1972 by Margaret Gary, Robert McAfee, Jr., and Carol L. Wolf, editors, American Geological Institute, Washington, D.C.) is one of the most often applied concepts in historical geology. "Uniformitarianism is the fundamental principle or doctrine that geologic processes and natural laws now operating to modify the earth's crust have acted in the same regular manner and with essentially the same intensity throughout geologic time, and that past geologic events can be explained by phenomena and forces observable today, the classical concept that 'the present is the key to the past.' The doctrine does not imply that change occurs at a uniform rate, and does not exclude minor local catastrophes." This principle was first proposed by James Hutton in 1795 and was popularized in 1830 by Charles Lyell.

To adequately interpret past geologic events, processes, or life forms, geologists must critically analyze and interpret modern analogues of these events, processes, and fossils. As an example of the application of this law, geologists have documented the types of rock deposition as well as the prevailing rates and directions of movement for various kinds of glaciers in Alaska, Greenland, and the Alpine regions of Europe. Geologists have used these modern data to substantiate geologic evidence of past glaciations during the Carboniferous period in South Africa, South America, and India. Similarly, by studying fossils, geologists base their interpretations of depositional and environmental conditions that affected bivalves (clams) that lived during Cretaceous times (66–144 million years ago) on the environmental conditions affecting living bivalves today.

Original Horizontality

The law of original horizontality was conceived by Nicolaus Steno (1638–1687). Steno observed that most sedimentary rocks are formed by particles that settle to the bottom of rivers, lakes, and oceans under the influence of gravity in essentially horizontal layers, which are approximately parallel to the earth's surface. Therefore, Steno reasoned, if sedimentary rocks are found in an inclined or folded attitude, they must have undergone movement *after* their deposition and lithification.

Superposition

The law of superposition was also proposed by Steno in 1669 after working in the mountains of western Italy. The principle of superposition was later effectively demonstrated by Hutton in about 1795. Hutton had observed that sedimentary rocks were formed by the accumulation of numerous layers. He concluded that in any undisturbed sequence of strata, the oldest (first-deposited) layer would be on the bottom and that the youngest (last-deposited) layer would be on the top. A necessary aspect of this principle is the assumption that the sedimentary strata have not been overturned or inverted by folding or faulting.

Lateral Continuity

Steno also proposed the law of lateral continuity, which states that most thick layers of sediment were originally deposited over geographically extensive areas and subsequently covered with overlying layers and lithified into sedimentary rocks. Uplift, erosion, faulting, and folding have in many cases disturbed the original lateral continuity of these layers. An example is found in the Grand Canyon. The same regional layers of rock on the north rim of the canyon are also found on the south rim, but the erosion of the canyon by the Colorado River has disrupted the continuity of these strata.

Cross-Cutting Relations

The principle of cross-cutting relations states that any rock unit or fault that cuts across other rock units is younger than the rock units through which it cuts. This definition is simple for faults that resemble the one shown in Figure 2.1, Example A, but relations must

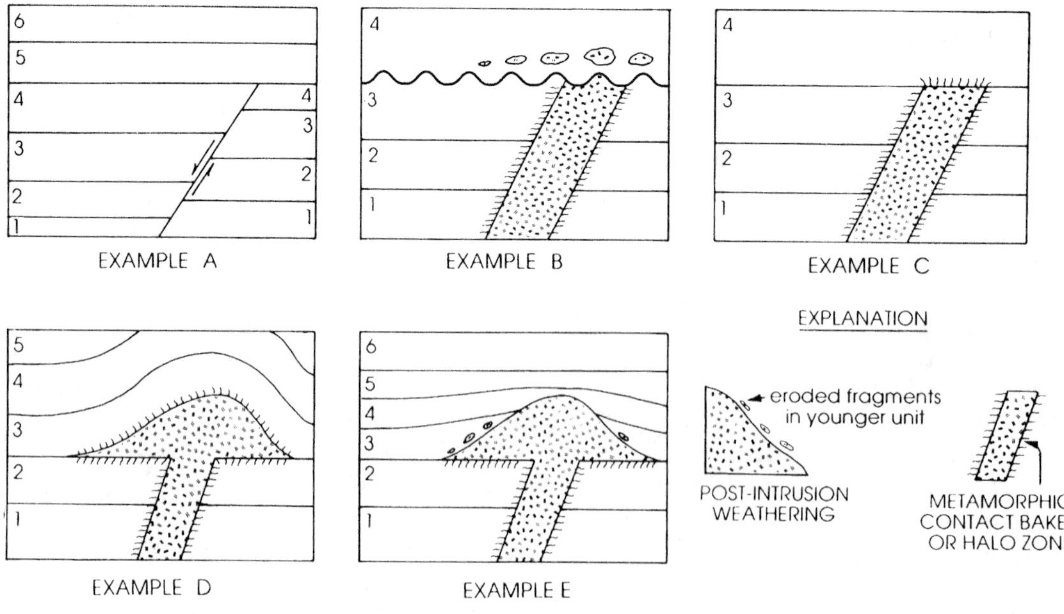

FIGURE 2.1 Examples of cross-cutting relationships.

be analyzed more closely when igneous intrusions are involved. In Examples B and C, igneous material has clearly intruded stratigraphic units 1, 2, and 3. The problem is the interpretation of when unit 4 was deposited. In Example B, the intrusion occurred before deposition of unit 4, as evidenced by the presence of eroded remaining components of the dike having been incorporated into the basal portion of unit 4. The wavy line indicates a period of erosion and is called an *unconformity*. In Example C the intrusion occurred after the deposition of unit 4. The evidence for that interpretive decision is the presence of the contact metamorphic zone that forms a halo around the intrusion (the fuzzy borders of the dike) including the basal portion of Unit 4 and the lack of eroded materials and the unconformity, as seen in Example B.

Another example of cross-cutting relations and their interpretation is an igneous intrusion of the type shown in Example D. Here, a laccolith has been emplaced after the deposition of units 1 to 5. This is evidenced by the cross-cutting of units 1 and 2 and the arching of units 3, 4, and 5, which have retained their uniform thickness and the metamorphic halo over the top of the laccolith. If the igneous material had cut units 1 and 2 and formed an extrusive body like a lava flow, as shown in Example E, one might have expected to find (a) a contact metamorphic zone only at the base, (b) eroded fragments of the igneous rock incorporated in the base of the overlying units 3 and 4; and (c) a gradual thinning of overlying units 3 and 4 due to a later, gradual depositional covering of the lava flow.

Inclusions

The principle of inclusion states that if a given layer of sedimentary rock contains inclusions or pieces of another kind of rock material within its matrix, these components or inclusions are older than that rock layer. This principle can be demonstrated by examining the composition, for example, of a conglomerate formed during the Eocene period 50 million years ago. Assume that at the time of the conglomerate's deposition, weathering and erosion had exposed an older, preexisting granite which was then 100 million years old. These clasts of weathered granite, having either stayed in situ or been transported, would eventually collect at a site of accumulation, where they would be lithified into the newer conglomerate. Thus the conglomerate, which present-day scientists would date at 50 million years old, would also contain inclusions of material from the much older granite (see Examples B and E, Fig. 2.1) that by now would be 150 million years old.

Similarly, sometimes a magma that intrudes into the earth's crust will break off pieces of preexisting crustal rocks as it fractures and intrudes upward through the crust. Some of these crustal fragments may not melt but become included in the solidifying rock as the magma cools. Thus pieces of 400-million-year-old schist may become included in granitic magma that intruded 40 million years ago and cooled slowly from a granitic intrusion or batholith. The schist inclusions are therefore older components of the younger granite surrounding them.

Fossil Succession

William "Strata" Smith (1769–1839), often regarded as the father of historical geology, carefully examined fossils that were embedded in the rock strata of England. He then documented that each layer in the succession of rock strata could be identified by the distinctive fossils it contained. On the basis of these observations, Smith proposed the law of fossil succession, which states that plant and animal fossils succeed one another in a recognizable order through the geologic record. Due to the rocks' being layered in stratigraphic sequence, and throughout that sequence each rock unit containing its characteristic group of fossils, Smith found visible evidence that fossils and groups of fossils change over time. As some groups of plants and animals were replaced by more successful groups, these changes were preserved in the rock sequence as changes in the overall assemblage of fossils. Consequently, the distinctive fossil groups and their host rock strata can be used to correlate strata from one geographic area to another.

REVIEW OF UNCONFORMITIES

In a succession of depositional environments occuring on the earth's surface, if all the deposited rock layers had been accumulated in a normal pattern without interruptions, geologists would say that they would be *conformable*. However, in reality there are many interruptions to the normal depositional patterns. These interruptions or breaks in the record are called *unconformities*. The presence of an unconformity implies that normal sedimentation had ceased and that the area had been uplifted to allow erosion to take place prior to the development of a new set of sediment accumulations covering the eroded surface. Therefore, an *unconformity* is a surface of erosion or nondeposition that separates younger strata from older ones. In a sense, this is the same definition as that of the law of superposition; i.e., younger rocks are overlying older units. However, the nature of the surface between the two older and younger units is the key. At the boundary between the two successive units in the definition of superposition, a normal sequence of deposition is considered to have taken place, like transgressive shale over a sandstone. However, in the case of an unconformity, a different sequence of events, and/or energy levels, within the depositional environment has taken place. During the time of non-deposition or uplift, the sequence of events could involve structural activity such as folding and faulting or erosion; both of which take time. How much time? One can only estimate. If the folded underlying strata are of Permian age and they are covered by Cretaceous sediments, then according to the geologic time scale, approximately 104 million years of geologic record could be missing. Alternatively, a continuous sedimentary sequence of rocks of Permian through Lower Cretaceous could have been deposited prior to Late Cretaceous erosion removing all the rock younger than the remaining Permian strata. In that case, the actual time for erosion could have been only a million years. Thus unconformities record not the duration of the erosion process but the apparent amount of geologic time missing and not recorded by the sequence of rock units in that area. Therefore, when studying unconformities, it is not the indicated time interval across the boundary that makes a unconformity a "great" unconformity.

There are three principal variations of this boundary called an unconformity. These variations are all based on significant changes of bedding attitude associated with tectonic activity at some level, changes in the sedimentary environments of deposition, fossils, energy within that environment, or introduction of igneous or metamorphic activity. All of these differences involve different intervals of time.

The three types of unconformities most commonly recognized in historical geology, along with their identifying characteristics, are described in the following discussions. It is assumed that any unconformable surface was, in general, originally initially horizontal. (The typical symbol used for unconformities in drawings is a wavy line, to indicate an irregular erosional surface.)

Nonconformity

A nonconformity is an unconformable erosional surface that involves and separates older igneous or metamorphic (crystalline) rocks from younger overlying sedimentary strata (see Figs. 2.2 and 2.3). Some of the criteria for identification of a nonconformable surface

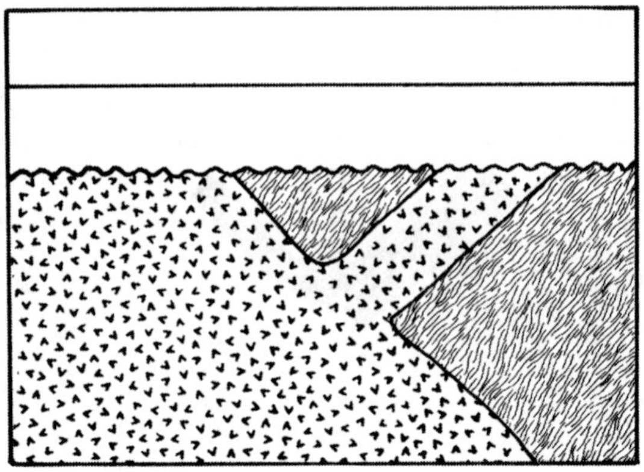

FIGURE 2.2 Nonconformity.

FIGURE 2.3 Nonconformity, Cambrian on Precambrian, Arizona.

include (1) incorporation of fragments of the igneous or metamorphic material within the basal portion of the overlying sedimentary units, (2) erosional truncation or abrupt termination of an underlying igneous rock mass at the unconformable surface, and (3) absence of contact metamorphism in the rocks immediately above the igneous or metamorphic rocks. Nonconformities imply that sufficient uplift has occurred to cause the erosion of any overlying strata as well as erosion of portions of the igneous and metamorphic units.

Angular Unconformity

At most unconformable bedding boundaries, the overlying stratigraphic units are usually in a nearly horizontal stratigraphic position. If the underlying units maintain a noticeable tilted, folded, or nonhorizontal attitude below the overlying unit, then a structural event is suggested to have taken place affecting the lower units.

This type of unconformity implies a definite sequence of geologic events. In the following example, units 1 through 6 are assumed to be sedimentary units that were initially accumulated in a horizontal position. Tilting or folding has occurred at a later time, with subsequent erosion and truncation of the upturned edges of units 1 through 6 before units 8 and 9 were deposited. Angular unconformities imply that an area has undergone uplift and that the uplift was accompanied by either folding or tilting, with erosion of the strata prior to later subsidence and continued deposition (Figs. 2.4 and 2.5).

Disconformity

A disconformity is an unconformable surface that separates essentially parallel sedimentary strata. Although disconformities are probably the most common type of

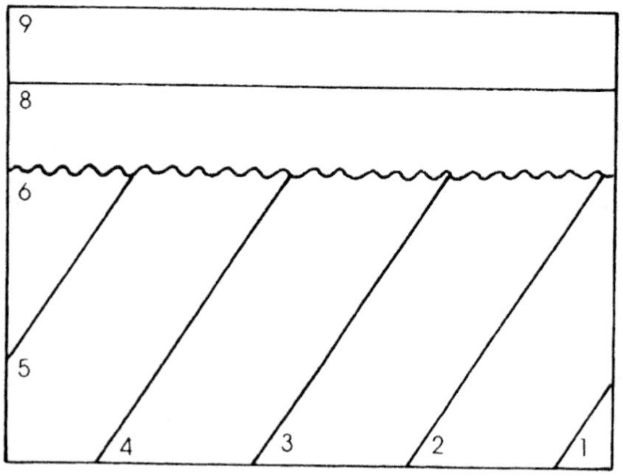

FIGURE 2.4 Angular unconformity.

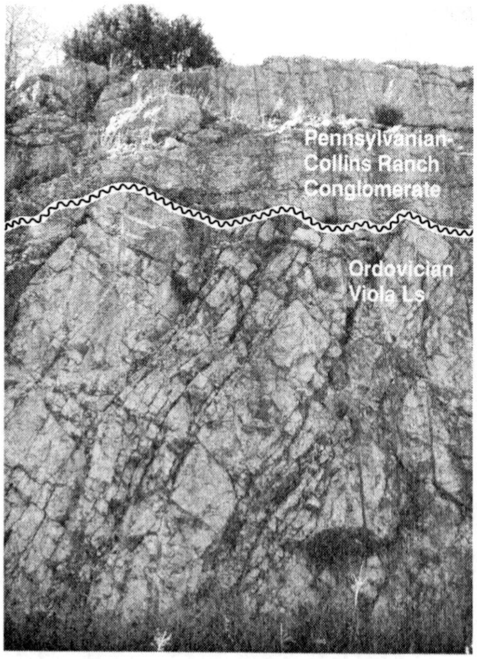

FIGURE 2.5 Angular unconformity, Pennsylvanian Collins Ranch conglomerate over Ordovician Viola Formation, Oklahoma.

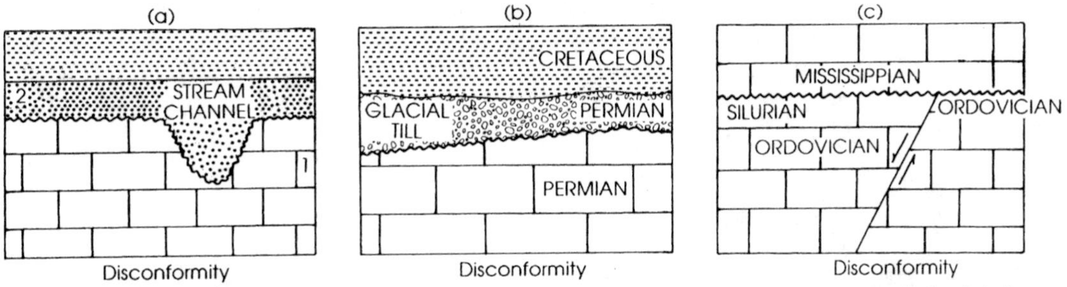

Erosion of preexisting sediments (unit 1) by a stream as evidenced by the channel sands and gravels (unit 2) and the incision of the channel.

Presence of a buried soil horizon (old glacial till) implies a hiatus before deposition of the overlying shale unit.

Parallel limestone units. Fossil dating evidence is necessary for recognition of the disconformity.

FIGURE 2.6 Disconformity examples.

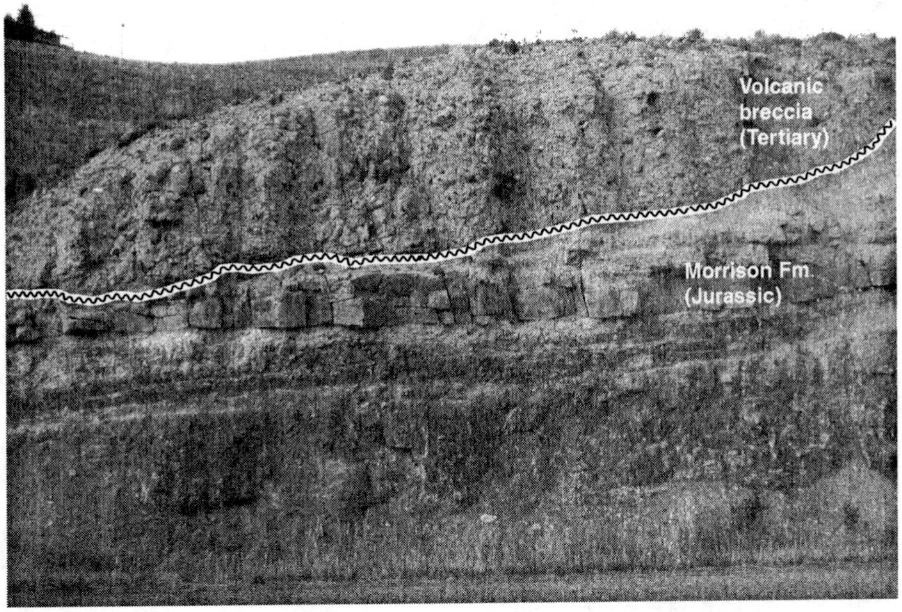

FIGURE 2.7 Paleocene volcanic breccia on Jurassic Morrison Formation, Colorado.

unconformities, they are often the most difficult to recognize, especially when rock types are similar above and below the erosion surface. Beds above and below a disconformity must be nearly parallel, but the contact (unconformable surface) may exhibit erosional relief of considerable magnitude. The presence of disconformities implies that regional uplift has occurred without severe deformation or a thermal event (see Figs. 2.6, 2.7, and 2.8).

REVIEW OF GEOLOGIC STRUCTURES

Crustal plate movements and other forces are known to create complex stresses (tension, compression, and shearing) within the earth's crust that produce strain in rock. Layered rocks at the surface are cold and brittle and will fracture in response to these stresses.

FIGURE 2.8 Navajo Formation and Chinle Formation on Precambrian igneous rocks, Colorado.

These layers at depth, heated and less brittle, can behave plastically and fold like the bellows of an accordion. Folds are produced by compressional stress; anticlines and synclines are the most common type of folds [see Fig. 2.9(a)]. Faults are caused by all three types of stress. *Normal* faulting is a result of tensional stress, a blistering or stretching of the crust. Compressional stress creates *reverse* and *thrust* faults; these are commonly found associated with folding and can signal a great deal of crustal shortening. Shearing stress produces lateral, or *strike-slip*, faulting. Figure 2.9(b) illustrates the several types of faults. The photographs in Figures 2.10(a), (b), (c), and (d) show examples of folding, faulting, and their associated surface crustal expressions.

Geologic structures are often illustrated using block diagrams, as shown in Figures 2.9 (a) and (b). These diagrams show something of the three-dimensional nature of the structures. The top surface of the diagram is called the *map* or *geologic map*. It illustrates what the structure looks like to us as we walk across the top of it or fly over it in a helicopter. It is the surface expression of the structure. The sides of the diagram are structural cross sections. They illustrate the structure as seen in a road cut or mountain side or river valley. This is a slice vertically through the structure. In reality, structures are often very large features and seldom can be confined to one hillside outcrop, so block diagrams are a convenient view of complex, regional geologic structures.

APPLICATION OF RADIOMETRIC GEOCHRONOLOGY

You should have a basic understanding of the rock cycle and, from your previous physical geology course, have an overview of igneous and metamorphic rocks. Igneous and metamorphic rocks are useful to geologists since they often contain associated minerals, such as uranium, potassium, rubidium, strontium, and thorium, that are used to help decipher the concept of geologic time. Geologists use two different concepts of geologic time. At times, a geologist may want to know as closely as possible when a specific catastrophic event in the earth's record has occurred, for example, a large preserved volcanic event. Alternatively, geologists use time in a relative sense, such as "Did this event come before or after that event?" or "Is this layer of rock older or younger than that series of

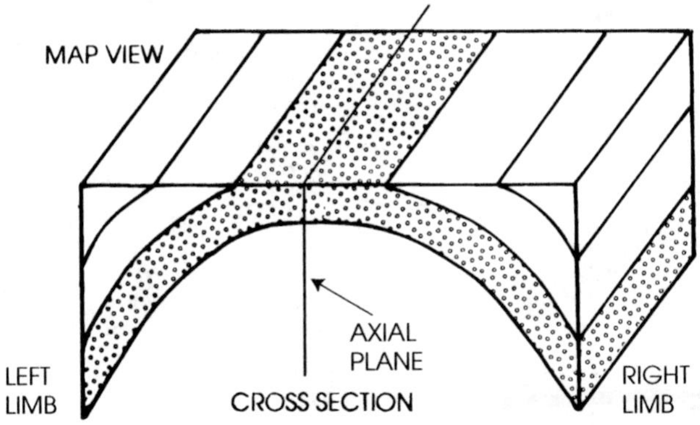

ANTICLINE

A fold where the strata have been bent into a convex upward arch. The core of the anticline will contain the stratigraphically *older* rocks (see patterned layer in diagram). See Fig. 2.10(b).

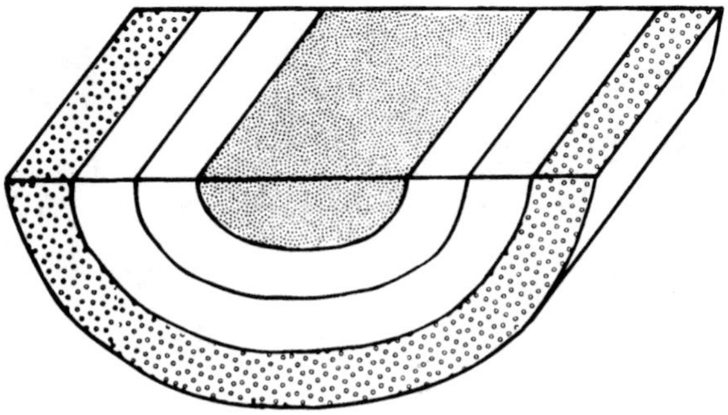

SYNCLINE

A fold that contains the stratigraphically *younger* rocks in the center or core of the structure (sand symbol on diagram). The strata are bent concave upward. See Fig. 2.10(b).

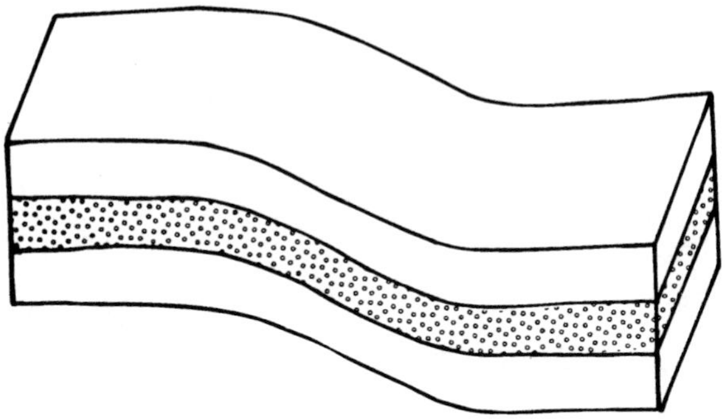

MONOCLINE

Strata that dip or flex from a horizontal attitude in one direction only. Generally a monocline will be a large feature of gentle dip. It should not be confused with the often more steeply dipping strata associated with one side or flank of a large anticline or syncline.

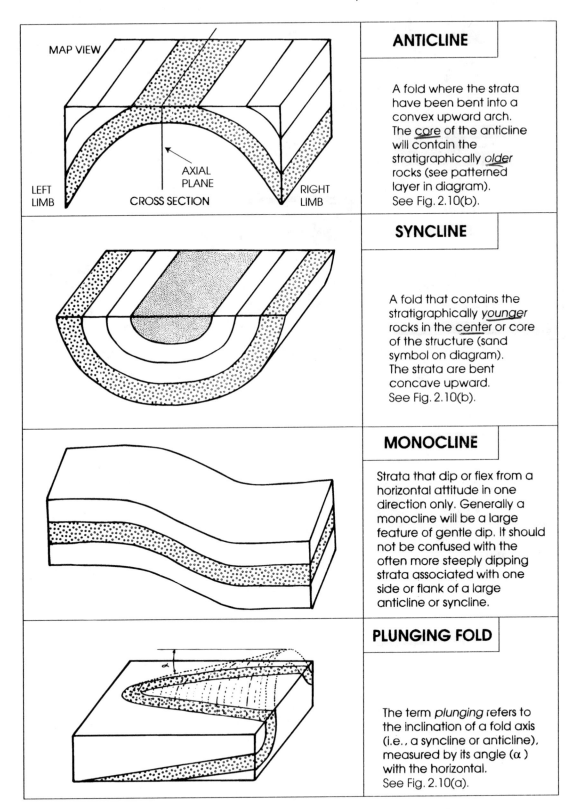

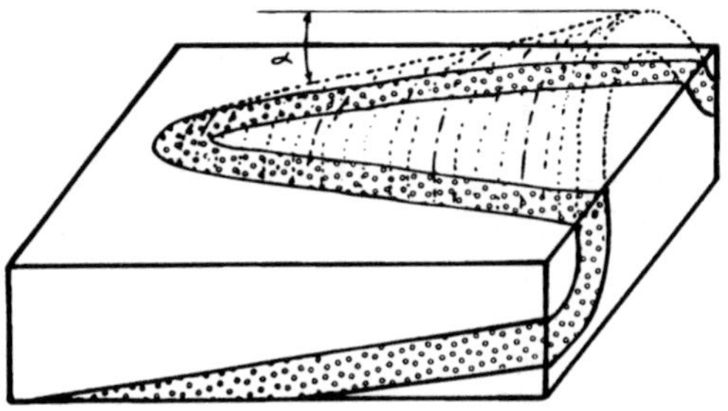

PLUNGING FOLD

The term *plunging* refers to the inclination of a fold axis (i.e., a syncline or anticline), measured by its angle (α) with the horizontal. See Fig. 2.10(a).

FIGURE 2.9(a) Major structures: folds.

NORMAL FAULT

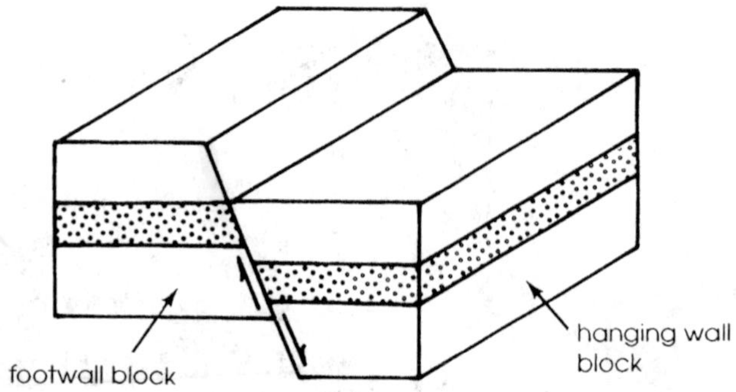

A fault in which the hanging wall block appears to have moved downward relative to the footwall block. The angle of the fault is usually 45°–90°. It is also called a gravity fault and is associated with tensional (pull apart) forces.

REVERSE FAULT

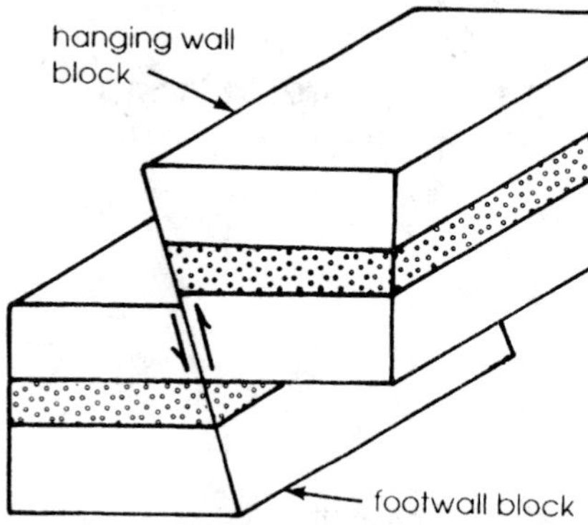

A fault in which the hanging wall block appears to have moved upward relative to the footwall block, showing vertical displacement. A reverse fault is the result of horizontal compression.

THRUST FAULT

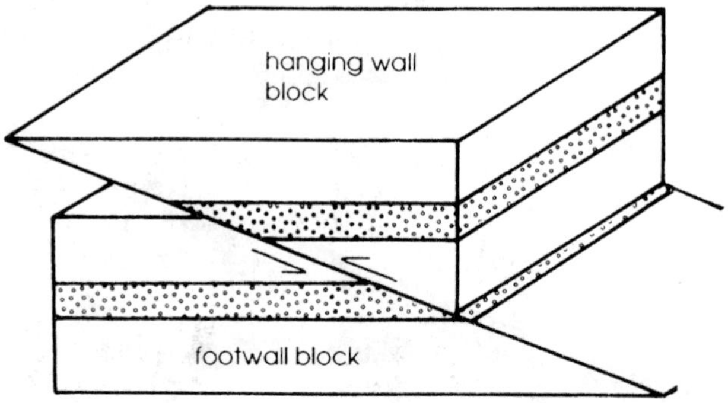

A low-angle reverse fault. See Fig. 2.10(c).

STRIKE-SLIP FAULT

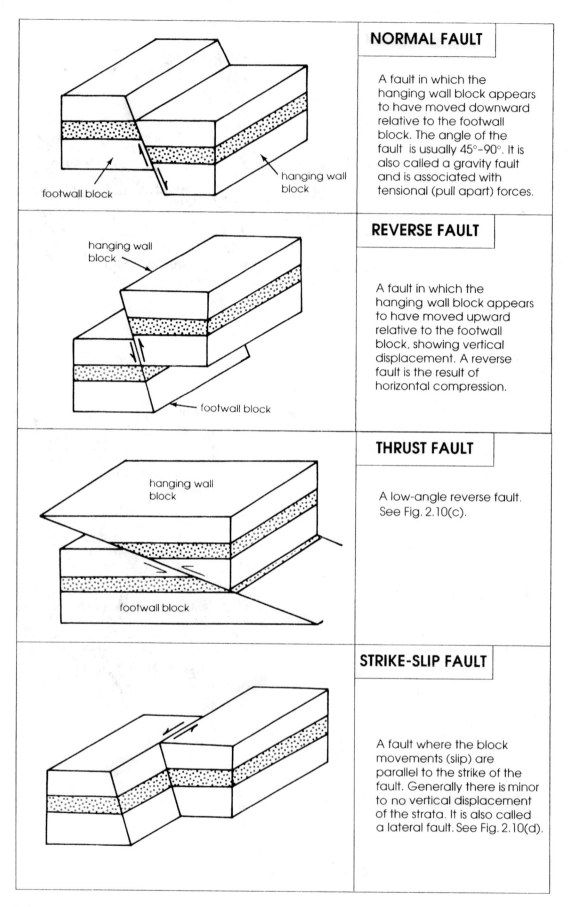

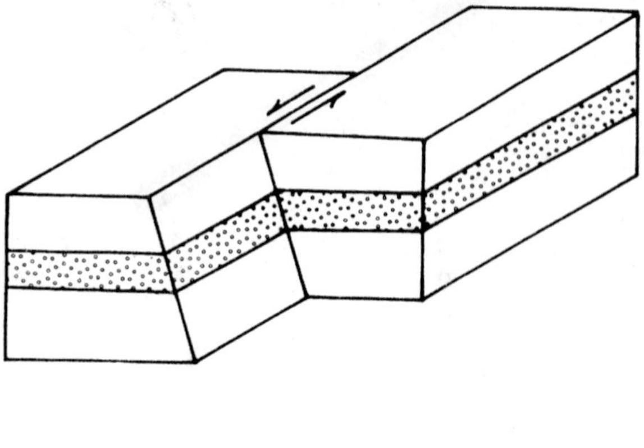

A fault where the block movements (slip) are parallel to the strike of the fault. Generally there is minor to no vertical displacement of the strata. It is also called a lateral fault. See Fig. 2.10(d).

FIGURE 2.9(b) Major structures: faults.

FIGURE 2.10(a) Virgin anticline, southwestern Utah.

FIGURE 2.10(b) Folded strata near Borah Peak, Idaho.

FIGURE 2.10(c) Voorhies thrust fault, San Bernardino Mountains, California.

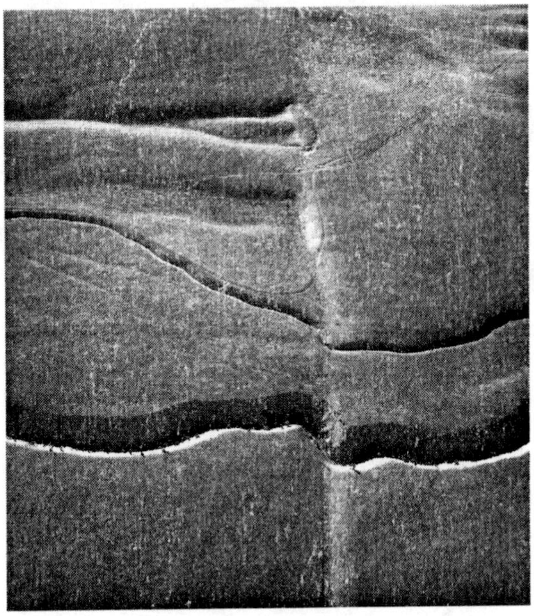

FIGURE 2.10(d) Right lateral offset of gullies, San Andreas fault, California.

Photos 2.10 (a) - (d) Courtesy of John Shelton

rock layers?" Relative geologic time will be explored in more depth in later chapters concerning fossils (Chapter 4) and correlation and mapping of rocks (Chapter 3).

Historical geology requires us to "read" a sequence of rocks to interpret the geologic history of an area. Part of the "reading" process involves the concepts of both measured and relative geologic time. In the following section we investigate the calculation and interpretation of measured geologic time using minerals primarily associated with igneous and metamorphic rocks.

RADIOMETRIC DATING

Radiometric dating is the process by which the absolute age of a rock or geologic event is determined. Elements such as uranium, potassium, rubidium, strontium, and thorium have radioactive isotopes. Radiometric dating uses radioactive isotopes that decay by emitting nuclear particles. These radioactive isotopes are called *parent* isotopes. In the decay process they are transformed into new isotopes of different elements called *daughter* isotopes. Some daughter products are themselves radioactive and decay to other daughter products. Eventually, all radioactive isotopes decay to a stable daughter isotope. For example, parent uranium isotopes decay through a series of steps to isotopes of lead, and parent potassium isotopes decay to calcium and argon.

Radioactive decay proceeds in various ways:

Alpha decay occurs when the nucleus of an atom emits an alpha particle, composed of two protons and two neutrons. This loss changes the atomic mass number (sum of protons and neutrons) and also the atomic number (number of protons). A change in the atomic number produces a new element. An example is the decay of uranium-238 to thorium-234 in the uranium decay series.

Beta decay occurs when a neutron loses a high-energy electron and is thus converted to a proton. This changes the atomic number but not the mass. The decay of rubidium-87 to strontium-87 provides an example of this type of decay.

Electron capture is the opposite of beta decay. An electron is incorporated by a proton in the nucleus. This results in one less proton and one more neutron. The atomic number changes but not the mass. Potassium-40 decays to argon-40 in this manner.

Radioactive decay is useful for dating rocks because experiments have shown that the rate of decay is constant and not affected by environmental factors such as temperature and pressure. Scientists determine the rate of decay from laboratory tests, and with additional

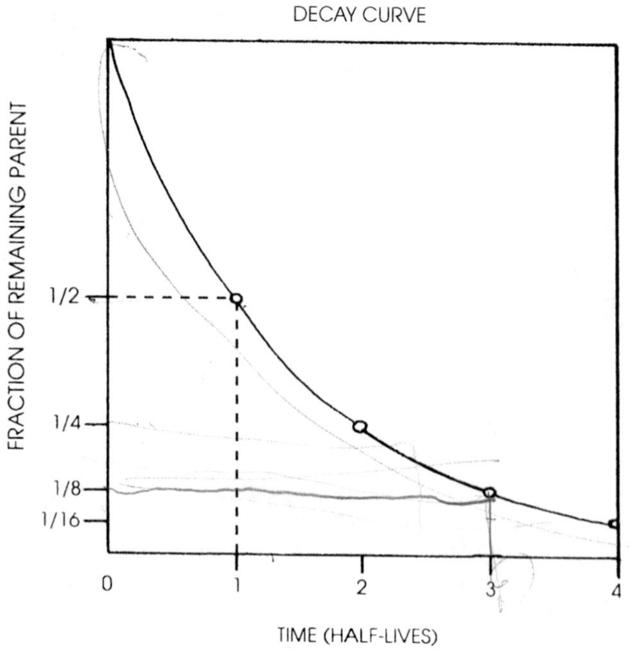

FIGURE 2.11 Decay curve.

mathematical computations they can construct a diagram called a *decay curve*, such as Figure 2.11, a generalized decay curve.

A radioactive decay rate is expressed in terms of a length of time called a *half-life*. A half-life is defined as the amount of time necessary for half of the original parent atoms to decay to daughter products or atoms. Thus, after one half-life, one-half of the original material is daughter products and the other half is the remaining parent isotope. After one half-life, the parent isotope abundance is reduced by one-half; after the second half-life, one-fourth of the original parent isotope remains. The progression continues as the remaining parent material is reduced during each successive half-life. Each radioactive isotope has its own specific half-life. Uranium isotopes (U-235 and U-238) have very long half-lives, approximately 713 million years and 4.5 billion years, respectively. Potassium-40 also has a long half-life, 1.3 billion years. Radioactive carbon found in organic material is very different; it is a short-lived parent isotope with a half-life of 5,730 years. Radioactive carbon is very useful in dating events that have occurred during the last 40,000 years of earth history.

An increasingly useful method of radiometric dating involves fission tracks. As a radioactive atom in a mineral crystal decays spontaneously, it emits a particle that travels through the crystal and disrupts its structure. This leaves a tiny trail called a *fission track*. By counting fission tracks, geologists can determine how many radioactive atoms have decayed. The crystal (commonly zircon) is then subjected to a neutron field, which causes the rest of the parent atoms to decay. Then the fission tracks are recounted, and a proportion of parent to daughter atoms can be established.

The general expression of the relationship between elapsed time and the radioactive decay of an isotope is

$$(N = N_0 e^{-\lambda t})$$

where N is the number of atoms of the isotope at time t, N_0 is the original number of atoms of the isotope, e is a mathematical constant equal to 2.718, and λ is the decay constant. The half-life of an isotope relates to the decay constant in this manner:

$$\lambda = \frac{\ln 2}{t_{hl}} = \frac{0.693}{t_{hl}} \qquad (t_{hl} = \text{half-life})$$

To calculate the age of a rock (t), the following formula is used:

$$t = \frac{\dfrac{\ln N}{N_0}}{-\lambda}$$

To illustrate radiometric dating in a simplified manner, Figure 2.11 can be used to demonstrate the technique.

Example: Assume you have a rock sample containing both a parent isotope X and its daughter isotope Y and that the parent isotope X has a half-life of 40 million years. You have sent the rock sample to a geochronology laboratory. The analysis shows the following: one-eighth of the total is X and seven-eighths is Y. What is the age of your rock?

Half-life = 40 million years
Number of half-lives that have elapsed = 3

This number is found by finding (1/8)X on the left side of the decay curve, drawing a horizontal line over to the curve, and drawing a vertical line to the bottom of the graph and reading the number of half-lives, which is 3:

40 million years × 3 half-lives = 120 million years

Igneous Rocks

Igneous rocks are formed at very high temperatures and therefore only very rarely contain discernible organic remains. However, both plants and animals can be preserved as fossils during burial by widespread volcanic ash deposits. Igneous rocks usually are not well stratified (layered) and, as a rule, are not geographically widespread when exposed at the earth's surface. The use of most igneous rocks in historical geology is threefold: (1) helping establish relative geologic time by various cross-cutting relationships, (2) delineating episodes of crustal unrest, and (3) measuring geologic time, since igneous rocks most often contain minerals with uranium or potassium that can be used for radiometric dating. Symbols for common igneous rock types used on geologic maps are illustrated on the inside front cover.

Intrusive (Plutonic) Igneous Rocks These igneous rocks are formed and cooled within the earth's crust. They may take the form of dikes, sills, laccoliths, or batholiths, depending on size and shape and whether they are concordant or discordant (cross-cutting) [see the basaltic dike in Fig. 2.12(a)], and have larger, intergrown crystalline *phaneritic* grains due to slower cooling.

Extrusive (Volcanic) Igneous Rocks Extrusive igneous rocks include lava flows, pillow lava, and fragmental deposits, including volcanic ash. These rocks have flowed or exploded out onto the earth's surface [see Fig. 2.12(b)]. This group of extrusive rocks has cooled rapidly, resulting in finer, or *aphanitic*, grain sizes. If cooled extremely quickly, they will appear glasslike with no apparent grains. Magma that has undergone both slow and rapid cooling at different times in its history may exhibit a *porphyritic texture* (mixed crystal sizes).

FIGURE 2.12(a) Intrusive (plutonic).

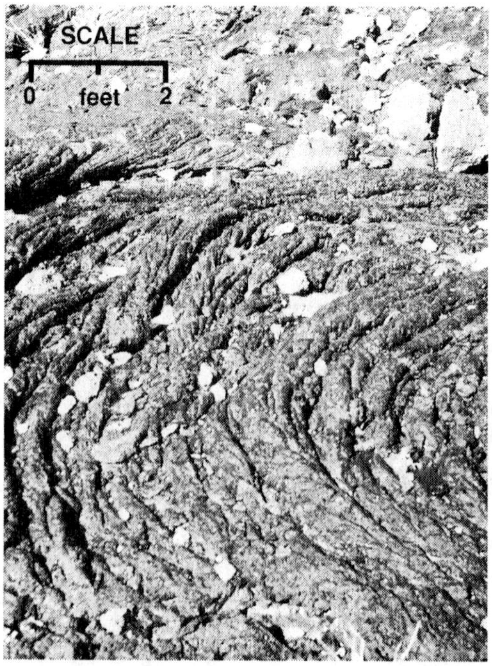

FIGURE 2.12(b) Extrusive (volcanic).

Metamorphic Rocks

As seen in Figure 1.1, metamorphic rocks are formed from preexisting sedimentary, igneous, or other metamorphic rocks. These rocks have been altered by some combination of increased temperature, increased pressure, or the introduction of chemically active fluids. The temperature and pressure can range from slightly above those existing at the surface of the earth to nearly those required to transform these rock materials into magma. Three major varieties of metamorphic rocks occur in nature.

Regional metamorphism: This is the most common and geographically widespread type of metamorphism. It is usually associated with a large thermal event or with an episode of mountain building as temperature and pressures are increased.

Contact metamorphism: This is important to the ordering of geologic events, even though it is often associated with a small local or specific igneous rock occurrence, like the emplacement of a dike. In this case, a narrow *halo* (baked zone) would develop surrounding the small dike cutting into sedimentary rocks near the earth's surface. If evidence of a baked zone or contact metamorphism is present, then the igneous rock has to have intruded into the preexisting rock layers and been the local heat source.

Dynamic metamorphism: This type of metamorphism is directly related to tectonic activity. As movement takes place along faults, intense grinding of the existing rocks occurs in these high-pressure zones, with new minerals like *mylonites* being formed at low temperatures. However, at higher temperatures, recrystallization can occur as the rocks are metamorphosed.

Metamorphic rocks are not used extensively in historical geology. The degree of metamorphism is directly related to the increasing pressure and temperature; thus different sets of minerals will be formed under these varying conditions. These mineral sets are called *index* minerals. Dating of index minerals within the exposed sequence of metamorphic rocks can yield a date for the thermal event. Occasionally, preexisting igneous or metamorphic rocks can be reheated by a younger thermal event, which can "reset" the older radioactive "clock" to a younger time. In some rare cases, low-temperature metamorphic rocks will be found to contain limited suites of fossils. Symbols for common rock types used on geologic maps are illustrated on the inside front cover.

CONSTRUCTION OF A GEOLOGIC CROSS SECTION

The location of cross section A-A′, stretching from Lake Winnebago, Wisconsin, to Toronto, Canada, is marked on the geologic map in Figure 2.13(a). First, place a sheet of paper adjacent to the line of cross section A-A′. Mark the position of points A and A′ as well as each point of contact between different geologic subdivisions. The marked sheet of paper should now look similar to the line of section shown on Figure 2.13(b).

In the geologic cross section A-A′, there are no strike and dip symbols shown because there is no room for such details on a regional map. Thus the mapped outcrop pattern of geologic units of differing ages must be used to infer or interpret the subsurface geologic relationships in the mapped area. Look at the geologic map and the line of cross section, and then study the structural interpretations on p. 61. Note that the youngest stratigraphic units in the cross section are in the center of the mapped structure (see Geologic Time Scale, inside back cover). From this information it follows that a synclinal structure is present; thus a generalized structural cross section could be drawn to show this interpretation (see Fig. 2.14). For more exercises on geologic maps, see Chapter 5.

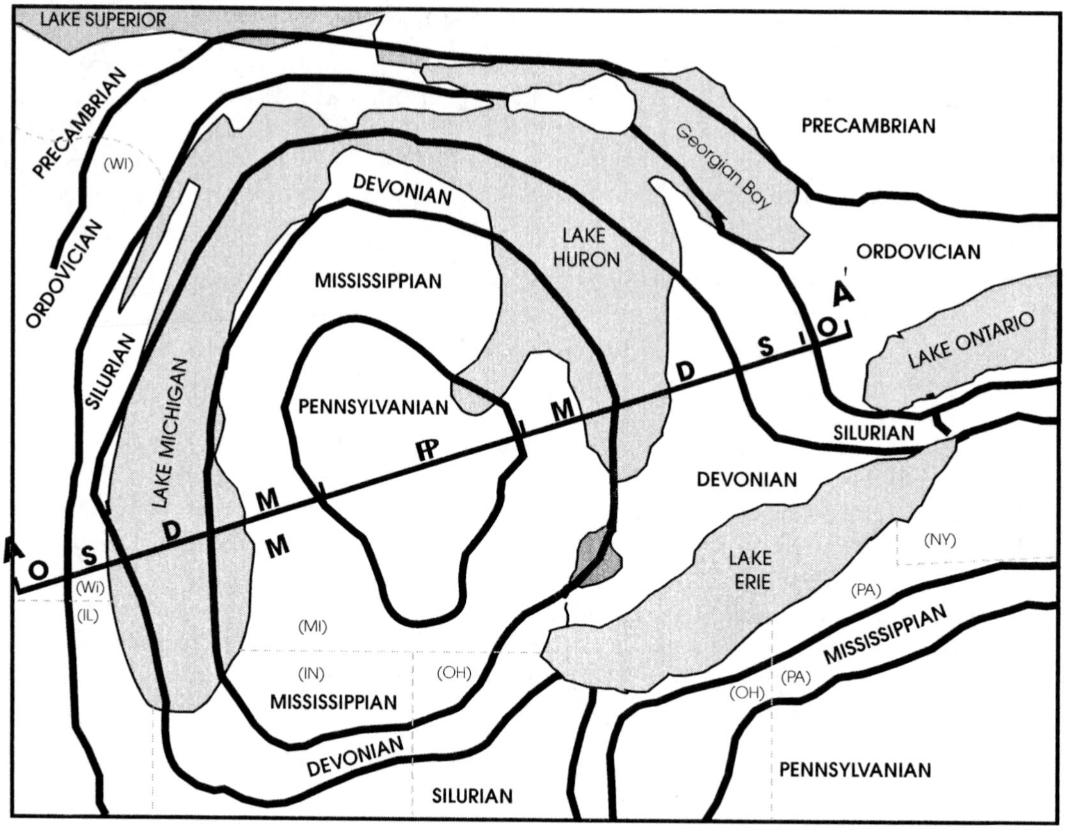

FIGURE 2.13(a) Geologic map of Michigan Basin, United States and Canada. Adapted from USGS, *Geologic Atlas of the United States.*

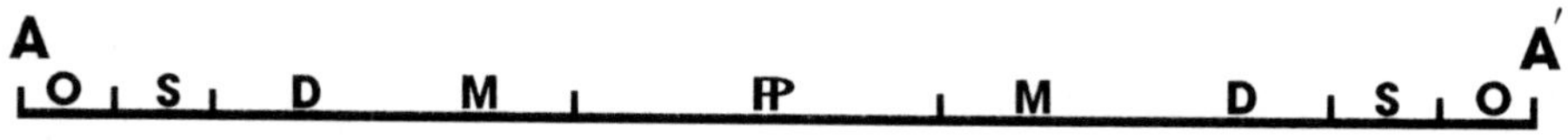

FIGURE 2.13(b) Cross section A-A′ with geologic contacts.

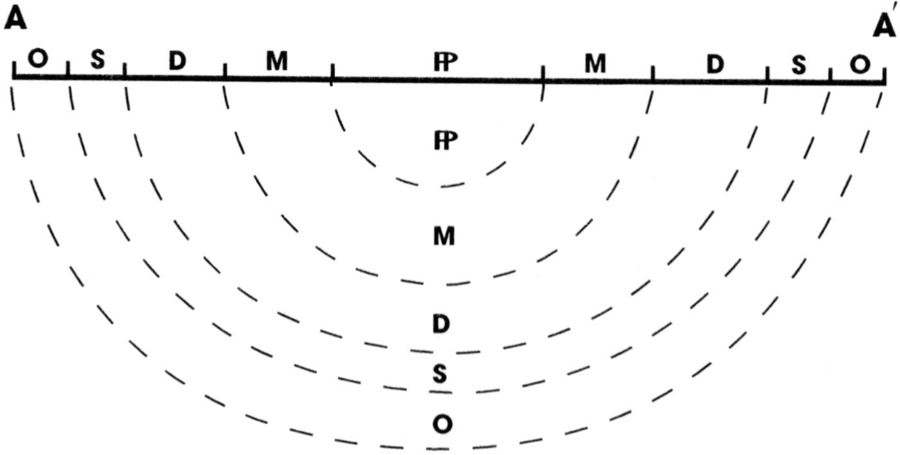

FIGURE 2.14 Generalized cross section.

RECONSTRUCTION OF A SEQUENCE OF GEOLOGIC EVENTS: THE GEOLOGIC HISTORY OF AN AREA

In this section, the procedure for organizing a sequence of geologic events into a logical order for the reconstruction of the geologic history of an area is developed and illustrated. To accomplish a historical reconstruction, a geologist must develop an ordered sequence for all of the geologic events that can be deciphered from the available data. Thus the geologist must be aware of many contributors, such as the overall layering of the rocks, fossils, cross-cutting faults or igneous rocks, structural activity, and other evidence that geologic events have taken place within the area being investigated. In the example provided in Figure 2.15, the first four significant features to notice are that some layers are horizontal, some layers are curved, an unconformity is present, and an intrusive igneous rock is seen crossing several of the layers. Along both sides of the geologic cross section in Figure 2.15, the sequence of events has been listed in a geologically correct order of occurrence. In the example, the approach to ordering a sequence of events that was followed placed the oldest geologic event at the base and the youngest geologic event at the top, as the geologic events occurred in geologic time. However, many geologists feel this approach is cumbersome and prefer simply to tabulate the event sequence as the normal chronologic events are determined to have taken place.

In the example (Fig. 2.15), it can be seen that at least four rock layers have been intruded by an igneous rock. As discussed earlier in the chapter, the law of original horizontality states that, under normal conditions, stratigraphic layers are initially deposited in a horizontal position. Then, a second law, the law of superposition, states that the successive superposed layers overlying the bottom layer are progressively younger. Assuming that these two laws have been met, layers 1–4 and maybe layer 5 were deposited horizontally prior to the injection of an igneous intrusion into the series of rocks.

Therefore, in Figure 2.15, the contact metamorphism seen associated with the dike cannot be seen affecting layer 6; across the unconformity. This implies that both folding of originally flat-lying layers 1–4 and the intrusion of the dike has occurred prior to the erosion of the area indicated by the unconformity. One unsolved question would be: What has happened to unit 5? Unit 5 may be absent because it was never developed; however, more likely, it was developed in sequence but has been removed during the interval of erosion associated with the development of the unconformity prior to the leveled surface and the subsequent overlying horizontal unit 6 and younger layers.

In any discussion of the reconstruction of a sequence of geologic events, when an unconformity, fold, or fault occurs, state the specific variety where possible. The exercises that

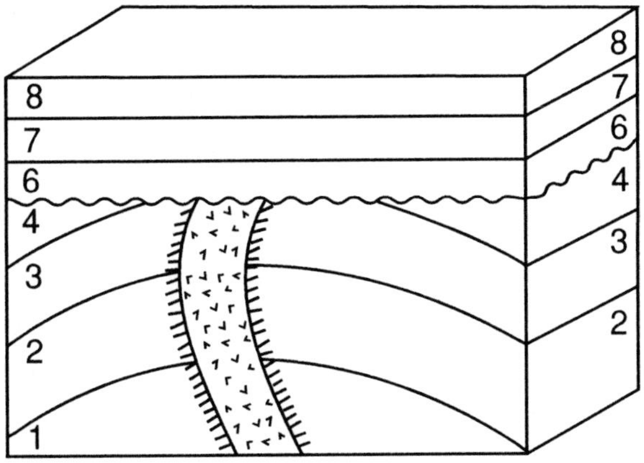

FIGURE 2.15 Example of sequence of events.

F. Deposition of units 7 and 8.

E. Submergence and deposition of unit 6, with the development of an angular unconformity over units 1–4 and a nonconformity over the dike. (In geology, a wavy line, like that between units 4 and 6, is used to signify the presence of an unconformity.)

D. Episode of erosion.

C. Regional uplift and folding of preexisting units.

B. Injection of dike with contact metamorphism of units 1–4.

A. Deposition of units 1, 2, 3, and 4 and perhaps unit 5.

follow involve the application of these basic geologic principles and laws and demonstrate the importance of ordering a sequence of events in the interpretation of geologic history.

EXERCISES

Exercise 2-1 ORDERING GEOLOGIC EVENTS

For each of the following cross sections (Figs. 2.16, 2.21, and 2.22), correctly identify the order of the geologic events.

Part A: Complete the ordering of geologic events in the four diagrams of Figure 2.16.

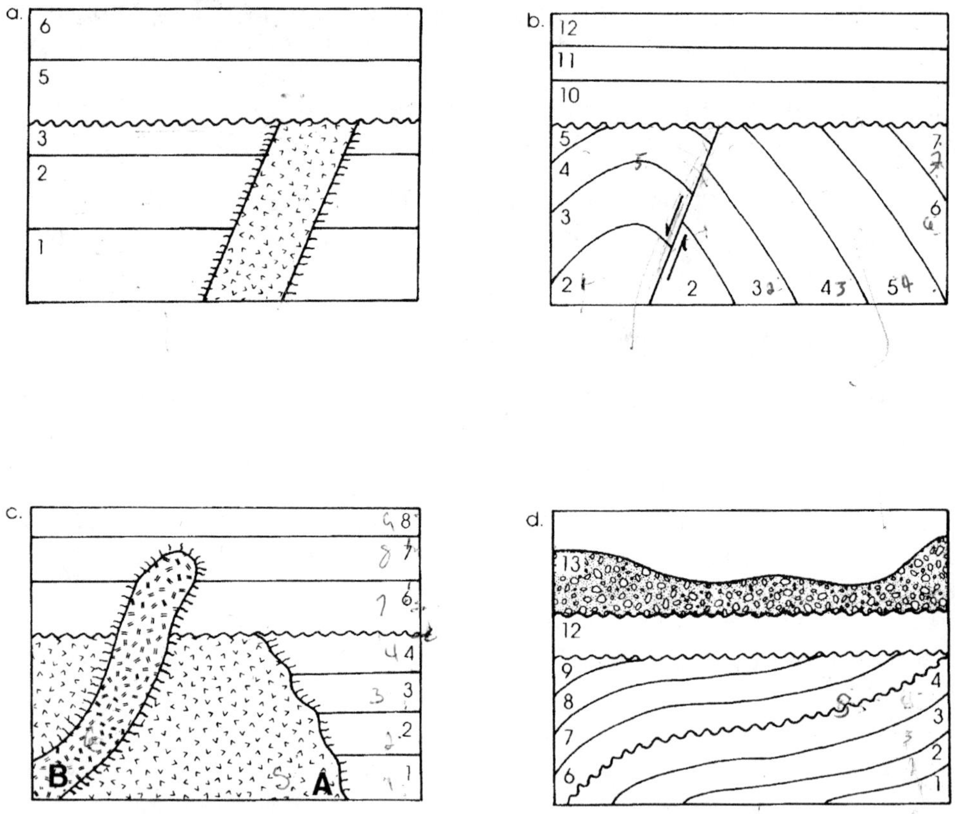

FIGURE 2.16 Series of block diagrams.

Part B: Draw in as many faults as you can find and then provide a general description for a sequence of geologic events for Figure 2.17.

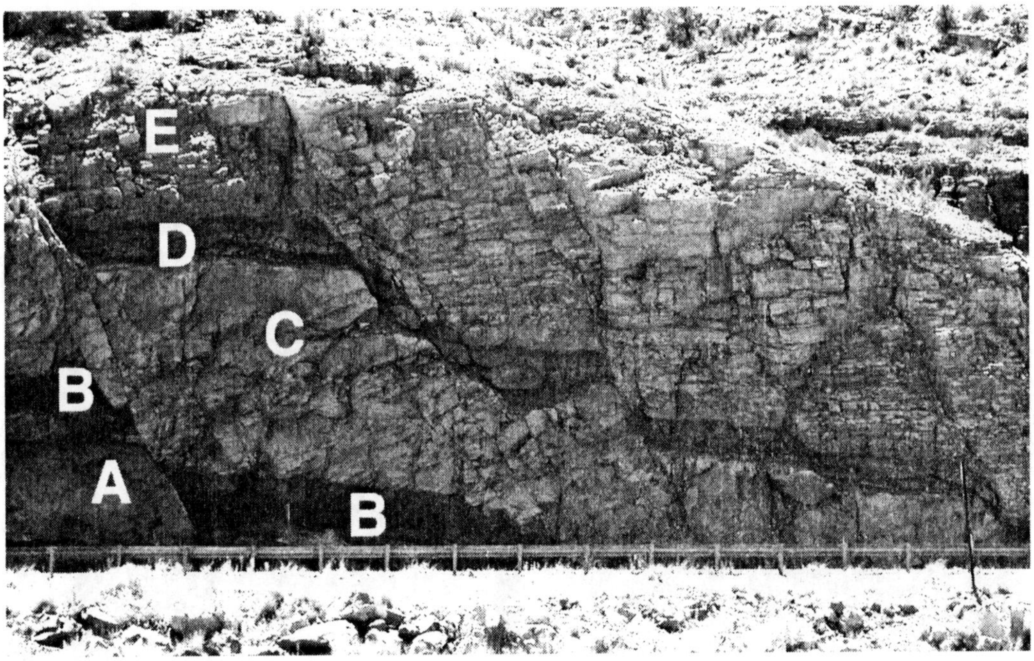

FIGURE 2.17 Faulted Entrada Formation, near entrance to Arches National Park, Utah.

Part C: Draw in the fault and describe a sequence of geologic events for Figure 2.18. What is the material called that appears to obscure the fault trace?

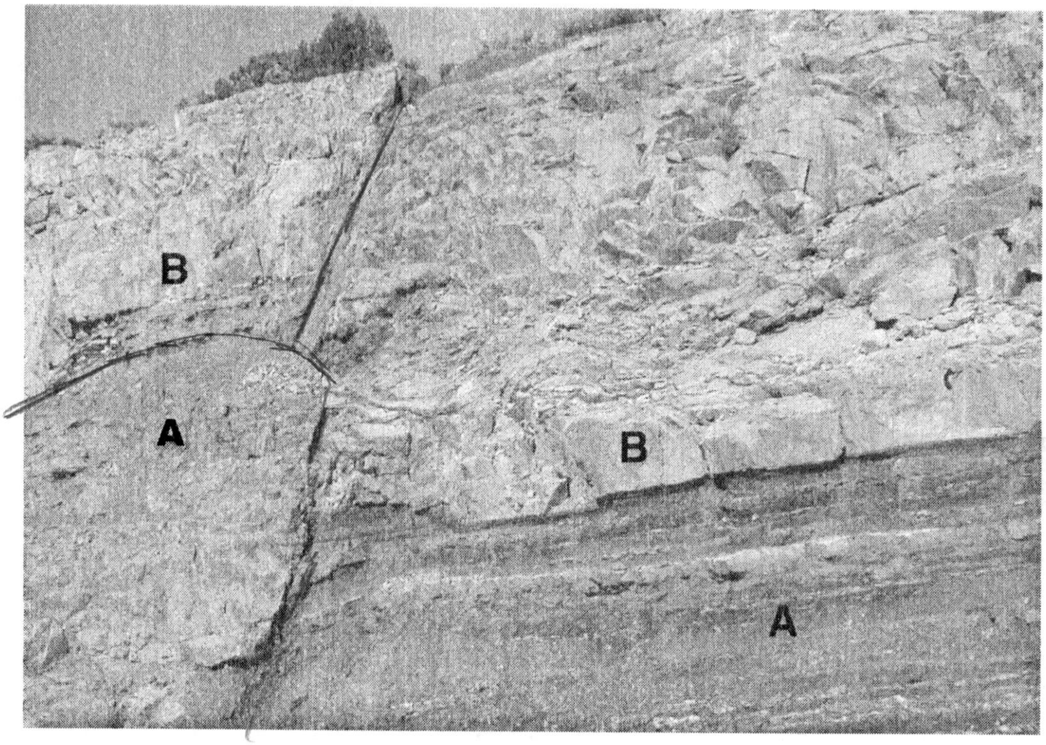

FIGURE 2.18 Entrada Formation, south of Moab, Utah.

Part D: Draw in four primary faults (there are several subsidiary faults). Describe a sequence of geologic events for Figure 2.19.

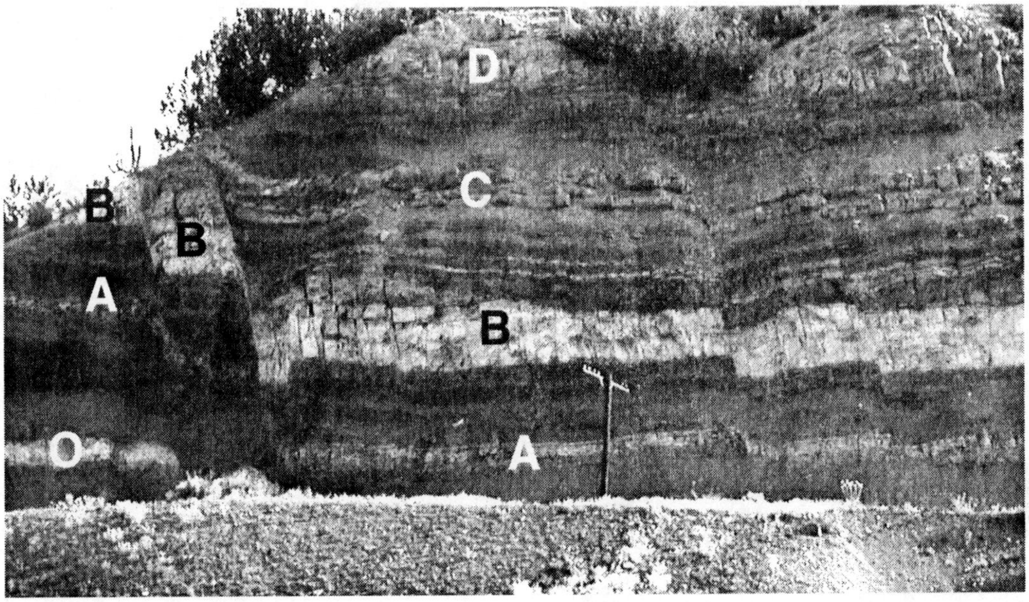

FIGURE 2.19 Twin Creek Formation, Hwy 6 near Thistle, Utah.

Part E: Roughly 1.1 billion years ago, a large continental plate began breaking up. Geological forces pulled the plate apart in three directions. The western break moved away from the eastern continental crust as the rifting helped form part of the Pacific Ocean. The sediments brought down by rivers and streams and an occasional flooding from the ocean hardened into the reddish Precambrian Unita Group of sediment, which ranges in age from 1100 to 925 million years old, reflecting a time when only a few early life forms such as bacteria, algae, and unicellular organisms existed.

Later numerous layers of continental and oceanic sediments such as the Madison (Mississippian), Glen Canyon (Triassic–Jurassic), Morrison (Late Jurassic), and Mancos (Cretaceous) formations accumulated on top of the filled-in rift. These thin formations contain an abundance of fossils, recording a vast progression of life forms from invertebrates to dinosaurs to mammals.

FIGURE 2.20 Sheep Creek Canyon, Flaming Gorge Reservoir area, south of Manila, Utah.

About the time that dinosaurs became extinct, another mountain-building event occurred. Roughly 65 million years ago, the North American continental plate began overriding the East Pacific Rise. These geologic forces crumpled the crust, creating the peaks and valley of the north–south running Rocky Mountains. This compression also reactivated the buried rift. The block forming the rift valley floor was forced up thousands of feet higher than the surrounding area, dragging the rock formations and forming the Uinta Mountains, the largest east–west mountain range in North America. Glaciation and erosion eventually swept the younger rocks away from the tops of the uplift, exposing the older structural relationships. (Descriptive geological data after informational road sign leading into the Sheep Mountain Geologic Trail in the National Forest on the Uinta Mountains, Utah.)

Draw in the location of the major thrust fault and describe the geologic sequence of events for Figure 2.20, citing geologic evidence in the photograph.

Part F: Complete the ordering of geologic events in the block diagrams within Figures 2.21 and 2.22.

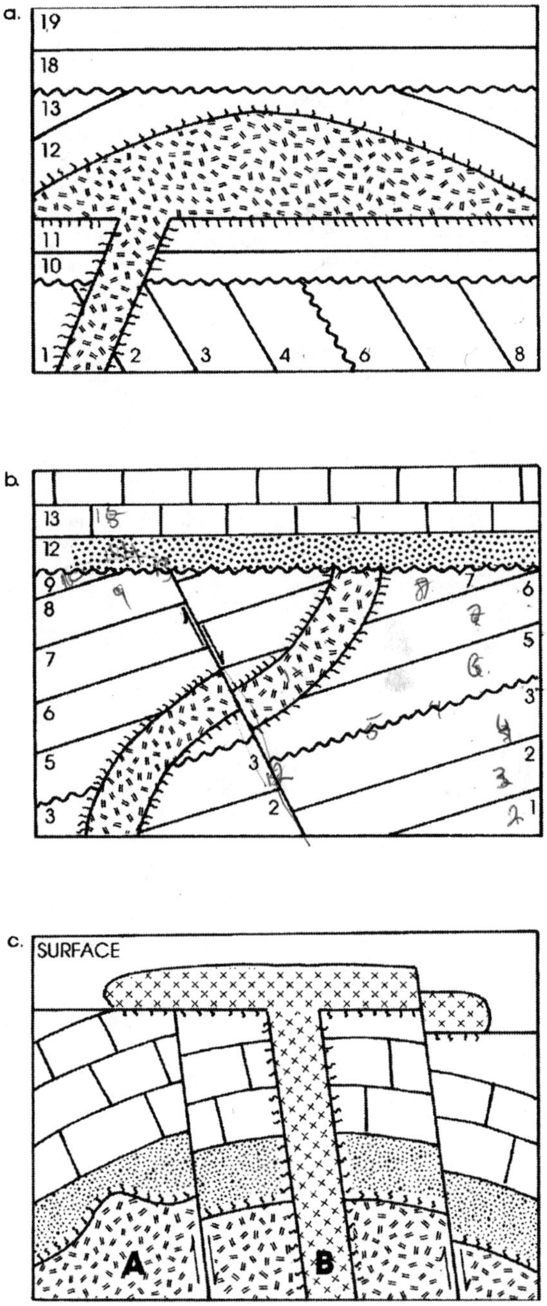

FIGURE 2.21(a,b,c) Series of block diagrams.

a.

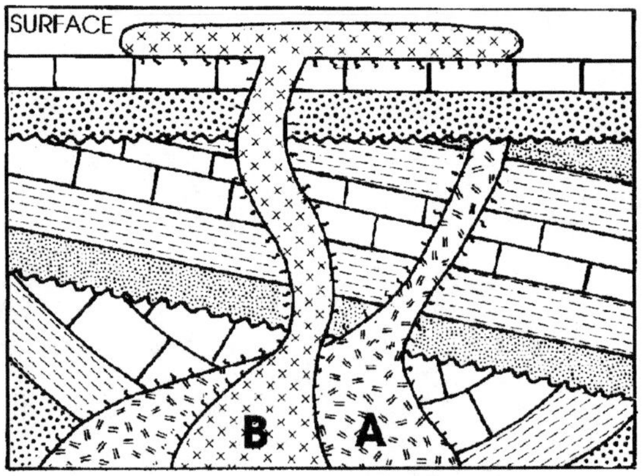

b.

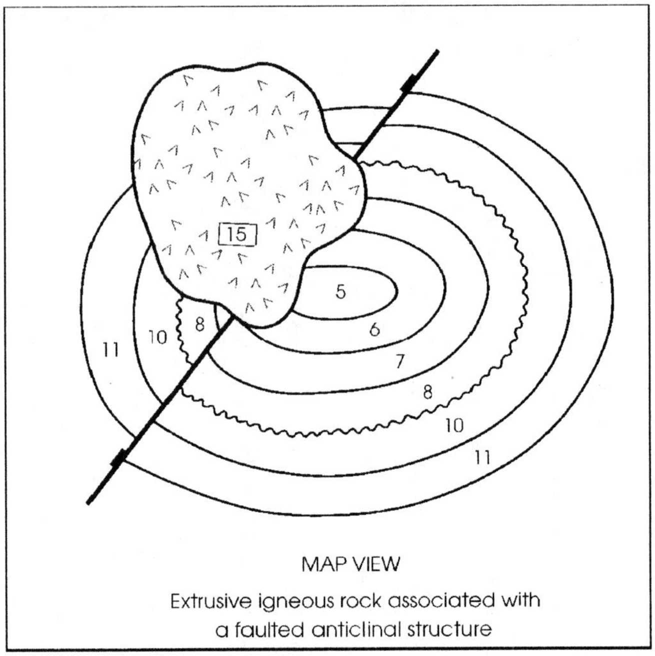

FIGURE 2.22(a,b) Series of block diagrams.

Exercise 2-2 RADIOMETRIC DATING

Use the radiometric decay curve shown in Figure 2.11 to answer the following.

a. Parent isotope Q has a half-life of 100 million years. Your rock sample has a ratio of $^{1}/_{16}$ isotope Q and $^{15}/_{16}$ daughter isotope R. What is the age of your rock?

b. Parent isotope L has a half-life of 20 million years. Your rock sample contains $^{1}/_{4}$ parent L and $^{3}/_{4}$ daughter isotope M. What is the age of your rock?

c. Your rock contains ½ parent isotope *A* and ½ daughter isotope *B*. Geochron Laboratories dates the rock specimen at 500 million years old. What is the half-life of isotope A?

---

d. Your rock contains ¼ parent isotope *S* and ¾ daughter isotope *T*. Geochron Laboratories dates the rock specimen at 200 million years old. What is the half-life of isotope *S*? Geochron has also told you that one of its lab technicians split the rock in half and accidentally crushed part of it and lost it down the sink drain. Does this affect the dating of your rock specimen?

Figure 2.26 (page 81) shows the decay curve for a specific isotope *X*. Refer to the curve and answer the following questions.

e. If a rock is 100 million years old, what percentage of isotope *X* is present?

f. If 35% of isotope *X* is present in a rock, what is the rock's age?

g. If a rock is 400 million years old, what percentage of the daughter isotope, *Y*, is present?

Exercise 2-3 MOHAWK VALLEY, NEW YORK

A generalized cross section of the Mohawk Valley to the Hudson River is shown in Figure 2.23. (Use the Geologic Time Scale, inside back cover.)

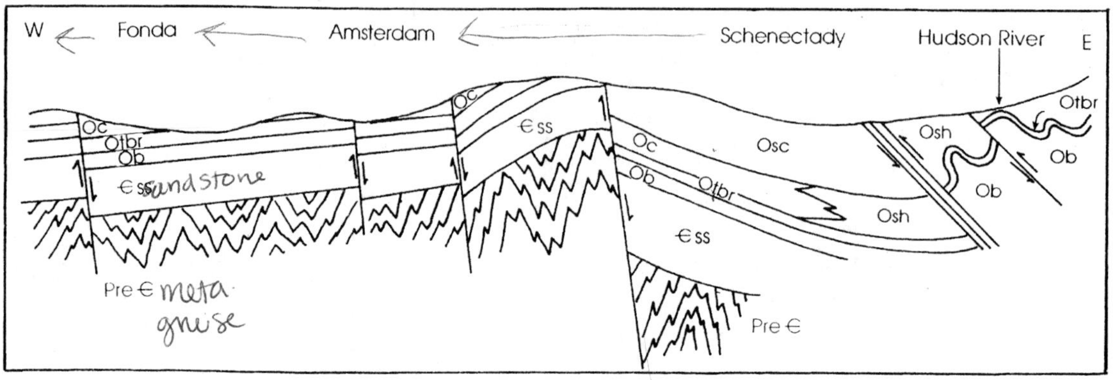

FIGURE 2.23 Structural cross section of the Mohawk Valley, New York. Adapted from USGS, *Geologic Atlas of the United States.* ~~Pre €~~, ~~€ss~~, ~~Ob~~, ~~Otbr~~, OC, Osh, OSC

a. Place all of the geologic units in order according to age, starting with the youngest _______, _______, _______, _______, _______, _______, _______ (oldest)

b. If the unit €ss is a fine-grained sandstone, and the Pre€ units are metamorphic gneisses, what kind of surface exists between these two units? ____________________.

c. Assuming that all of the faulting west of Schenectady occurred at the same time in the cross section, what is the age of that faulting? (circle one)

~~**1.** before €ss~~ **2.** before Osh **3.** after Otbr ~~**4.** after Osc~~

Exercise 2-4 CHESTER VALLEY, PENNSYLVANIA

In the Philadelphia area, the Chester Valley contains a series of rock units that are shown diagrammatically in Figure 2.24. (Use the Geologic Time Scale, inside back cover.)

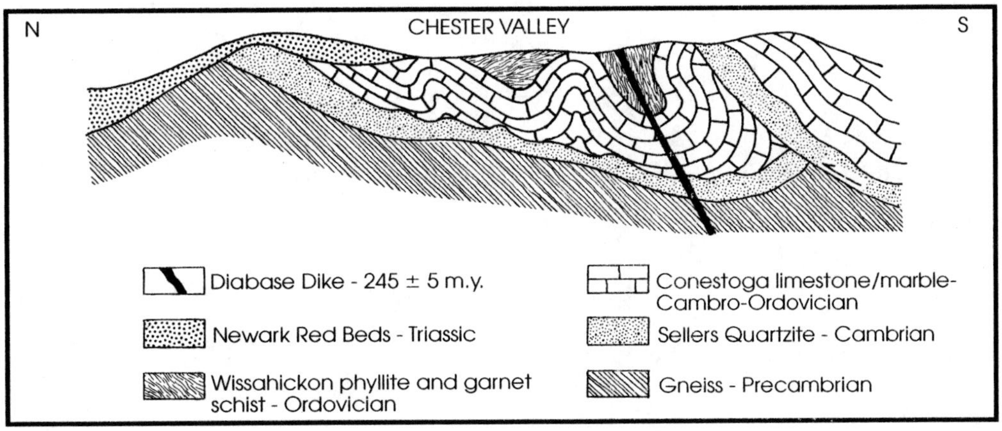

FIGURE 2.24 Structural cross section of Chester Valley, Pennsylvania. Adapted from USGS, *Geologic Atlas of the United States.*

a. Name all of the Paleozoic units.

b. Along the north side of the Chester Valley some of the Newark beds are mapped. An obvious unconformity exists between these Triassic beds and the underlying units. What type(s) of unconformity exist(s) along the old, erosional surface?

c. What evidence shows that the Chester Valley area has undergone compression and metamorphism in the geologic past?

Exercise 2-5 SEQUENCE OF RADIOMETRIC EVENTS IN NEW MEXICO

Figure 2.25 shows a cross section of an area in northern New Mexico where two dikes cut across numerous sedimentary units. The dikes are composed of igneous rock that contains a very small but significant amount of radioactive isotope *X*. For the hypothetical isotope contained in the dike rocks and the schist in Figure 2.25, a decay curve has been constructed (see Fig. 2.26).

a. Using Figures 2.25 and 2.26 determine the following.

(1) The date of metamorphism of the schist: _______ (million years)

(2) The date of the intrusion of *A*: _______ (million years)

(3) The date of the intrusion of *B*: _______ (million years)

(4) The half-life of isotope *X*: _______ (million years)

(5) What is the age of beds 1–4? _______ (Periods) (See the Geologic Time Chart, inside back cover.)

(6) What is the age of beds 5–9? _______ (Periods)

(7) Bed 10 could not be any older than _______ (Period)

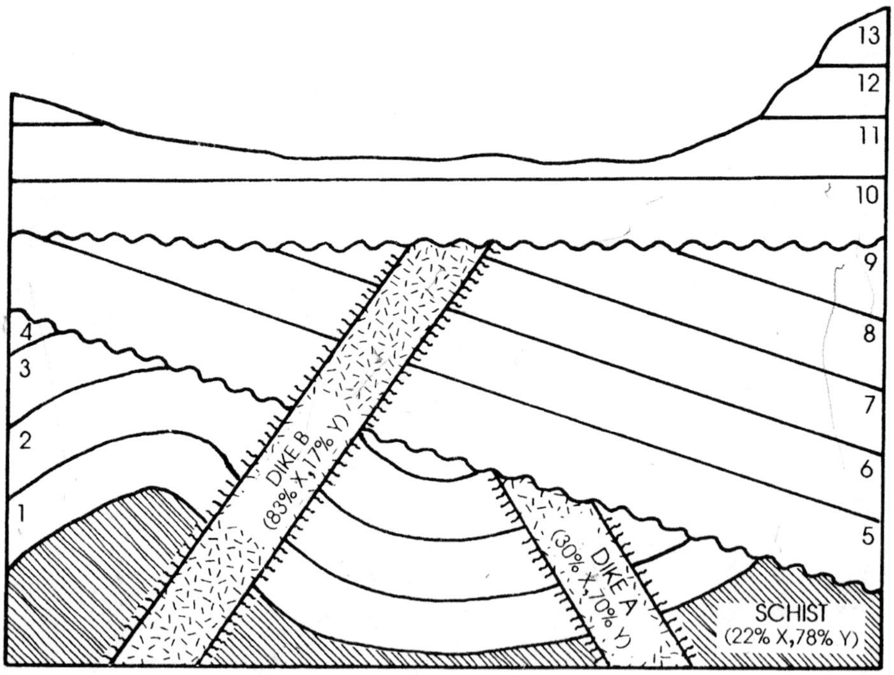

FIGURE 2.25 Cross section, New Mexico.

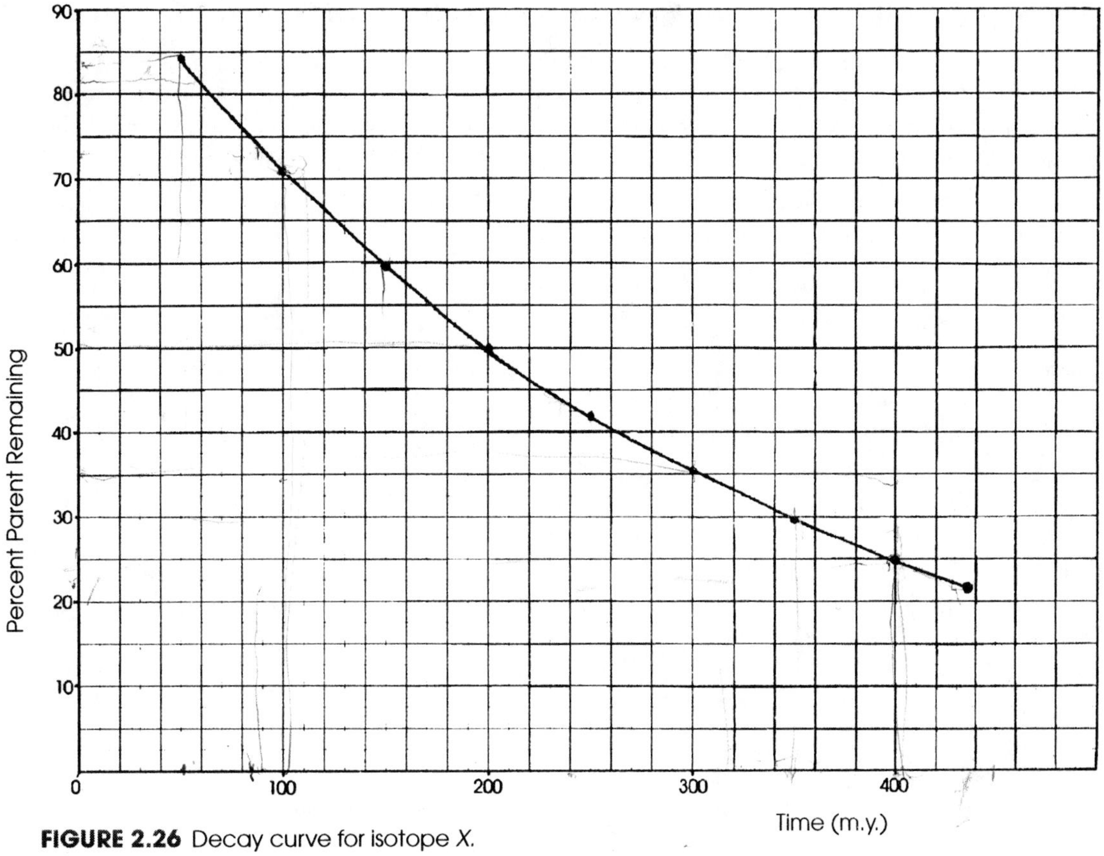

FIGURE 2.26 Decay curve for isotope *X*.

b. Write a brief narrative description of the geologic history for the cross section.

Exercise 2-6 INTERPRETING THE GEOLOGIC HISTORY OF THE GRAND CANYON

The Grand Canyon in northern Arizona is one of the most spectacular examples of stream erosion in the United States. The walls of the canyon are nearly 100 miles long and 1 mile deep. These excellent exposures of Paleozoic strata in the canyon make this is one of the best areas in the world for illustrating geologic history (see Figs. 2.27 and 2.28).

Answer the following questions by studying the geologic cross section of the Grand Canyon (Fig. 2.28).

a. What is the name of the oldest rock unit in the cross section?

b. Which is older, the Zoroaster granite or the Grand Canyon supergroup?

c. For the rocks exposed in the Grand Canyon, number each of the unconformities present and label its type. For each of these unconformities, list any geologic systems that may be missing and make a calculated estimate as to how much geologic time (in years) is missing. (Use the Geologic Time Scale, inside back cover.)

d. What is the name of the youngest Precambrian unit shown in the cross section?

e. List the units that were deposited during the Paleozoic era.

f. What is the youngest unit of rocks shown in the cross section?

g. Based only on the evidence in the cross section, during what geologic period did the Colorado River begin to carve the Grand Canyon?

h. In brief narrative form, describe the geologic history of this region, beginning with the Precambrian and ending with the Cretaceous. Be as detailed as the evidence permits. (Use a sheet of notebook paper.)

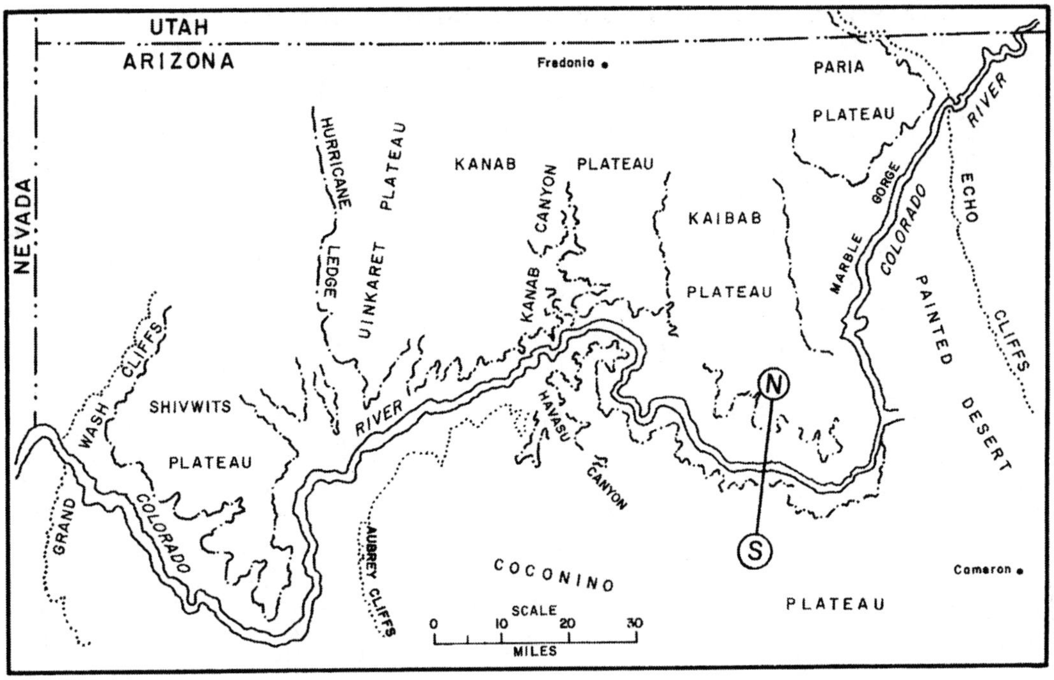

FIGURE 2.27 Location map of Grand Canyon cross section.

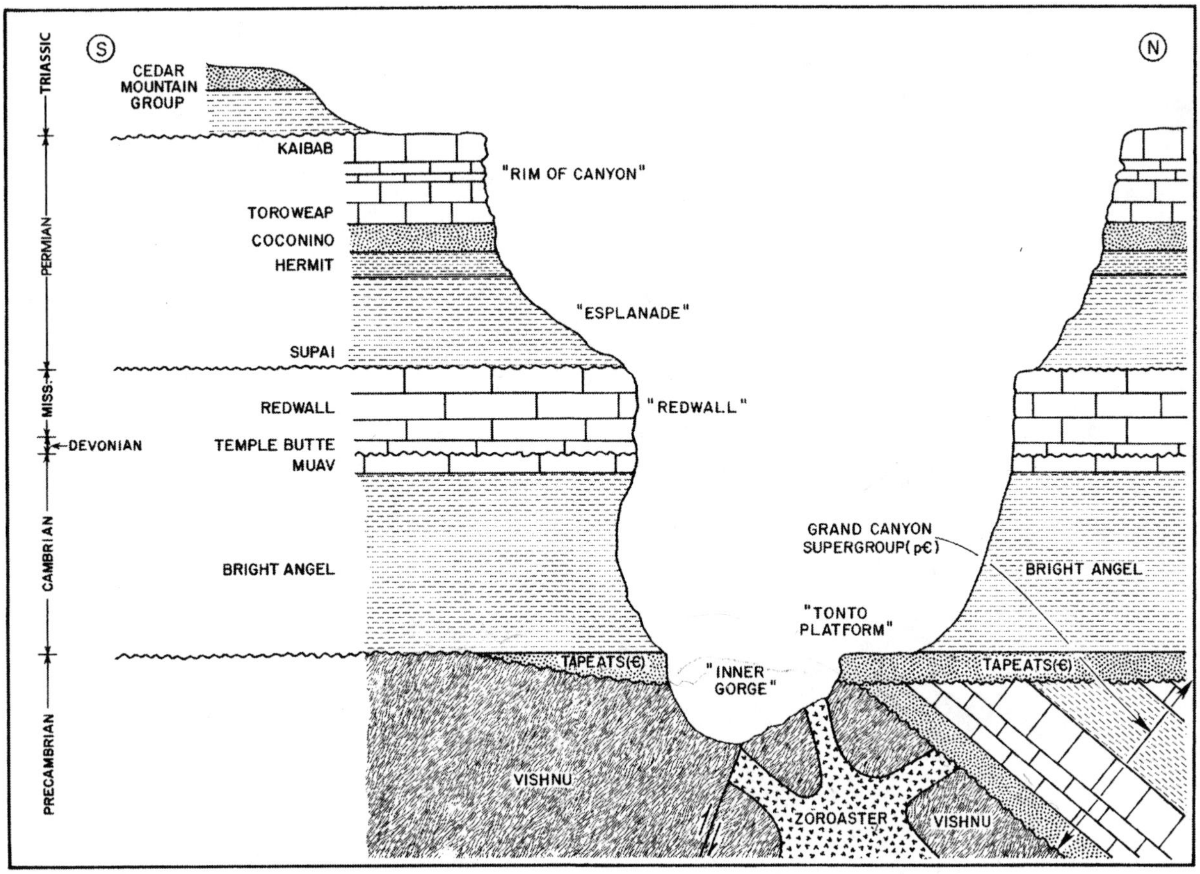

FIGURE 2.28 Cross section of Grand Canyon.

Exercise 2-7 INTERPRETATION OF THE VALLEY AND RIDGE PROVINCE IN NORTHWESTERN GEORGIA

The Valley and Ridge Province is one of the major subdivisions of the Appalachian mountain system. This province extends in a long linear belt from New York to Alabama (see the physiographic map, Fig. 5.1 on p. 213) and is dominated by parallel folding and low-angle faulting of Paleozoic sedimentary rock. In Georgia, the Valley and Ridge Province (see Fig. 2.29) is confined to the northwestern corner of the state. To the south and east of this region in Georgia, the folded Paleozoic sediments are bounded by the igneous and metamorphic rocks of the Piedmont and Blue Ridge provinces.

a. There are two faults in the cross section. What type are they? Cite your evidence.

b. Project the Silurian and Devonian strata from Lookout Mountain to Taylor Ridge.

(1) Circle the correct structure:

(a) syncline **(b)** breached anticline
(c) monocline **(d)** graben

(2) From the available evidence, what is the earliest geologic period this structure could have formed?

c. The Pennsylvanian rocks form a conspicuous ledge-forming stratigraphic unit at the top of Lookout Mountain. Why are these rocks more resistant to erosion than others in the cross section?

d. What type of unconformity is found here?

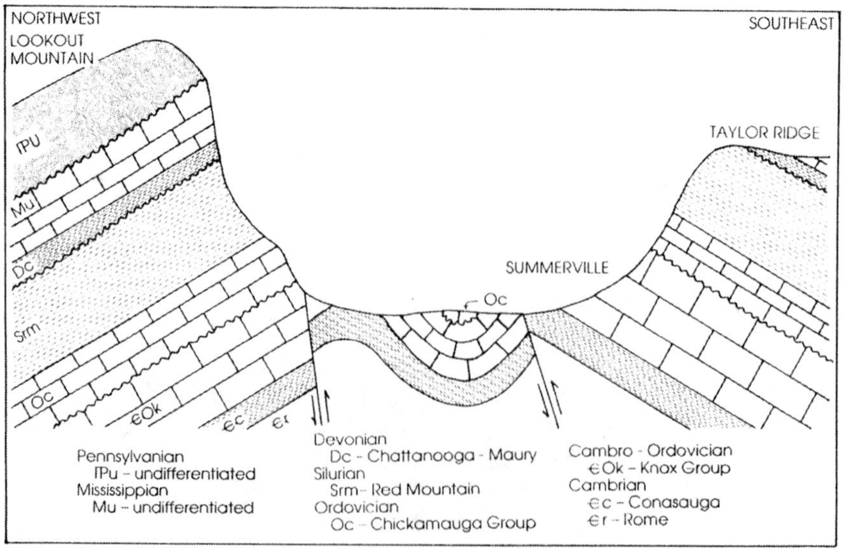

FIGURE 2.29 Structural cross section, Georgia Valley and Ridge Province. Adapted from USGS, *Geologic Atlas of the United States.*

Exercise 2-8 TEN MILE RIVER MINING DISTRICT, COLORADO

The Ten Mile River Mining District is a lead, zinc, and silver mining area in north central Colorado near the famous Leadville Mining District. The mines were opened in the late 1870s and are still being worked today. Use the decay curve, Figure 2.26 to determine the age of the igneous intrusions in this area. Then answer the questions that follow.

Intrusion	% parent X	Age
EM	83%	_____
LP	76%	_____
R	60%	_____
GN	xxxx	1.8 billion years

a. Using the foregoing radiometric ages, superposition, and cross-cutting relationships, determine the relative age relationships for the rock units present in Figure 2.30. Enter the symbols in the correct space on the cross section.

b. Using Figure 2.30, list the types of folds and faults present in this cross section. Describe the forces involved.

c. The Lincoln Porphyry intrudes the Maroon formation along a bedding plane. What is the name for such a concordant intrusion?

d. From the information given, how can you determine the age relationship between the Elk Mountain Porphyry and the fault?

e. Write a brief geologic history of this area.

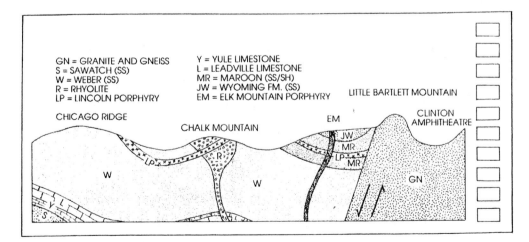

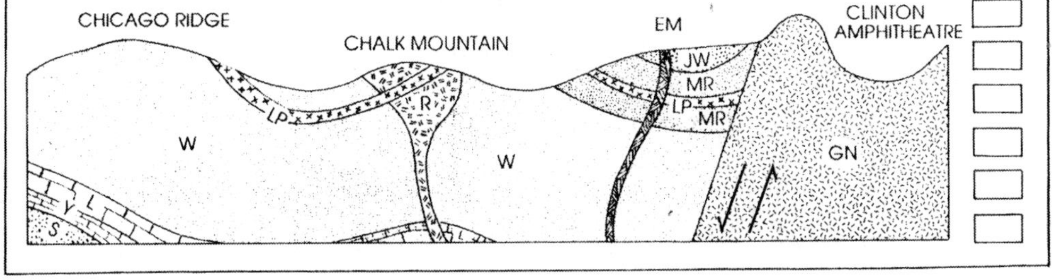

FIGURE 2.30 Cross section of Ten Mile River Mining District, Colorado. Adapted from USGS, *Geologic Atlas of the United States.*

Exercise 2-9 INTERPRETATION OF THE GULF OF SUEZ

The Gulf of Suez is a northern extension of the Red Sea, west of the Sinai Peninsula (see Fig. 2.31). It contains the largest oil-producing area in Egypt. Much of the oil is brought ashore at Ras Shakeir in Egypt and is carried by pipeline to Cairo.

Until the relatively recent petroleum exploration, little was known about the subsurface sediments or structures within the Gulf of Suez. With all the seismic and drilling data currently available, geologists have been able to determine that the Gulf of Suez is a large graben, with much subsidence and internally complex, smaller-scale horst and graben faulting of the main downdropped block. The majority of oil fields in the Gulf of Suez have a structural configuration as shown in Figure 2.32. Geological interpretations based on drilling indicate that the first major episode of faulting occurred during the Late Paleocene. Prior to the Early Tertiary, the depositional patterns and structures in northeastern Egypt were tied to the overall development of the eastern Mediterranean area; the Red Sea had not yet developed. As the marine waters filled the developing Gulf of Suez basin during Eocene, Oligocene, and Early Miocene times, thick accumulations of evaporites were formed. More recently, Late Miocene normal faulting cross-cut the Early Miocene and older clastics and

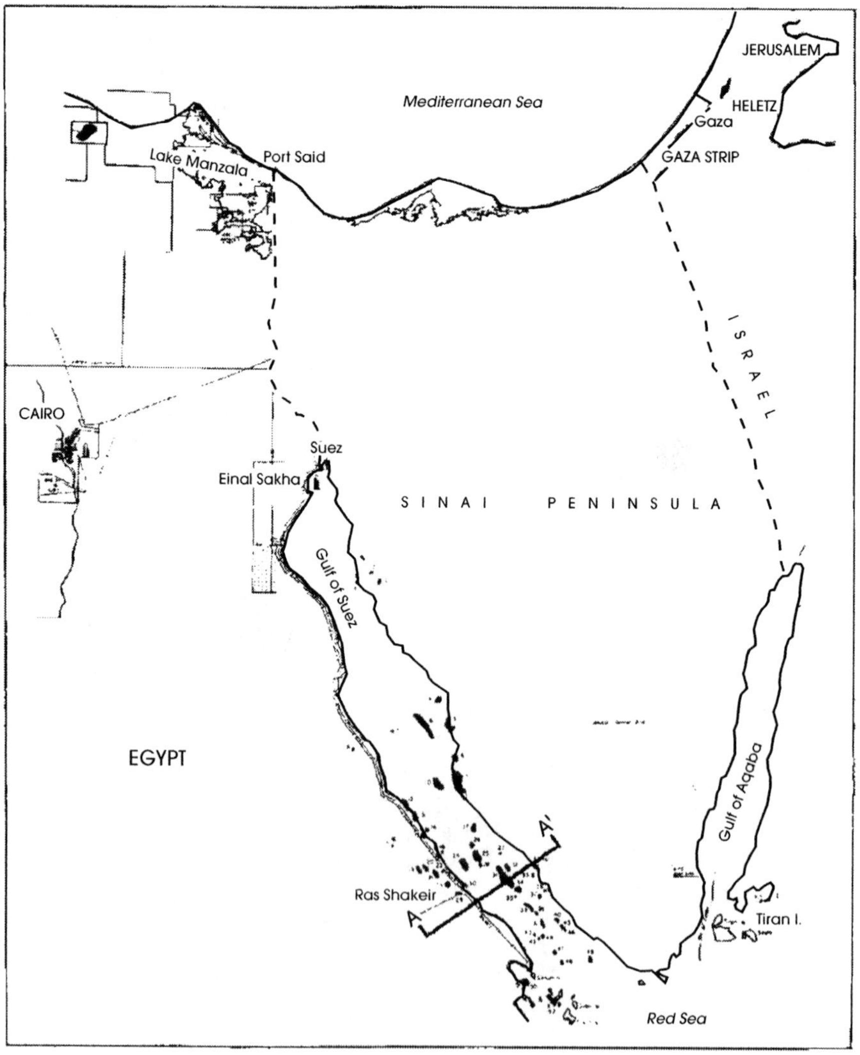

FIGURE 2.31 Location map, Gulf of Suez.

evaporites. The Miocene faulting reactivated some of the older Paleocene faulting and was responsible for forming most of the petroleum traps associated with the present geologic structures in the Gulf of Suez. Subsequently, the Gulf of Suez joined the Red Sea and became part of the world's developing rift system. Use Figure 2.32 to answer the following questions.

a. What types of unconformities are shown in Figure 2.32?

b. Why do the faults seem to terminate within the thick layer of evaporites?

c. What geologic events have occurred in this area, in what sequence, and during which geologic periods or epochs? Discuss briefly.

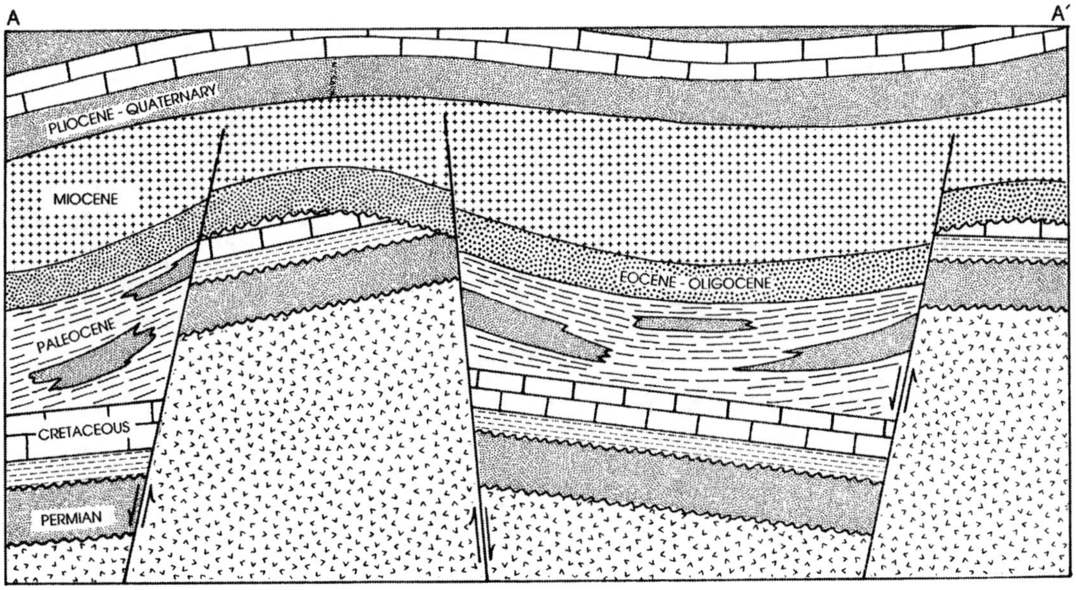

FIGURE 2.32 Generalized geologic cross section, Gulf of Suez, A-A′.

Exercise 2-10 SUBSURFACE PAKISTAN

The area in Figure 2.33 is located in the foothills of the Himalaya Mountains in northern Pakistan. The headwaters of the well-known Indus River are in this area. Northern Pakistan has been involved in divergent and later convergent plate tectonic movements (see Chapter 6). The cross section represents a section across the northern edge of a large late Tertiary sedimentary basin to the south that has been filled with erosional materials from the rising Himalaya Mountains further to the north. Refer to Figure 2.33 to answer the following questions.

a. **(1)** What type of faulting is found in the Mesozoic sediments?

(2) What force produced this faulting?

b. Did the faulting episode end before or after the deposition of the Paleocene sediments? Cite your evidence.

c. **(1)** What type of unconformity exists between the Mesozoic–Paleocene strata and the Pliocene sediments?

(2) Approximately how many million years are missing in this unconformity? (Refer to the Geologic Time Scale, inside back cover.)

d. Write a brief geologic history of this area.

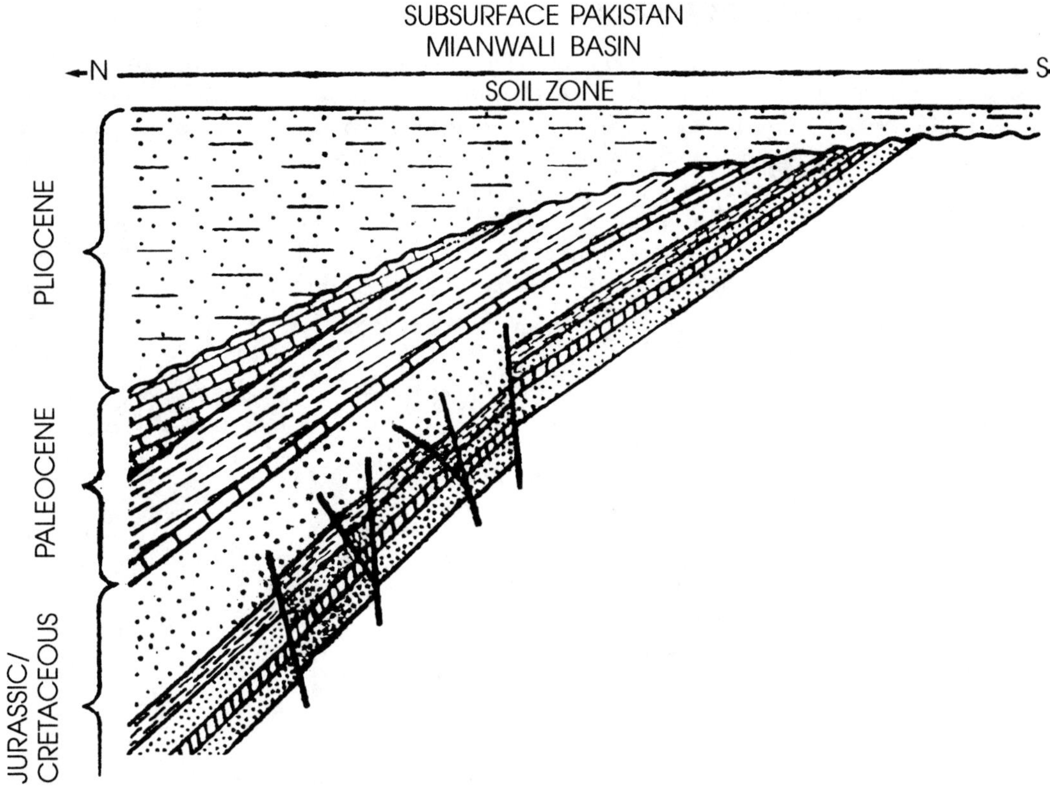

FIGURE 2.33 Cross section of Mianwali Basin, Pakistan.

Exercise 2-11 GEOLOGIC CROSS SECTION AT BISBEE, ARIZONA

The Bisbee, Arizona, area was an unimportant lead-mining district until copper ore was discovered at the Copper Queen Mine in 1880. The ore was extremely rich (23% copper) and sulfur-free. Before the richest ore was exhausted, in 1884, the main shaft had been dug more than 300 feet down an incline into the hillside as it followed the ore body. The owners bought the nearby Atlanta Claim and soon discovered another major new ore body that connected the two mines. By the early 1900s, all of the richest oxidized copper ore had been mined. The mines operated into the 1970s by economically using ores with copper content as low as 0.25–0.75%, thanks to advances in mineral extraction and refining technology. Many of the world's finest copper mineral specimens of malachite, azurite, cuprite, and others were collected from the Bisbee Mining District, in particular the Copper Queen Mine.

a. Using Figure 2.34, examine the cross section and determine the superpositional relationships among the rock units present. Enter the symbols for the rock units in the stratigraphic column to the right of the figure.

b. What types of faults are evident in the area of the cross section? What forces were involved in producing them?

c. What stratigraphic relationship probably exists between the Pinal schist and the Bolsa quartzite?

d. List and name the types of unconformities in this cross section and determine the missing periods of geologic time. (Use the Geologic Time Scale, inside back cover.) Add the unconformity symbol to the cross section in the appropriate places.

e. Why does limestone create hills in this desert area of Arizona?

f. During which geologic period was the granite porphyry intruded into the Bisbee area?

g. What are the host rocks for the various copper mines?

h. Hot copper-bearing fluids penetrated the crust toward the surface. What rock type was preferentially replaced by these fluids in the Bisbee Queen Mine area?

i. Determine the geologic history of this area.

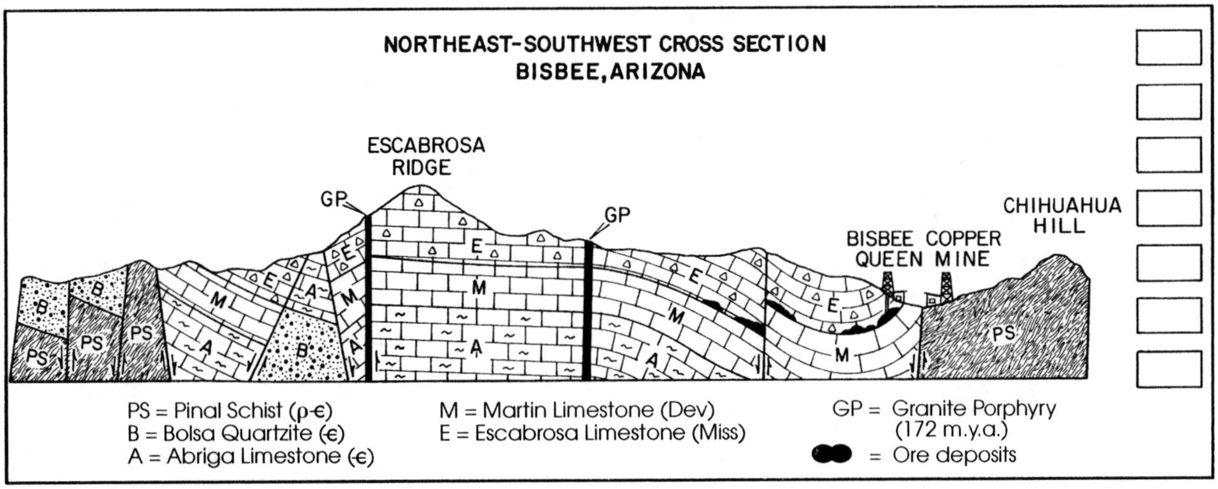

FIGURE 2.34 Cross section, Bisbee, Arizona. Adapted from USGS, *Geologic Atlas of the United States.*

CHAPTER

3

Physical Stratigraphy

STRATIGRAPHY AND CORRELATION

Overview

In the last chapter you learned to place a series of rock units in sequential order in terms of both relative and measured time by determining the relationship of a series of geologic events one to another. In this chapter you will be examining the relationship of those rock units in stratigraphic terms by correlating the units' lateral and vertical relationships through time.

The chapter covers two primary aspects of physical stratigraphy. First, the discussion will focus on how geologists differentiate and interpret individual or sequences of rock units in terms of being defined as rock-stratigraphic or time-stratigraphic units. Next, the discussion centers on how geologists define and use multiple correlation techniques to develop an equivalency among multiple outcrops of rocks. Correlation is an important concept to geologists. Correlation is one of the main tools used to establish local, regional, or worldwide similarities or changes of rock units within a given relative time frame. The establishment of a correlation between rock units is important when interpreting regional environments of deposition, the presence or absence of unconformities, facies changes within rock units, and a region's geologic history.

STRATIGRAPHY

Physical Stratigraphy

To begin the complex process of conducting an interpretation of the geologic history of a given area, such as the Black Hills of South Dakota or the Catskill Mountains of New York, students and professional geologists alike must first collect geologic data from old maps, outcrops, or drilling information, followed by analysis and interpretation. Typically rock outcrops are available along riverbanks and railroad and highway cuts. Subtle changes in the soil types of forests or fields yield information on the underlying rocks. Additionally, detailed subsurface stratigraphic data can be obtained through exploration drilling data associated with water, petroleum, or mining activity. By measuring, describing, and interpreting these scattered pieces of data, geologists can create a broad, detailed interpretation database to develop a region's geologic history. The correlation process of either matching the similarities or noting the differences in the sequence of rock strata from area to area can often be complex. Successful correlation can be accomplished only if the strata have been carefully grouped and environmentally analyzed for evidence that can indicate equivalency. Once correlations are complete, the resulting analysis will lead to an interpretation of an area's geologic history.

Time and Time-Rock Units

Geologists use two basic stratigraphic methods to subdivide groups or sets of various rock strata. The first involves relative geologic time. In this method, the overall geologic time scale (see inside back cover of the manual) is progressively broken down into smaller, more detailed time subdivisions (largest to smallest): *eon, era, period, epoch,* and *age*. The second method involves designation of which groups or sets of rock strata have been deposited during any given interval of geologic time. The three main subdivisions are (largest to smallest) *system, series,* and *stage*. Those rock strata designated to have been deposited during a specific unit of time are called *time-rock* units. For example, the Ordovician *system* (a time-rock term) consists of whatever rock strata were deposited during the 70 million years designated as the Ordovician *period* (a time term). The equivalent time and time-rock terms are given in the following chart.

GEOLOGIC TIME UNITS	TIME-ROCK UNITS
Eon	*
Era	*
Period	System
Epoch	Series
Age	Stage

*The terms *Eonothem* and *Erathem* have been proposed but are not considered as practical and are thus not commonly used.

Rock-Stratigraphic Units

Within the realm of time-rock units, geologists must describe and interpret all of the rock units individually and in groups to determine and place these stratigraphic units into the time framework. Thus, geologists have a third set of defined terms, called *rock-stratigraphic* terms, which are not related to time. Individual or small sets of rock layers are correlated locally and regionally to determine the lateral boundaries of those units. Then specific rock-stratigraphic terms are given to these individual or groups of rock layers. The smallest rock-stratigraphic unit that a geologist usually designates on a geologic map is a *formation*. A particular rock unit must meet at least two criteria for a geologist to term it a formation: (1) It must be widespread and thick enough to be mappable, and (2) it must be readily distinguishable from adjacent units. A formation's strata were usually deposited during a restricted time interval under a certain set of environmental conditions within a given area.

ROCK STRATIGRAPHIC UNITS
Group
Formation
Member
Bed

Group—Two or more associated formations.
Formation—A distinctive and mappable rock unit.
Member—Subdivisions of a formation (at least two must be present for such subdivision to be valid).
Bed—The smallest recognized rock-stratigraphic unit (a distinctive portion of a member; *examples:* coal bed, iron-rich bed, and oil-bearing sand bed).

No set limits clearly establish whether formations are mappable. Some individual rock units that are treated as formations are only 4–5 feet thick, whereas other units, such as the Mancos and Pierre formations (Cretaceous) in Colorado and Wyoming, are over 1,000 feet thick. In most cases a formation must be approximately 10–20 feet thick before it is considered readily mappable. Formations commonly are composed of rock units that accumulated under the same environmental conditions or under conditions that were uniformly changing. Therefore all the parts of a formation should be developmentally related. A formation consisting primarily of one rock type usually incorporates that rock type in its name (Pierre Shale). A formation including two or more related rock types will be named more generally (Morrison Formation).

The boundaries of formations, called *contacts*, usually represent major changes of lithology or sedimentary environment. A formation may start to form first in one area; but then, as a depositional environment expands, the resulting formation will also expand. Consequently, some parts of a formation may have been deposited at different times in different places.

In Figure 3.1(a), the stratigraphic column shows some of the varying formations that can make up a group. Figure 3.1(b) indicates how one formation from this group, the Oread Limestone, is subdivided into members after detailed local mapping. For each named formation (or any rock-stratigraphic unit), a key location and outcrop exists, for example, a creek bank or a canyon. The locality where the formation is well exposed and first defined is called the *type locality*. If it is in the proximity of a well-known geographic landmark, the formation name may be derived from that landmark. For example, the type locality for the Morrison Formation of Jurassic age is near Morrison, Colorado.

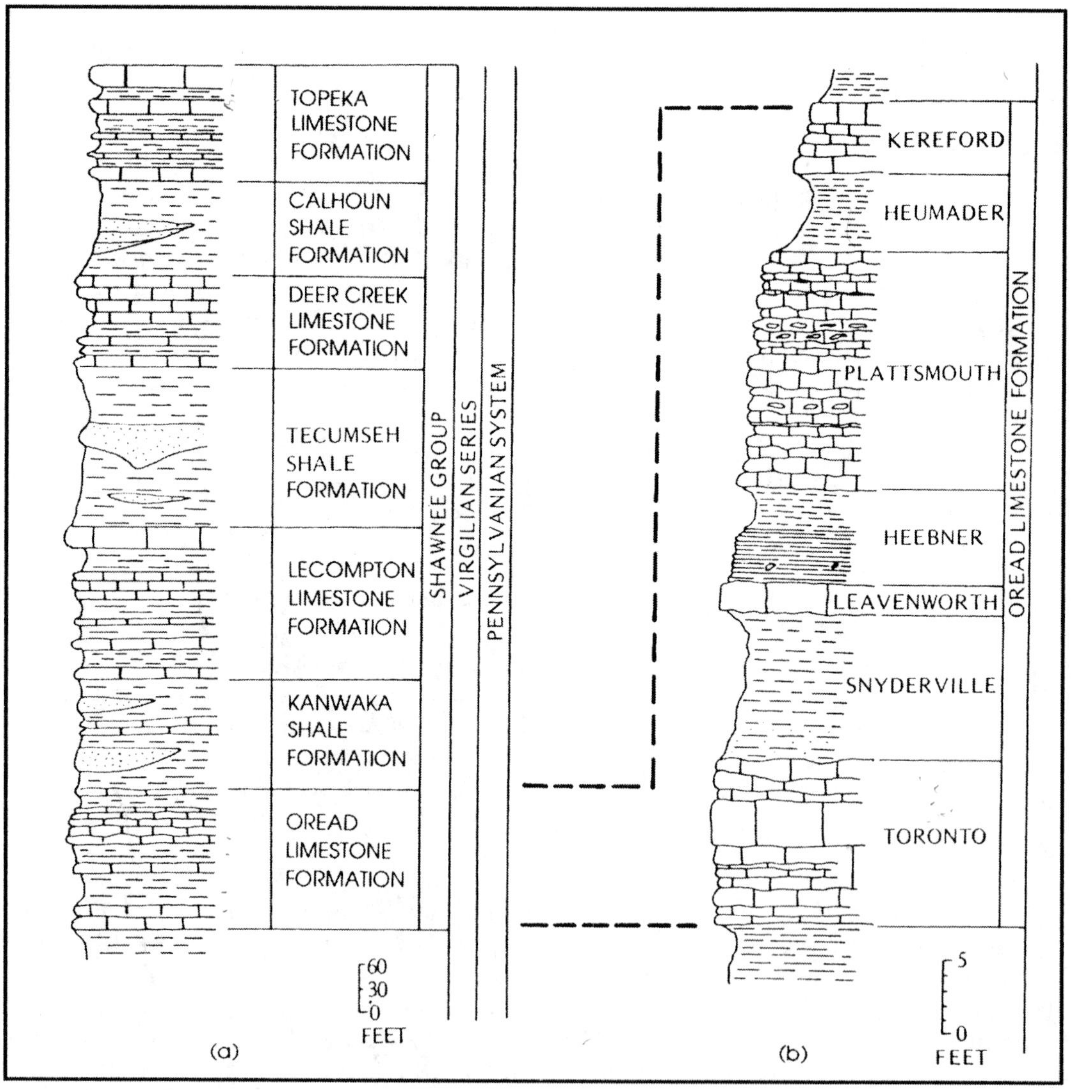

FIGURE 3.1 Measured stratigraphic section of Pennsylvanian strata, Kansas.

In addition to the formal rock-stratigraphic units just discussed, there are some other less formally defined stratigraphic units, such as *biostratigraphic* units (which are identified by a distinctive assemblage of fossils), *paleogeographic* units (which are determined by regional depositional patterns), and *economic* units (which are used to describe the specific occurrence of a mineral-rich zone). The most important and widely used of these is the biostratigraphic unit. In the Gulf Coast region of the United States, for example, there is a widespread and distinctive index fossil, the foraminifera *Heterostegina,* which appears in the subsurface and marks the famous "Het" zone. For geologists and oil companies, the Het zone serves as an important marker for regional correlations.

EXERCISES

Exercise 3-1 STRATIGRAPHIC SECTIONS, COLORADO

The information in Table 3.1, the data set table, provides thickness measurements and descriptions for the geologic units present in two Paleozoic outcrop sections in Colorado. These data are used to construct and fill in the two correlation columns shown in Figure 3.2. Use

TABLE 3.1 Data Set—Colorado Stratigraphic Columns

Southeast Colorado (Column A)	*Feet*	*South central Colorado (Column B)*	*Feet*
Leadville Formation			
(Middle Mississippian)			
Limestone, nodules of chert	110		
Quartzite	5		
Limestone, medium bedded, unfossiliferous	25		
Chaffee Formation		Leadville Formation	
(Upper Devonian–Mississippian)		(Middle Mississippian)	
Limestone, medium bedded, fossils	20	Limestone, thick bedded, crinoids	120
Dolostone, thick bedded	30	Limestone, fossils	10
Quartzite, conglomeratic	25	Limestone with chert	15
Covered	20		
Manitou Formation		Ouray Formation	
(Lower Ordovician)		(Upper Devonian)	
Shale, thin bedded	10	Limestone, thick beds fossiliferous	55
Dolostone, thick beds, fossils	15	Shale, fossils	10
Limestone, chert nodules	45	Sandstone and shale	35
Sawatch Formation		Ignacio Formation	
(Upper Cambrian)		(Upper Cambrian)	
Sandstone, massive	10	Quartzite with lenses of conglomerate	75
Quartzite, massive	5		
Quartzite, irregular cross-bedded	90		
Precambrian		Precambrian	
Total	410	Total	320

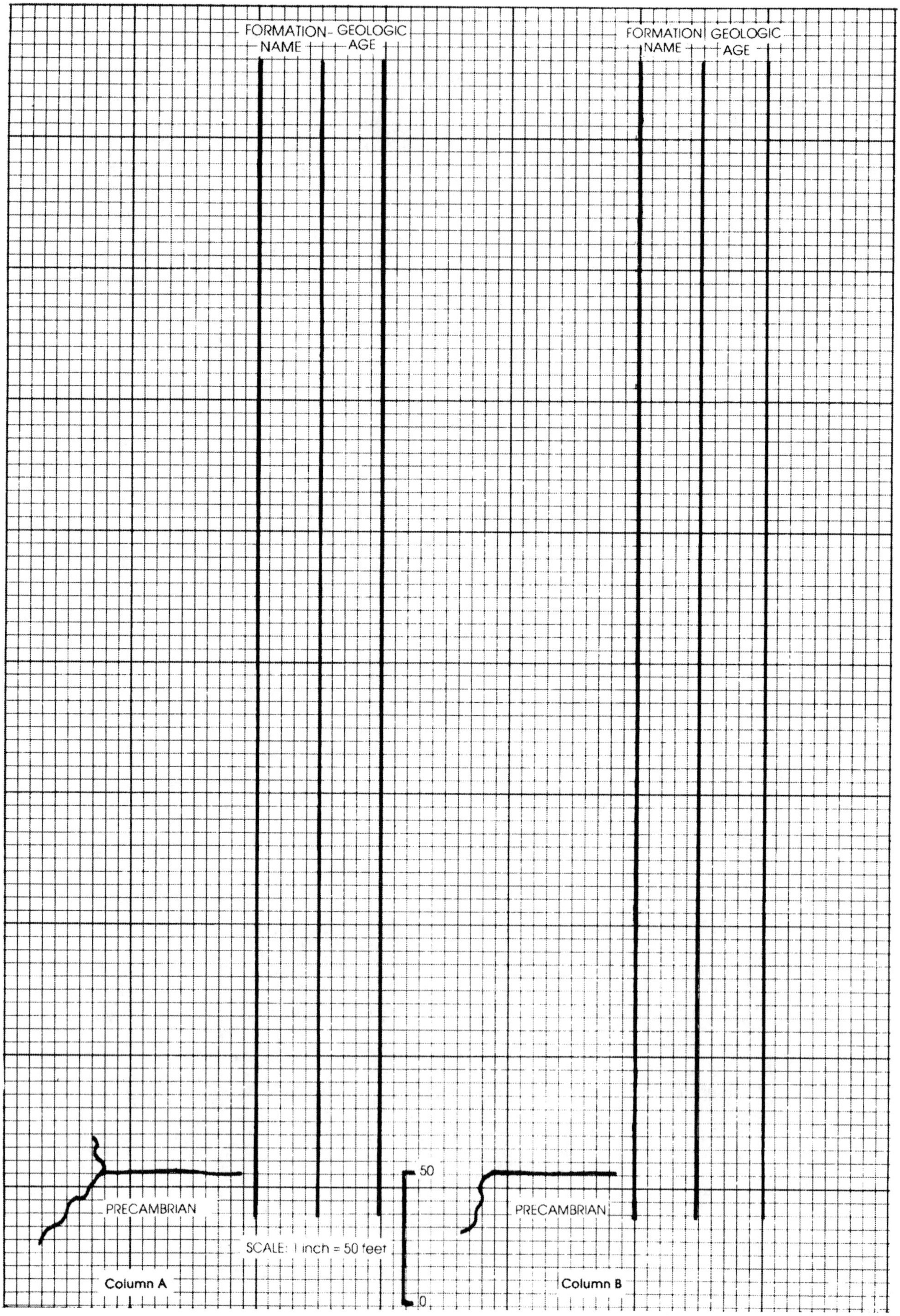

FIGURE 3.2 Graph for construction of columns using Table 3.1 data set.

the column profile shown in Figure 3.1 as a model for a graphical presentation of how a measured section should look along with its weathering profile. Remember that limestones in most climates resist weathering relative to the softer erosion-prone shales; thus, the carbonates are generally ledge formers and the shales tend to have a concave inward pattern in profile. The graphic scale for the thickness of the units is provided on Figure 3.2. When you are finished building the columns, like that shown in Figure 3.1, correlate them by age and lithology and answer the following questions.

a. Which formations are equivalent by age and lithology even though the formation names are different?

b. Which formation is present in one area but not in the other?

c. Which formation changes both its name and thickness from one area to the other?

LITHOSTRATIGRAPHIC ANALYSIS

Geologists ultimately need to synthesize all the geological data that have been accumulated in the field and the laboratory. One of the most useful ways to synthesize this information is to construct *lithofacies* and *paleogeographic* maps of the study area. The accurate construction and interpretation of these maps depend on an adequate understanding of the principles of correlation, sedimentary environmental analysis, and facies relationships.

A concept fundamental to the interpretation of a sequence of sedimentary rocks is that of the relative position of a shoreline. When compared to the reference position indicated in Figure 3.3(a), the shoreline in Figure 3.3(b) has moved toward the center of the continent. This landward change of the shoreline with time (called *transgression*) can be caused by a subsidence of the land or by a rise in sea level or by a combination of both processes. Conversely, if the sea level is lowered or the continental margin is raised, the shoreline will be forced away from the continent, as shown in Figure 3.3(c). This process is generally known as *regression*.

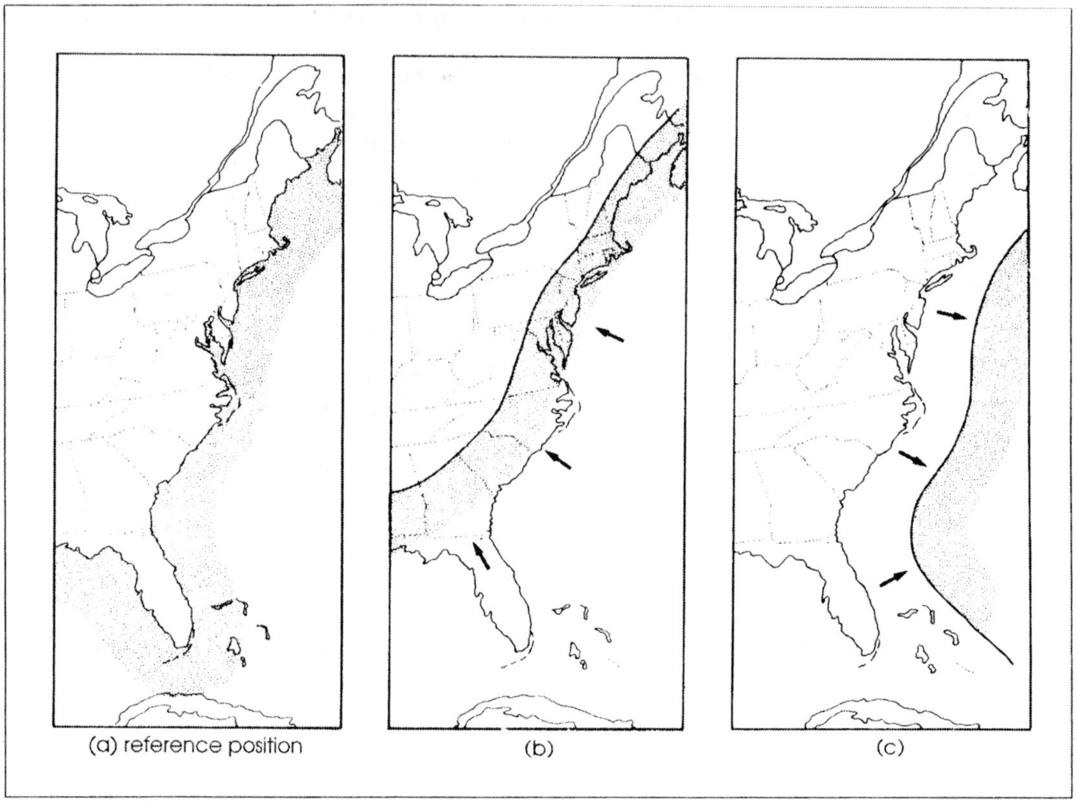

FIGURE 3.3 Relative shoreline positions.

Development of Sedimentary Facies

Assuming the existence of stable conditions, a profile that extends from a coastal plain onto a continental shelf might show a system of lithologic variations, indicated by the patterned zones in Figure 3.4. Under normal-energy conditions, the coarsest material transported by a stream will be deposited at the first major gradient change, probably on the coastal plain or in the beach area. Deposition offshore will consist of progressively finer and finer sediments. Further out on the offshore shelf, beyond where clastics are

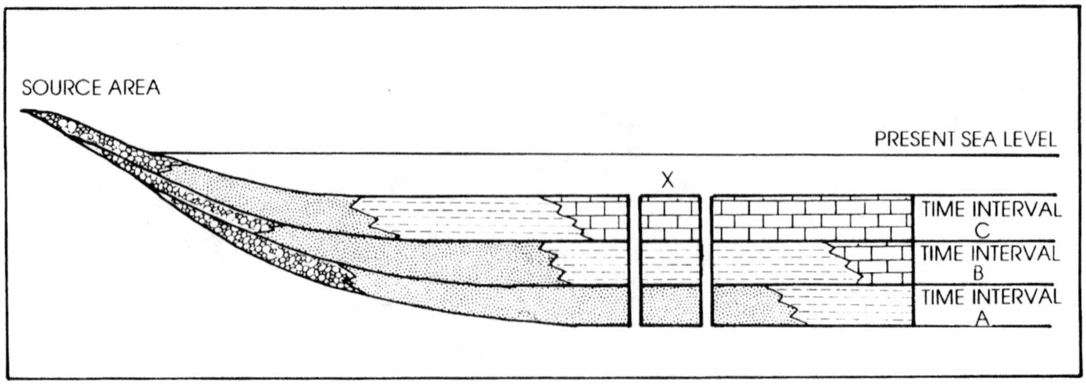

FIGURE 3.4 Transgression.

deposited, precipitated minerals will be dominant. Figure 3.4 shows a hypothetical distribution of particle sizes.

The distribution of sediments on a continental shelf is a complex process that depends on the shelf's proximity to nearby streams, current systems, and wave action. Each distinctive type of sediment (sand or limestone, for example) is called a *sedimentary facies* or *lithofacies*. A *facies change* is the transition zone where one of these rock lithologies grades into another. For example, in time interval C in Figure 3.4, four rock facies and three facies changes are shown.

Movement of Facies Patterns: Transgression and Regression

Geologists have documented numerous intervals throughout geologic time when relative sea level has changed. The transgression of the sea onto a continent produces a phenomenon known as *onlap*. Onlap occurs when a series of sedimentary facies (one combination set of sands, shale, and limestone: see time interval B in Fig. 3.4) moves landward or shifts to a position further onshore (compare time interval C in Fig. 3.4 to time interval B). In a cross section (Fig. 3.4), evidence of a transgression can be found by looking at the vertical sequence of rocks. For example, assuming that column X in Figure 3.4 represents a sequence of strata exposed in an individual outcrop, a transgressive series of sedimentary facies would begin with a sand unit and subsequently be overlain first by a shale unit and finally by a carbonate unit. This sequence of strata also indicates a decrease of energy with the depositional environment. The vertical column X represents a fixed point in space along the line of the cross section and thus "sees" the successive movement of facies past this point. It is important to note that the order of the units in the column is also the same as the hypothetical lateral distribution of sedimentary facies at any one time (see time interval B in Fig. 3.4). This stratigraphic concept of the succession of sedimentary facies is known as *Walther's law*.

During a regression the reverse sequence of events takes place. As the continent becomes elevated (emergent) or as sea level drops, more sediments are transported toward the coastal depositional environments and the distribution of facies is reversed. Thus deposition occurs progressively further offshore, away from the old shoreline (see Fig. 3.5). Again, using column X in Figure 3.5 as an example, the vertical section begins with a carbonate unit at the base and proceeds to a sand unit at the top. Thus the stable low-energy depositional environment represented by the carbonate unit is gradually replaced by the higher-energy clastic facies. Therefore, when a given outcrop is examined, a geologist is able to interpret whether the strata in the outcrop represent a transgressive or regressive series of

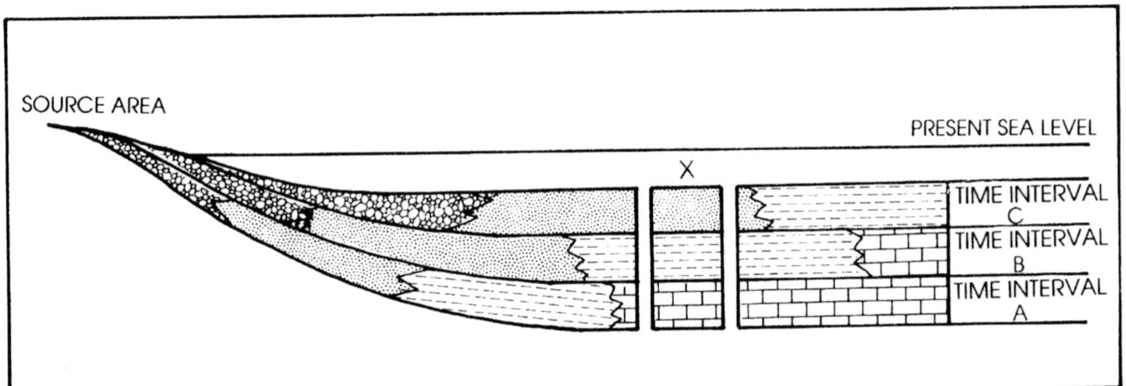

FIGURE 3.5 Regression.

units by noting whether the rock strata in progression from bottom to top (older to younger) become increasingly finer or increasingly coarser. Thus, the order of the rock units in the outcrop may yield information about the relative position of the shoreline at that time.

Time-Transgressive Stratigraphic Units

A geologist often attempts to trace a given lithology or formation over a large region. In doing so, the geologist must resolve whether a specific lithology (for example, the coarse sandstone shown in Fig. 3.6) accumulated during a single-time interval (such as time-rock interval B) or whether the formation is a combination of units that have accumulated in different time-rock intervals (such as intervals B and C). In the example pictured, if all the sandstones were correlated (regardless of their time of deposition) as one distinctive rock unit, they would be time-transgressive. In other words, when correlating the sandstone of time-rock interval B with the sandstone of time-rock interval C, two different periods of deposition are involved, and the correlation thus crosses or transgresses the time line dividing the two units (see Fig. 3.6). Fossils can be valuable tools in recognizing time-transgressive units.

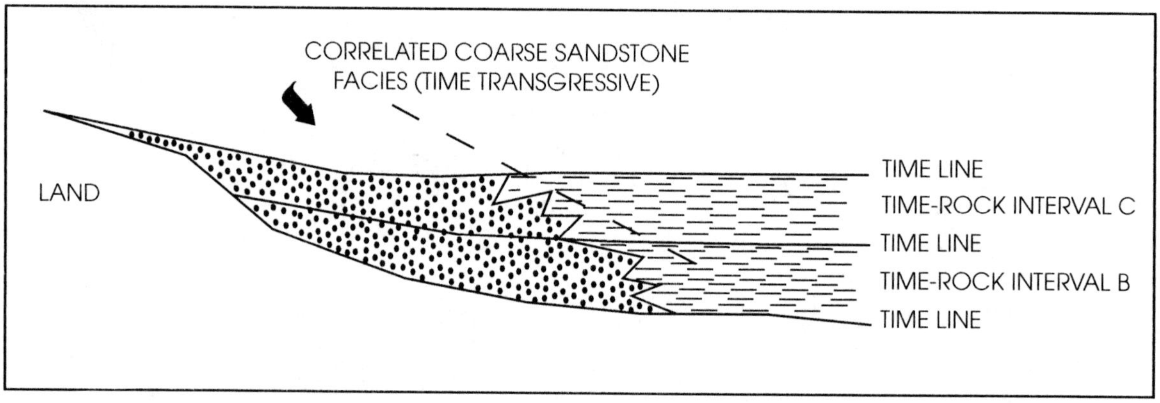

FIGURE 3.6 Time-transgressive facies.

Construction of Lithofacies Maps

Lithofacies maps are used to define the geographic distribution of similar facies or rock types. Lithofacies maps may be constructed for modern environments of deposition, such as coastal regions, river systems, or deserts, or for almost any specified environment. Such a map can also be made for the reconstruction of the depositional patterns during a specific interval of the geologic past. The only difference is that in a modern environment the samples are easily collected on or near the earth's surface, but compilation of a lithofacies map for any time in the geologic past (for example, the Mississippian of Kentucky) requires collecting both surface and subsurface geological data.

Method of Completion

Step 1. First, scan the sample locality map to mentally differentiate the distributions of the various rock types involved [see Fig. 3.7(a)].

Step 2. Next, isolate all samples of the same facies. Remember that the procedure is very similar to that of drawing contour lines, but do not connect the points of similar

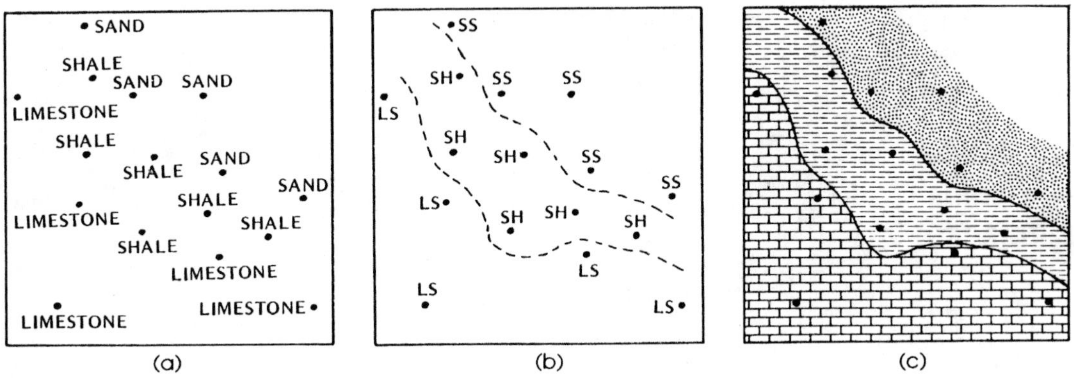

FIGURE 3.7 Construction of a lithofacies map.

lithology directly. Each point represents only an individual rock sample; the actual facies change occurs between two adjacent locations of differing rock types [see Fig. 3.7(b)]. Sketch in a line that weaves between adjacent points of differing rock types and separates the two rock types. The completed lithofacies pattern is shown in Figure 3.7(c).

EXERCISES

Exercise 3-2 DEVONIAN PALEOGEOGRAPHY

The following problem will illustrate the technique used by geologists to interpret lithofacies maps.

a. Using the technique shown in Figure 3.7, construct a lithofacies map for the Devonian depositional environments in part of the northern Appalachians using Figure 3.8. Devonian rocks are absent in the shaded portion of the data map.

b. Write a synopsis of the geology during the Devonian in this region of the United States.

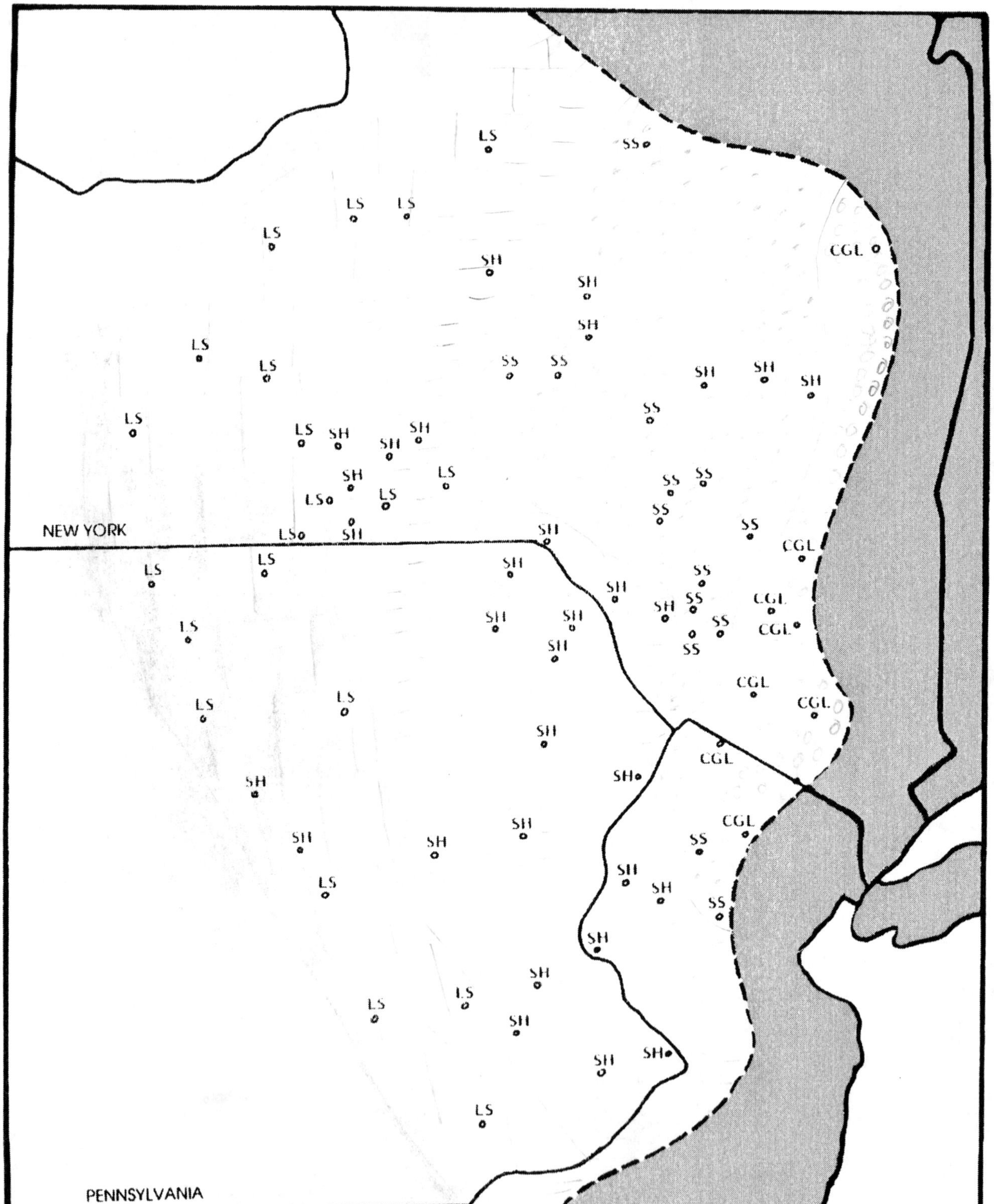

FIGURE 3.8 Devonian field sample locations. – Deeper waters to the west.
– Source zone to the east.

Exercise 3-3 CAMBRIAN IN THE "FOUR CORNERS" AREA OF THE UNITED STATES

The paleogeographic map of the Cambrian system (Fig. 3.9) shows the preserved Cambrian strata in the "Four Corners" area of the southwestern United States. In Arizona, the names of the formations are the same as those Cambrian units found in the Grand Canyon (see Fig. 2.28). In the upper corner of the map, a correlation chart gives the names of the Cambrian formations in western Colorado and eastern Utah. The following tabulation lists the lithologic samples collected.

1. limestone	**13.** sandstone	**25.** shale	**37.** sandstone
2. shale	**14.** absent	**26.** limestone	**38.** limestone
3. limestone	**15.** absent	**27.** limestone	**39.** shale
4. dolostone	**16.** sandstone	**28.** dolostone	**40.** sandstone
5. dolostone	**17.** shale	**29.** dolostone	**41.** absent
6. dolostone	**18.** absent	**30.** dolostone	**42.** absent
7. shale	**19.** sandstone	**31.** dolostone	**43.** absent
8. limestone	**20.** sandstone	**32.** dolostone	**44.** absent
9. limestone	**21.** shale	**33.** limestone	**45.** absent
10. shale	**22.** sandstone	**34.** limestone	**46.** limestone
11. shale	**23.** sandstone	**35.** limestone	**47.** dolostone
12. sandstone	**24.** sandstone	**36.** shale	**48.** limestone

Questions

a. Plot the listed lithologies at the localities marked on Figure 3.9. Construct a lithofacies map for the Cambrian system (method shown in Fig. 3.7). Be sure to stop your lines at the boundaries bordering the two uplifts.

b. What geologic explanations could account for the following observations about the map you have constructed?

(1) The Cambrian facies stop against the sides of the Zuni and Uncompahgre uplifts and are absent on the uplifts.

(2) Cambrian facies are absent in the southeastern portion of the mapped area.

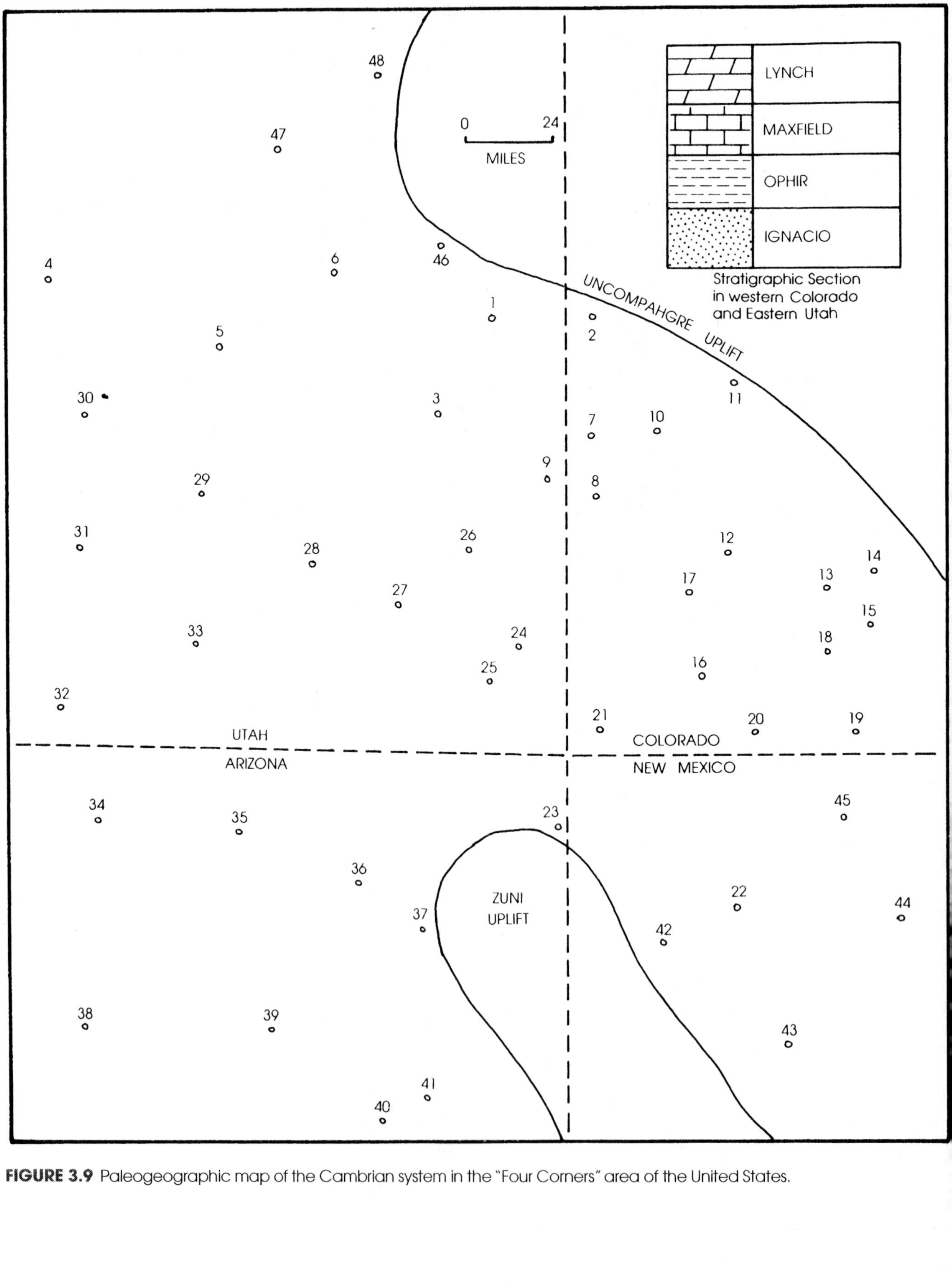

FIGURE 3.9 Paleogeographic map of the Cambrian system in the "Four Corners" area of the United States.

Exercise 3-4 LITHOFACIES OF DOGGETT FORMATION

The data in Figure 3.10 were derived from cores while drilling for petroleum. From these data, construct a lithofacies map for the Mississippian Doggett Formation.

DATA

a. Data points 1, 2,11,15,16, 26, and 33 revealed no Doggett Formation but showed granite instead.

b. All the cores were analyzed, and a dominant lithology was determined for the Doggett Formation at each point. These lithologies are located in the ternary diagram of Figure 3.10.

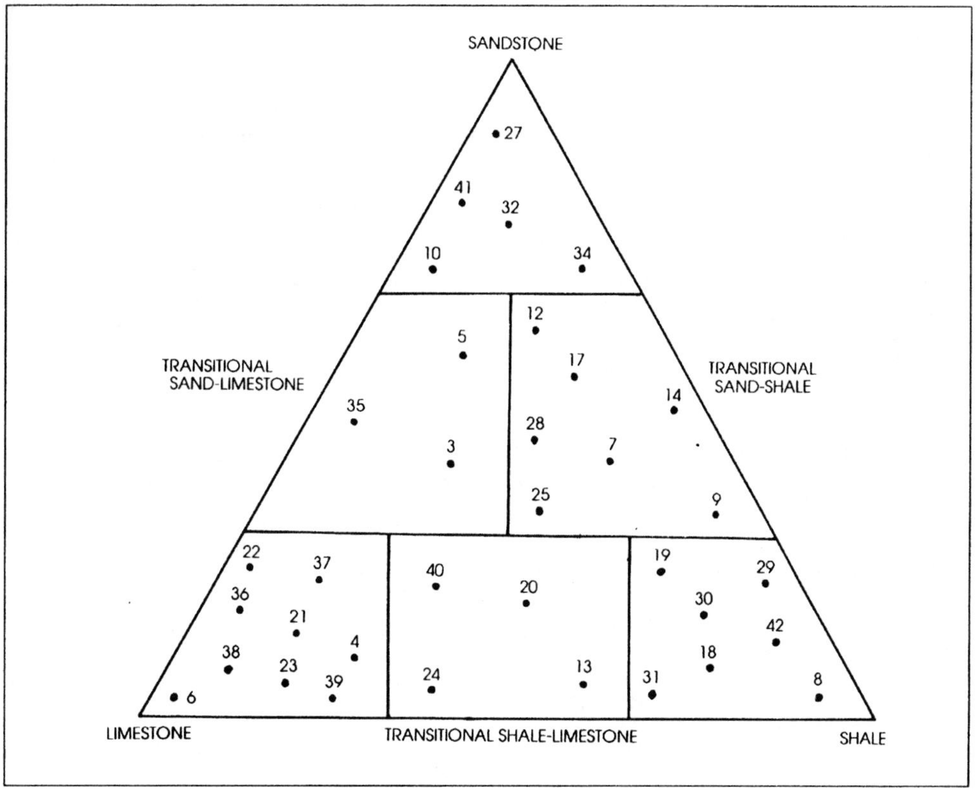

FIGURE 3.10 Ternary diagram of lithologies, Doggett Formation.

Procedure

Find the lithology representative for each data point and transfer it to the location map (Fig. 3.11). Next, sketch in the lithofacies boundaries on the lithofacies map. (Remember, both the end members and the transitional lithologies must be included on the map.)

Interpretation

Briefly discuss the paleogeography and distribution of lithologic patterns within the mapped area.

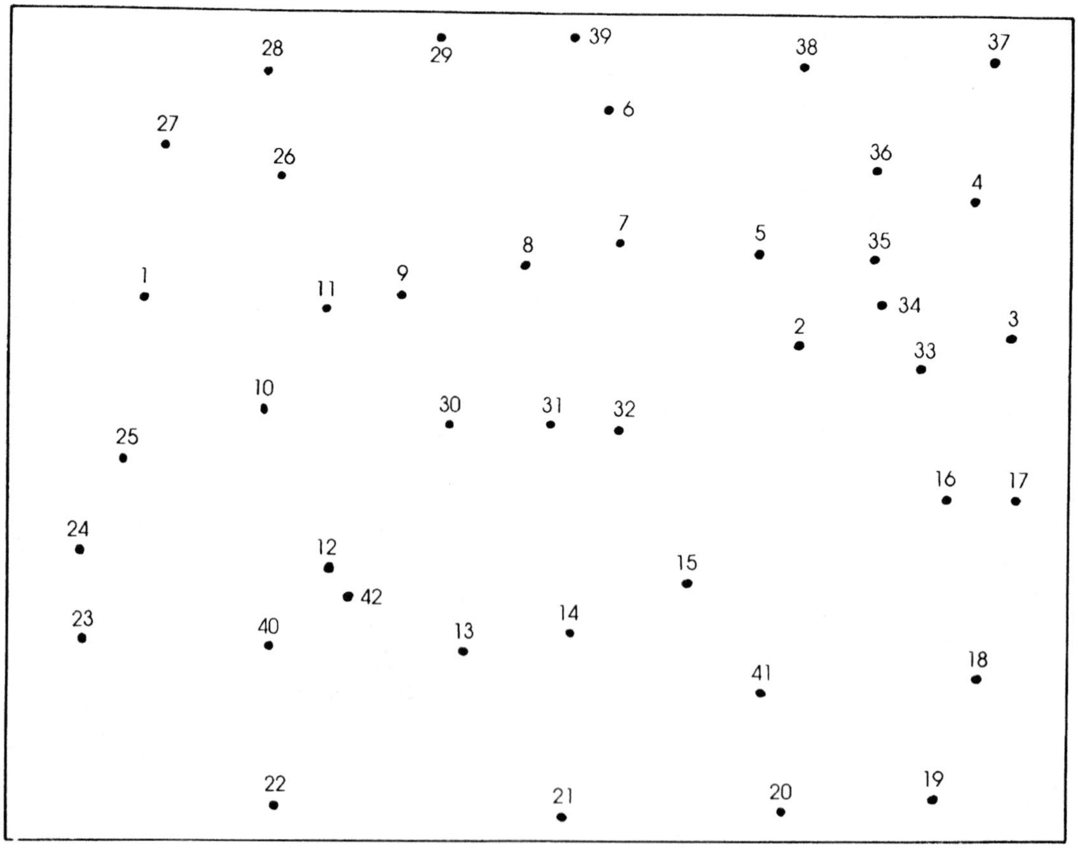

LITHOFACIES MAP

FIGURE 3.11 Lithofacies map, Doggett Formation.

CORRELATION

Correlation is the matching of rock units in terms of space or time and is very important to geologists. A geologist examines numerous outcrops along rivers, road cuts, and fields and then integrates the collected data into a composite stratigraphic column, such as that of the Oread Limestone shown in Figure 3.1. Several similar sections are then correlated on the basis of either time or lithology. From these data, a geologic map and a geologic history can be constructed for the region under study. There are many different approaches to the process of correlation, but usually one or more of the methods listed in the following discussion are used. These basic correlation techniques were used by William "Strata" Smith in England in producing some of the first geologic maps.

Lithologic Similarity

Rock units are often distinguished from one another on the basis of color, composition, sedimentary structures, and texture. When initial correlations of a series of measured stratigraphic sections are attempted, even within a relatively small area, only the most reliable lithologic characteristics of the strata should be used. After the key lithologic properties of the rock units are defined, correlation of an expanded group of measured sections throughout a larger area can then be attempted. Some lithologies, such as a zone of red clay or an intraformational conglomerate, may be very distinctive on a local scale but of less importance in regional correlations. Widespread stratigraphic units, such as a massive coarse-grained grey limestone unit, can probably be correlated confidently over quite a large geographic area. The presence of distinctive key beds or marker beds, such as thin widespread beds of volcanic ash, black shales, or fossil-rich shelly layers, are often desirable for both local and regional correlations.

Physical Tracking (Walking Out the Outcrop)

Tracing a marker bed by walking along the unit from one outcrop to another is a reliable method of correlation. Lateral tracing of strata appearing continuously in surface outcrops is one of the fundamental approaches used in constructing geologic maps from field data and in constructing maps from aerial photographs. Obviously, mapping using this approach can be severely limited by rugged, faulted, or eroded terrains, deep soil zones, or the presence of abundant vegetation. Such conditions commonly can lead to structural and terrain interpretation problems.

Stratigraphic Sequence and Thickness

Once marker beds have been defined and correlated within a series of strata, the sequential order of rock units between these marker beds should be noted. The thickness of the individual beds or the thickness of the total interval between the marker beds may vary, but the sequence of rock units between the marker beds is often regionally similar and thus very useful in establishing correlations. This approach must be taken with caution, however, if unconformities or facies changes are suspected within the area.

Paleontological Criteria

Correlations of distinctive assemblages of fossils contained in rock units usually produce one of the most reliable regional correlations. Index fossils, which are defined as having a wide geographic distribution and short vertical time range, are useful for local correlations and essential for most regional correlations. In detailed paleontological studies, researchers use range zones, concurrent range zones, and assemblage zones for correlation (see the Glossary).

Remote and Subsurface Stratigraphic Tools

Electric Log Characteristics In areas between outcrops, information on the sequences of rock units usually comes from wells that have been drilled into the subsurface. Data concerning subsurface rock sequences can be obtained by studying such well cuttings or core samples or by utilizing electrically generated records called *electric logs*. Electric logs provide a rapid and inexpensive way to obtain a continuous profile of the subsurface layering and basic rock types.

The exact nature of subsurface rock sequences on the basis of electric log characteristics is determined by lowering a logging (recording) device (an electronic capsule called a *sonde*) to the bottom of a well. As the device is raised, an electric current (or sometimes sound waves) generated at the base of the capsule is transmitted through different rock layers and received at the top of the device. Since the ability of a rock to resist or conduct electric current (or the ability to transmit sound frequencies) is a function of the rock's permeability, porosity, and fluid content, fluctuations in the transmission of electricity are recorded as the capsule is raised. These fluctuations produce the transmission profile of the subsurface rock layers called an electric log. Most sedimentary rocks have a predictable range of electrical properties that greatly facilitates subsurface interpretation and correlation from well to well. There are numerous varieties of well logs, including ones that measure the dip of the rock strata, rock densities, and natural radioactivity.

Note that the electric log in Figure 3.12 has two distinct curves, spontaneous potential (S.P.) on the left and resistivity on the right, which are both scaled from zero to positive readings. The three basic sedimentary lithologies (sandstone, shale, and carbonates) are associated with a fairly diagnostic set of electric log characteristics. Sandstone has a positive "kick" on both the S.P. and the resistivity curves. Shale exhibits relatively neutral or nearly zero S.P. and resistivity curves, whereas limestones evidence a nearly neutral S.P. curve but a highly positive resistivity "kick" (see Fig. 3.12).

Seismic and Radar Exploration Advances in seismic and radar technologies have greatly increased scientists' ability to investigate the subsurface configurations of stratigraphic sequences. By tuning the frequency of a radar signal to the shallow, low-velocity sediments at the earth's surface, geophysicists can use radar to "see" objects or structures below ground or in shallow water. This remote sensing technique has been successfully used for military purposes, for mining exploration, and in archaeological studies.

Early seismic analysis was restricted to recording energy imparted to the earth by shooting dynamite in individual shot holes. Modern analyses have progressed to sophisticated three-dimensional surveys at sea with continuous series of acoustic shot points on 75-foot grids. Modern "shaking" machines substitute ground vibration for explosions in seismic surveys. With the help of modern computers, these seismic data can be greatly enhanced by noise and signal filtering. Ultimately, geologists use these seismic interpretations to produce detailed subsurface correlations that enhance the interpretation of structural folding and faulting and stratigraphic sequence configurations both on a detailed local level and on a regional basis. Deep and costly specialized seismic profiles have also been conducted on a crustal scale to investigate large-scale tectonic systems, such as has been done for the Appalachian Mountain system by COCORP (Consortium for Continental Reflection Profiling).

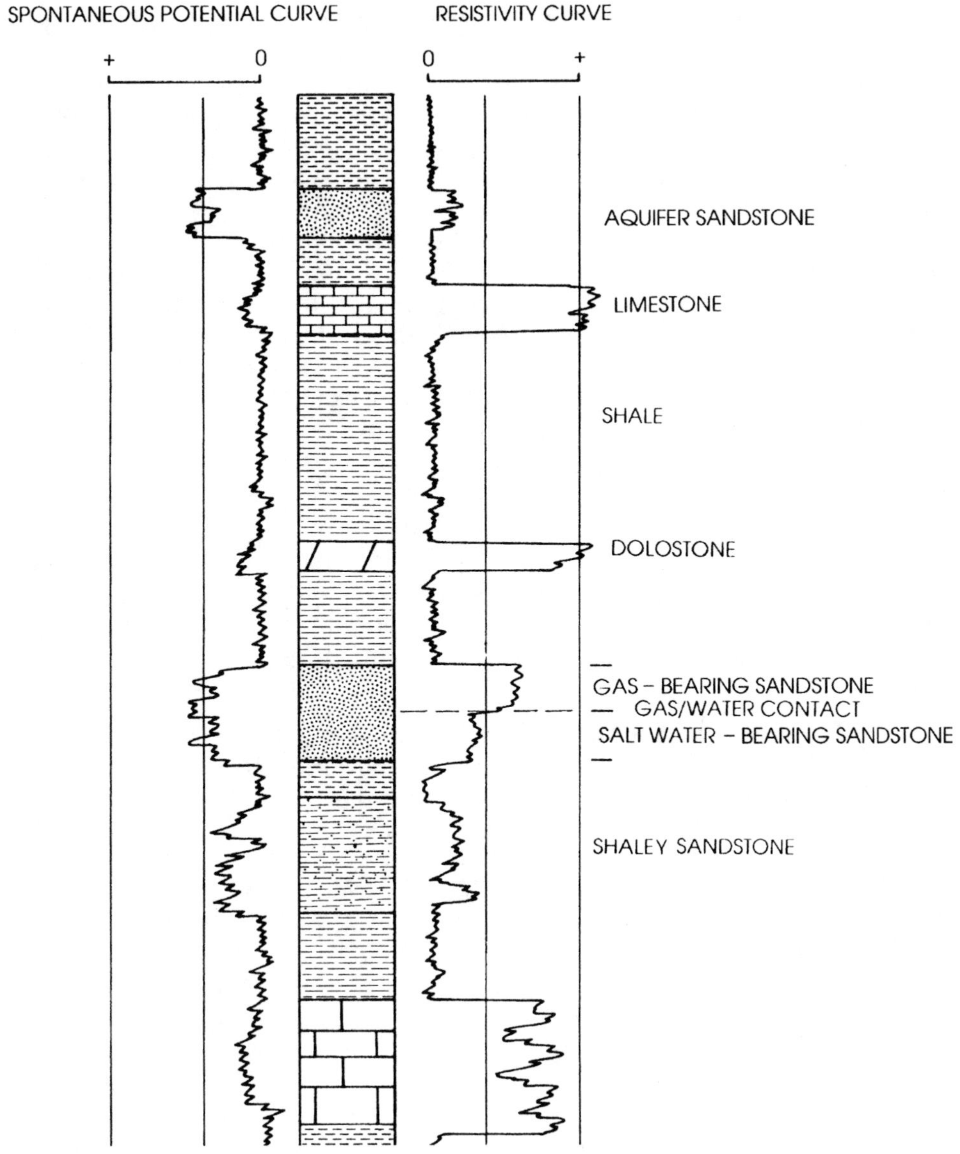

FIGURE 3.12 Electric log.

Common Correlation Problems

All of the correlation techniques just discussed are very useful to field geologists. However, ideal conditions often do not exist for many outcrops, nor do the lithologies or fossils remain consistent from one area to another. Outcrops are subject to erosion, growth of vegetation, human activity, and different degrees of structural changes. Any or all of these activities can alter or disguise all or portions of outcrops and geological correlations.

Facies Change Even though distinctive stratigraphic units may be present in several correlative sections, other rock units may change their lithologic appearance over relatively short geographical distances. As discussed previously, these physical changes, called *facies changes*, can complicate correlation. Another danger in correlation is the reliance on apparently persistent colors of rock units. There are numerous color variations possible in common minerals such as calcite, quartz, and limonite; therefore, color can be misleading when used to interpret rock units.

Another facies change that can cause correlation problems involves the presence or absence of a certain type of fossil within a formation. Plants and animals that could tolerate only a very limited range of environmental conditions are often associated with specific lithologies or facies and are called *facies fossils*. A sudden change from an open-marine to a brackish or freshwater environment could yield a marked local or regional change in the fossils contained within a formation; such a change would probably produce only a very subtle lithologic change in the rock unit itself. Therefore, if the correlations are primarily based on facies fossils, they would not necessarily properly establish the regional time equivalence of rock units. Use of well-established index fossils avoids this problem.

Structural Deformation Structural deformation often can complicate regional correlations. Folding and faulting may have disturbed the normal sequence of stratigraphic units. Any sequence of rocks from such an area should be examined in detail before it is used for regional correlation. A previously established regional correlation may help to determine the proper sequence of such disrupted strata.

Incomplete and Covered Sections Very often outcrops of rocks appear incomplete because they are covered by debris from overlying, more resistant rock units tumbling downslope; slumping of unstable hillsides (causing rotational displacement of the rock units downhill); or erosional materials of mass wasting (such as earth flow) (see Fig. 3.13, A, B, and C, respectively).

Where subsurface correlations are involved, incompleteness of rock data can result from faulting, unrecognized facies changes or unconformities, or from lost portions of well cores or samples. Careful analysis of the stratigraphic relationships in adjacent areas, additional drilling, or outcrop trenching may minimize any problems of incompleteness of geologic data.

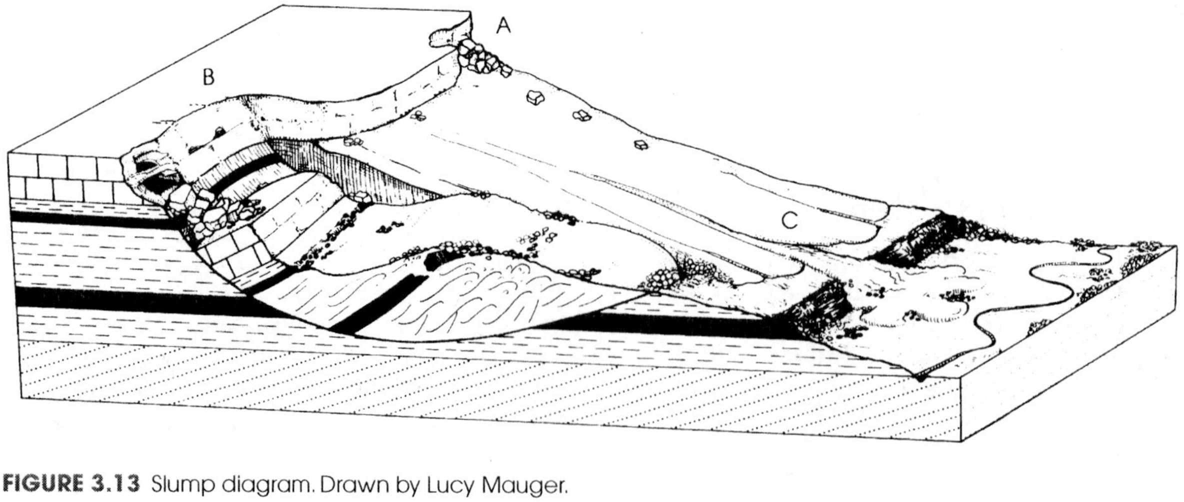

FIGURE 3.13 Slump diagram. Drawn by Lucy Mauger.

Correlation and Stratigraphic Interpretation

The following paragraphs outline the typical methodology and thought processes involved with correlation. Two outcrop-based columns of rock to be correlated are shown in Figure 3.14. The steps for correlation are as follows.

Step 1. First correlate the arkosic unit. The arkosic material is near its basal granitic source; thus, correlation is achieved by using the similar lithologies and same thickness of the unit in both sections. (*Note:* When one is confident about a correlation, one usually draws a solid line between the correlated points.)

Step 2. Above the correlated arkosic unit are several sedimentary units that appear to be in the same order but that vary in thickness or content. The next unique lithologic unit that can be correlated with confidence is the pyroclastic unit. Therefore the strata between these two correlated units should also be correlative, but one might be less confident about their correlation. If Walther's law (see page 99) is used to interpret the correlated units in the interval between the pyroclastic layer and the arkosic layer, it is apparent that these units tend to become finer as they progress upward and that a transgressive sequence is in evidence. The limestone unit below the pyroclastic layer shows that a facies change is present as the massive limestone in outcrop A is thinner and contains algae in outcrop B. Since outcrop B contains a higher percentage of clastics and shallow water algae in the limestone, it is interpreted to have been closer to the source area or shoreline than outcrop A. The thickness variations of individual lithologic units could

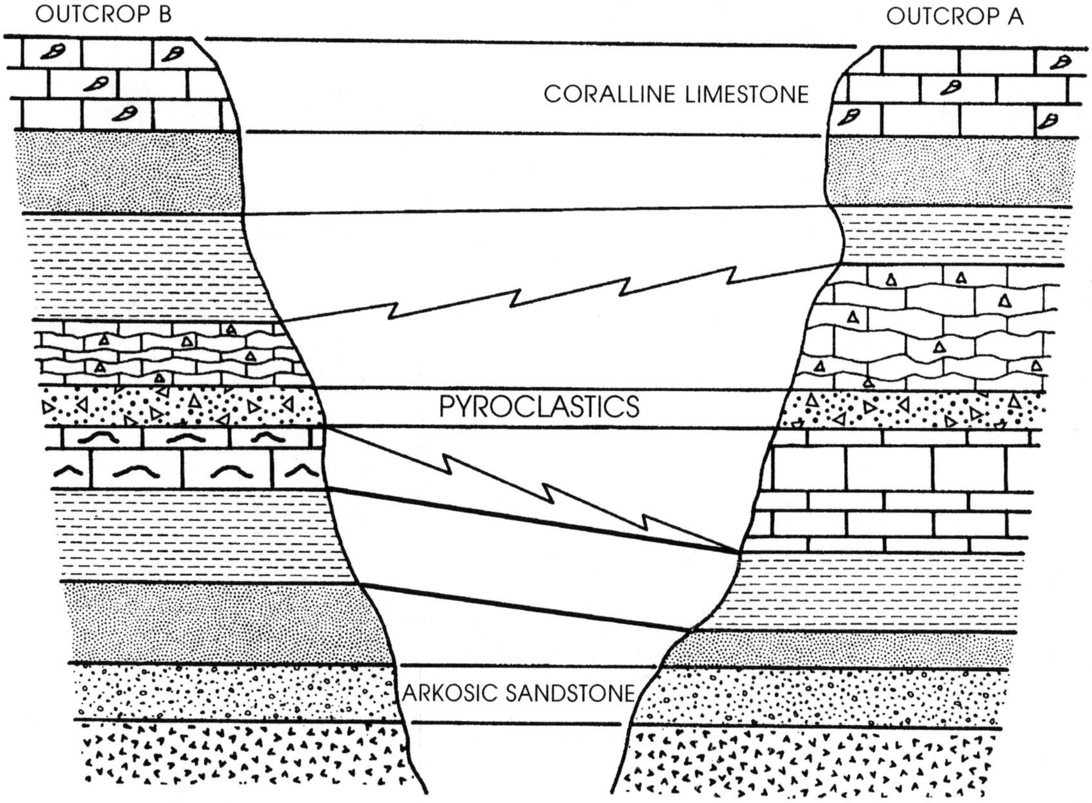

FIGURE 3.14 Correlation example.

mean that the depositional environment began to change because of transgression or regression or that it remained stable longer in one area than it did in another.

Step 3. The next major marker unit above the pyroclastic unit is the uppermost distinctive coralline limestone. Again, each of the lithologic units between the marker beds must be correlative. One interpretation of this sequence of strata could point to progressively more energy in the depositional environment, thus the gradual upward coarsening of the sediments. According to Walther's law, a regressive sequence of sediments is preserved from the low-energy wavy-bedded cherty limestone to the sandstone unit directly underlying the coralline limestone. Therefore the overall interpretation of the two correlated outcrops would be that a transgression occurred when the shoreline moved to the left, and then a terrestrial pyroclastic event was followed by a regression when the relative position of the shoreline returned nearly to its original position. The variation in thickness of the wavy-bedded cherty limestone from outcrop A to outcrop B can be indicated by a facies change.

EXERCISES

Exercise 3-5 CORRELATION EXERCISES

In the next several parts to this exercise, there will be two or more columns of lithostratigraphic units to correlate by drawing in the lines at points of equivalence.

Part A.

1. Correlate the sections in Figure 3.15(a) and (b) on the basis of similar lithologies.

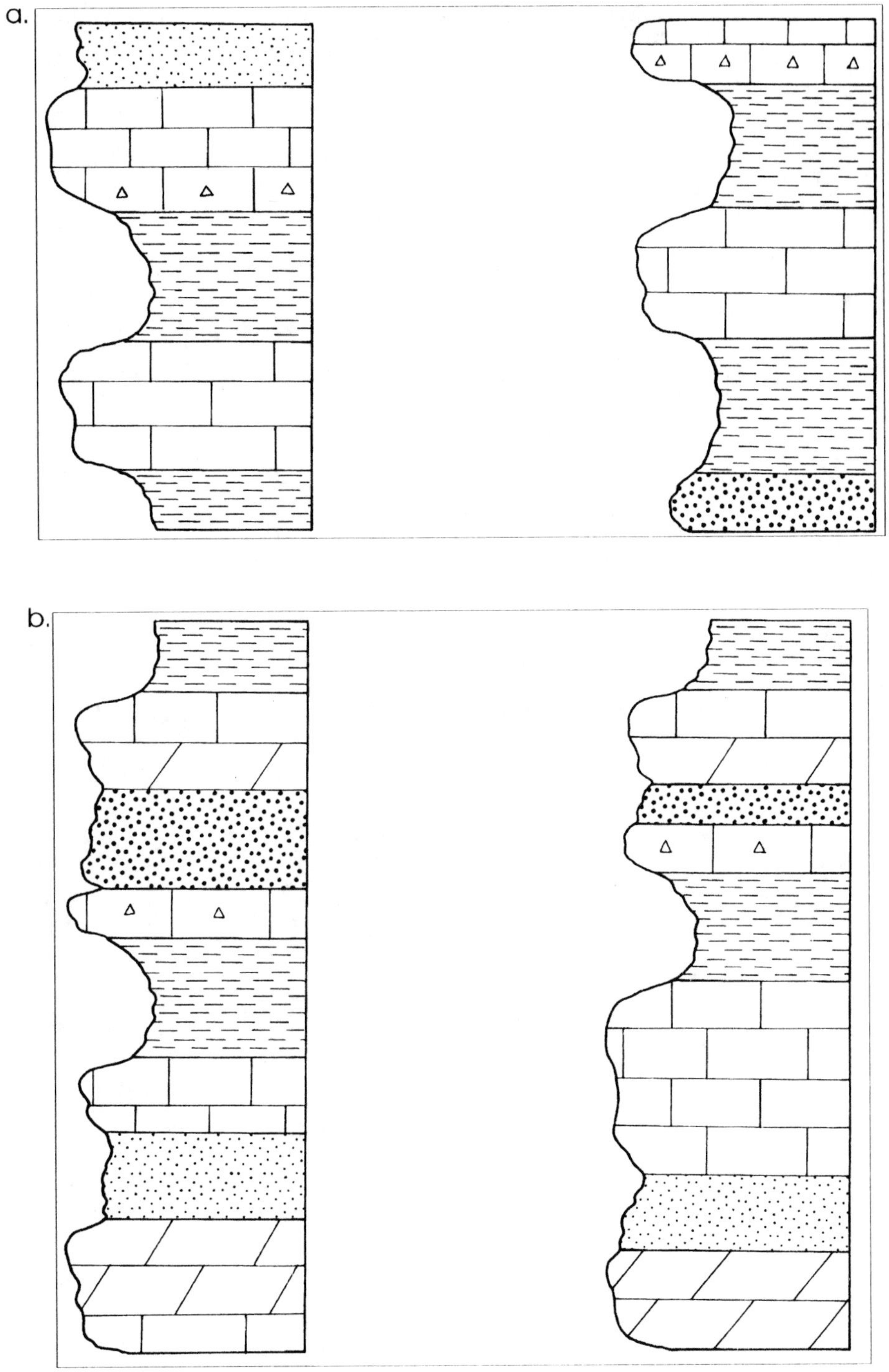

FIGURE 3.15 Correlation columns for Problem 1.

2. In the three columns in Figure 3.16, explain what you think happens to the sandstone unit between Section C and Section A.

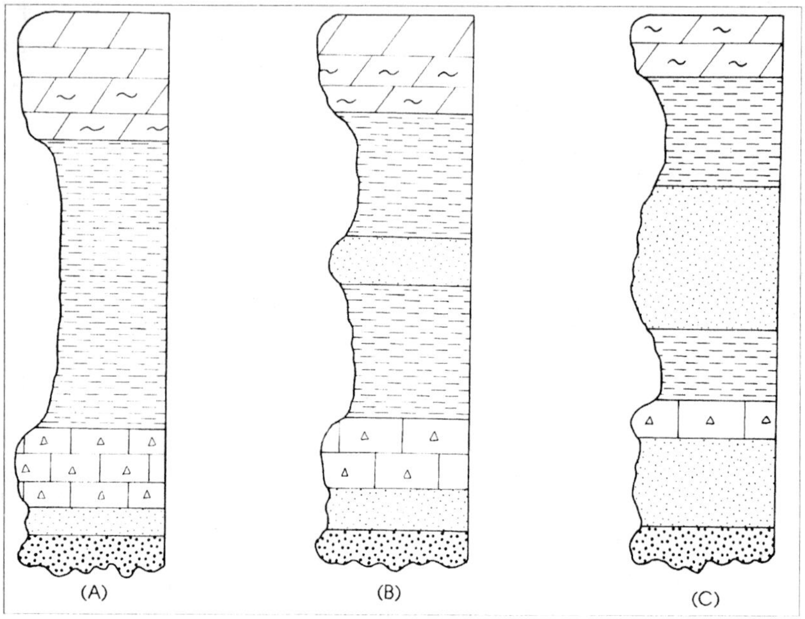

FIGURE 3.16 Correlation columns for Problem 2.

3. In Figure 3.17, why is the limestone in column B thicker than that in column A?

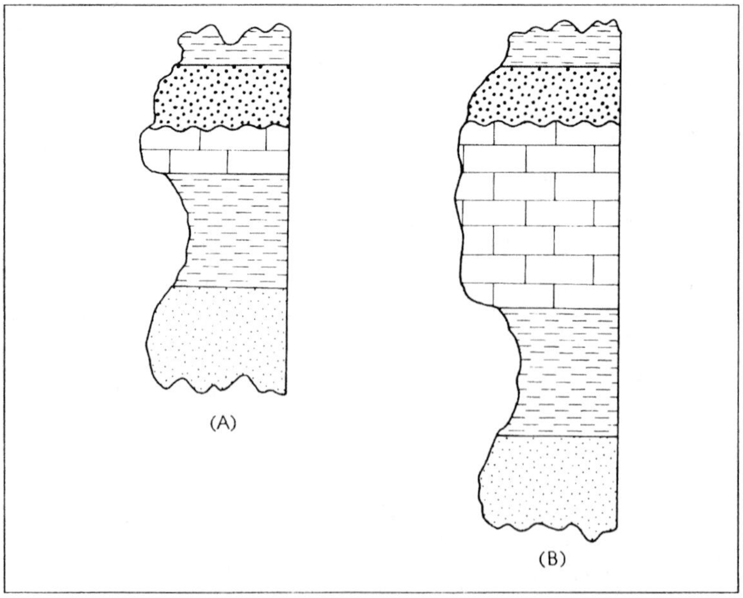

FIGURE 3.17 Correlation columns for Problem 3.

4. Correlate the three columns in Figure 3.18, and answer the following questions.

(a) Explain what is happening to the limestone unit at time 3 in the column sequence A-B-C. Can you suggest a reason for your observation?

(b) If all the rocks were deposited in a marine environment, which column is the one most likely to have been deposited farthest from land?

(c) Which one was deposited closest to land?

(d) In which direction did the land lie? Explain.

(e) What was happening to sea level from time interval 1 to time interval 3?

(f) What was happening to sea level from time interval 3 to time interval 7?

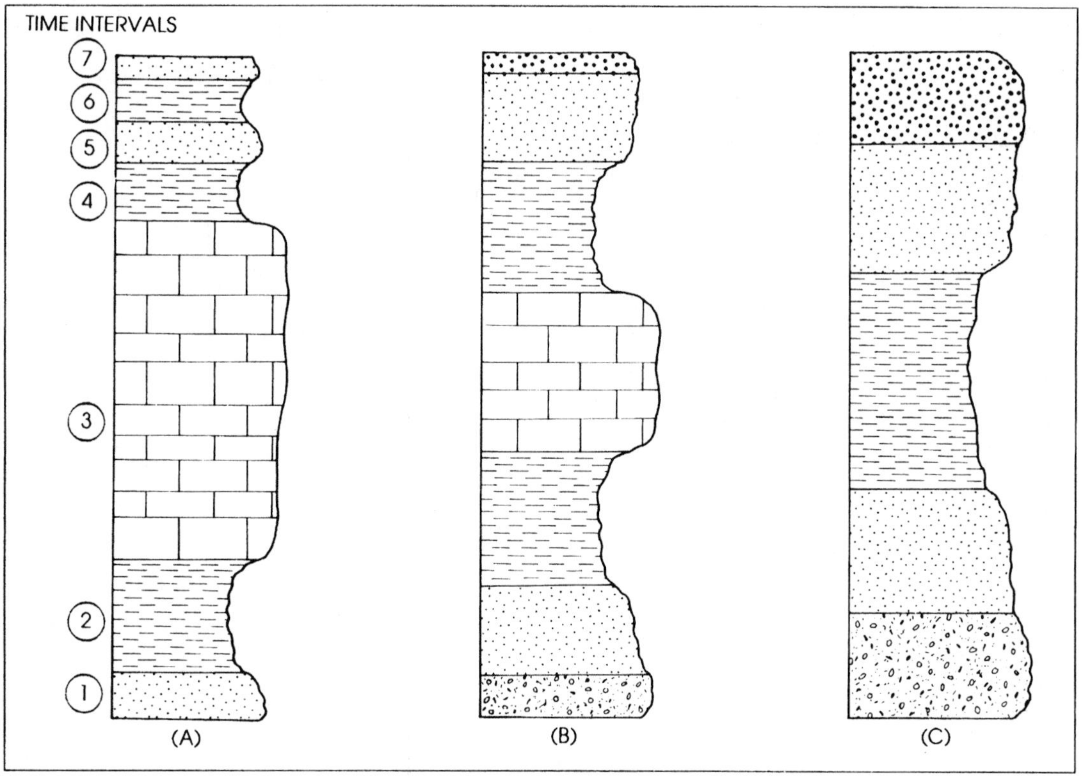

FIGURE 3.18 Correlation columns for Problem 4.

5. Examine the stratigraphic columns in Figure 3.19. Correlate them appropriately, and then answer the following questions.

(a) What kind of unconformity is present in these two sections?

(b) For the sequence above the coal, does it most likely represent a transgressive or a regressive sequence? Cite your evidence.

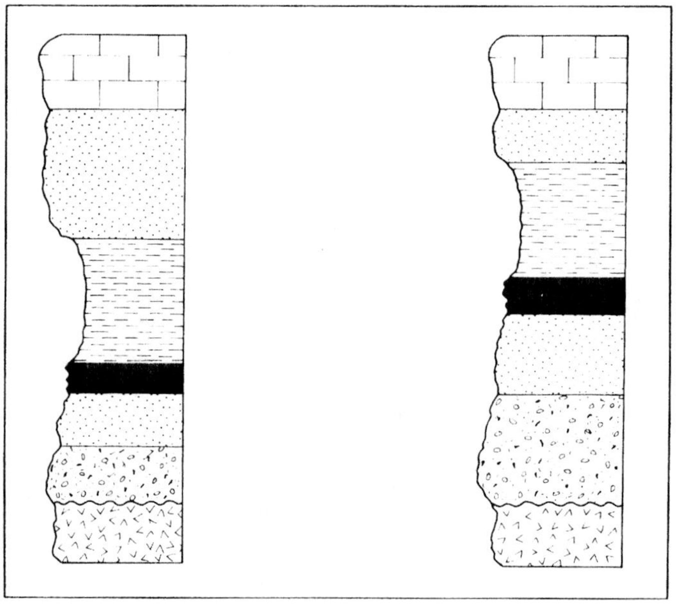

FIGURE 3.19 Correlation columns for Problem 5.

Part B. Correlation Exercise

Correlate the stratigraphic sections in Figure 3.20. (Note the presence of the unconformity and some fossils in the limestone. In these columns, use the fossil symbols only as correlation indicators.)

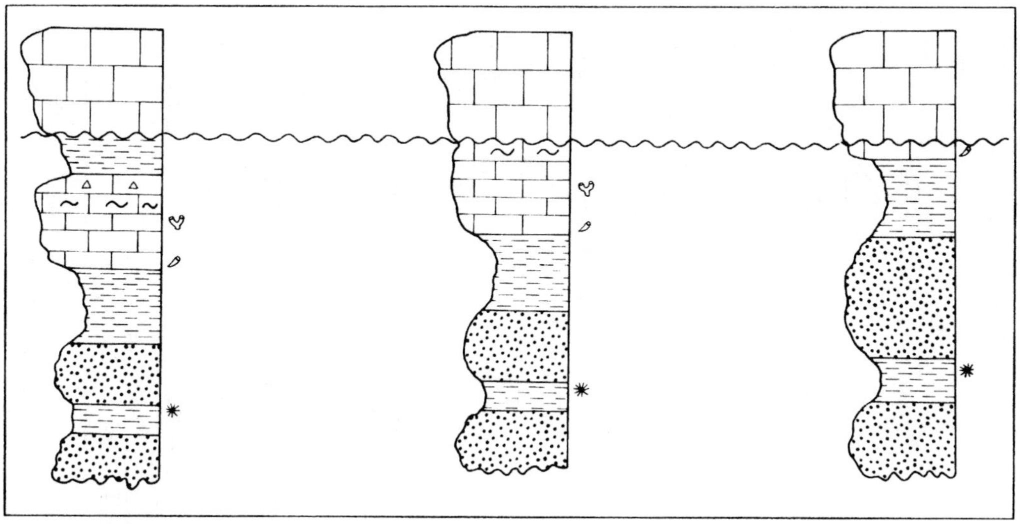

FIGURE 3.20 Correlation columns for Part B.

Part C. Correlation and Faulting

Correlate the stratigraphic sections in Figure 3.21. Indicate facies changes with appropriate symbols. Also locate and name the types of any unconformities. Correlate the intrusive rocks among all sections and label the type of intrusive structure shown.

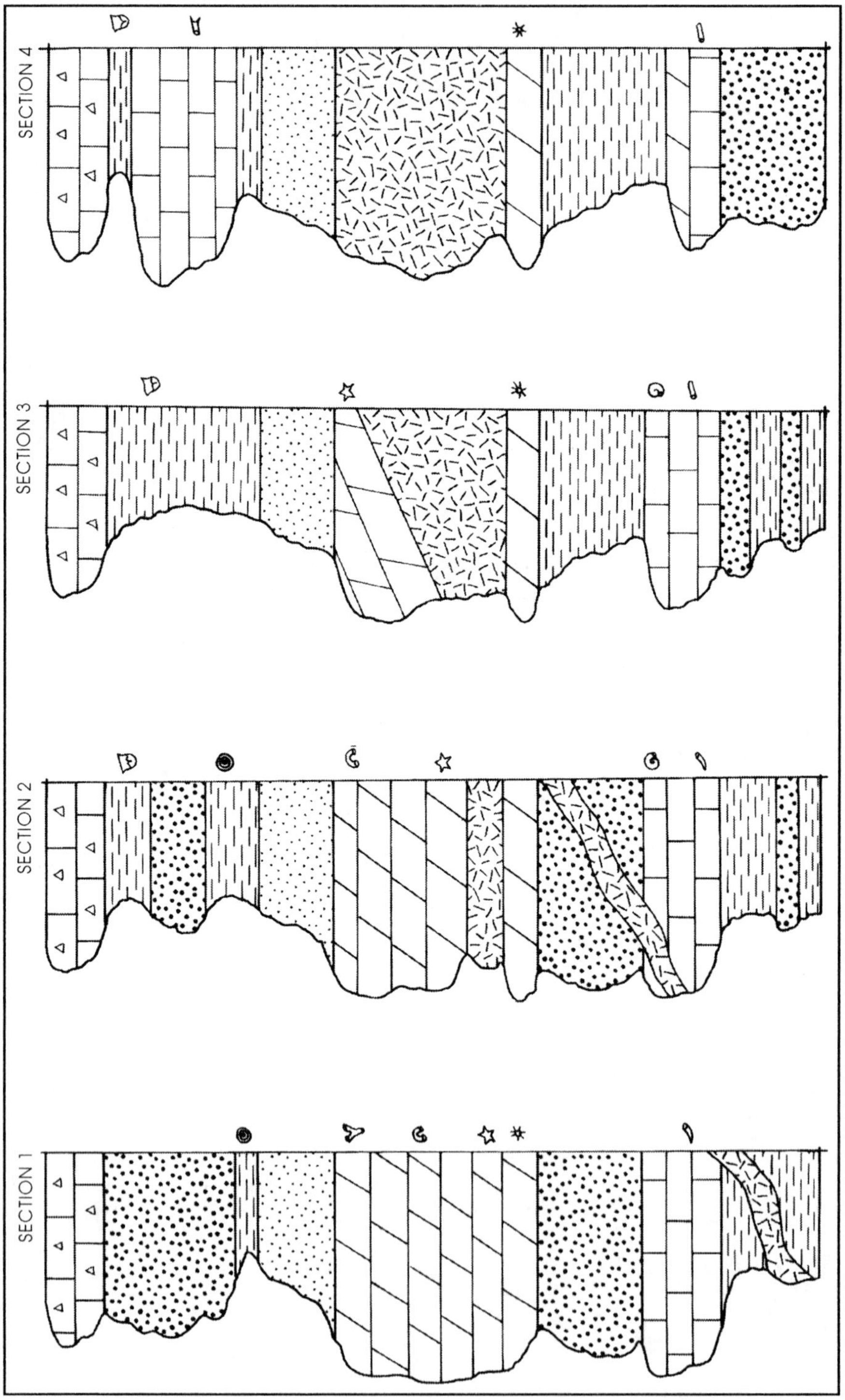

FIGURE 3.21 Correlation columns for Part C.

Part D. Correlation and Faulting

a. Correlate columns A to E in Figure 3.22. The two key marker beds are the cherty limestone and the evaporite units (see the lithology key on the inside front cover). *Note:* There *is* an unconformity present and a nearly vertical fault between column B and column C. (Assume that no beds are curved due to faulting.)

b. The fault has a displacement or throw of ___________ feet.

c. What type of unconformity is present?

d. Was the faulting prior to or after the unconformable units were deposited?

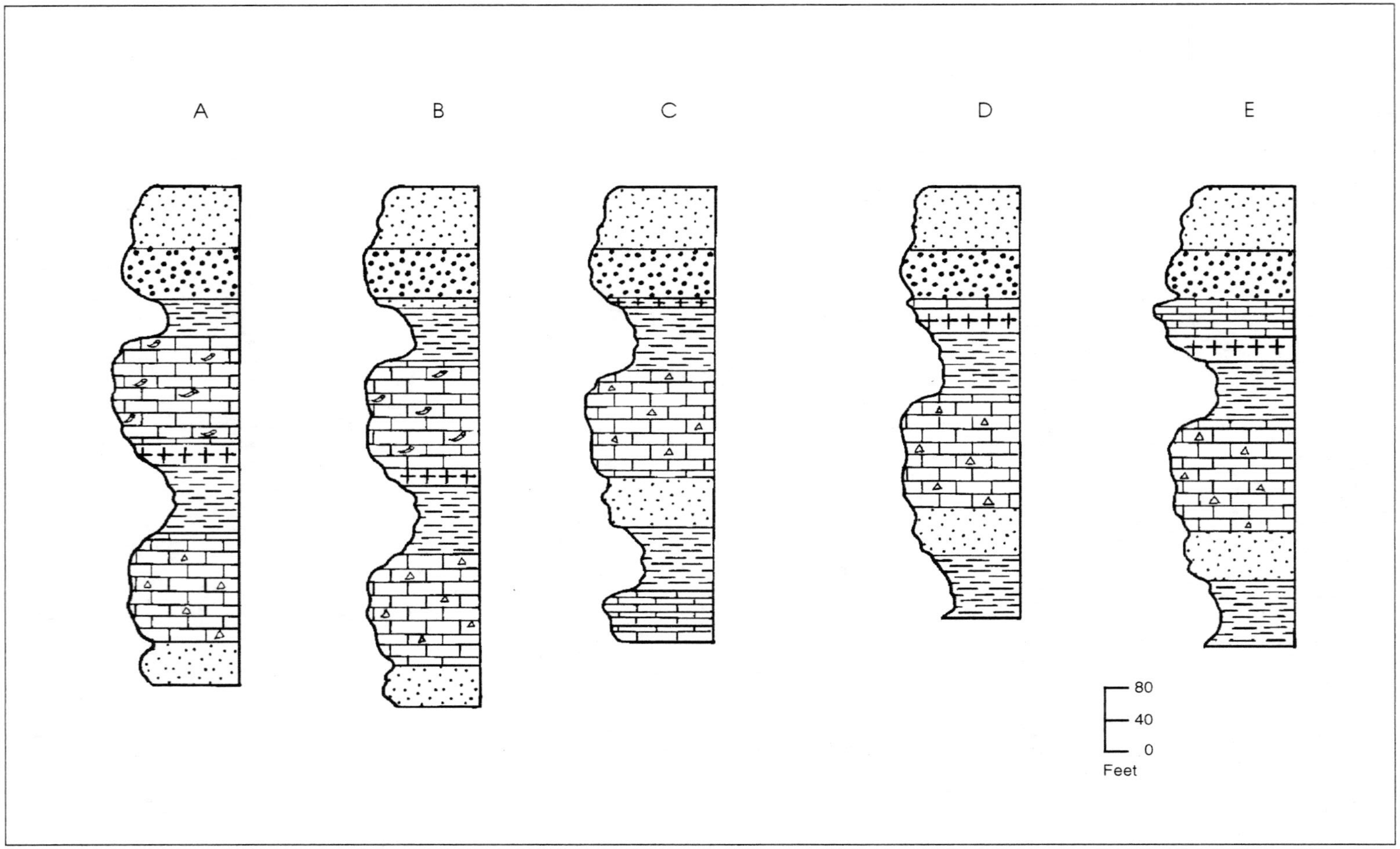

FIGURE 3.22 Correlation columns for Part D.

Part E. Electric Logs

From the given electric logs and lithology, correlate the subsurface geologic data from portions of the three oil wells in Figure 3.23 (see log example, Fig. 3.12, page 105).

The "pay zone" is the conglomeratic sandstone in the center of well B. From information derived from these wells (refer back to the interpretation of well logs, pp. 103–106), should a fourth well be drilled to the east or to the west of these three wells? (Explain why.)

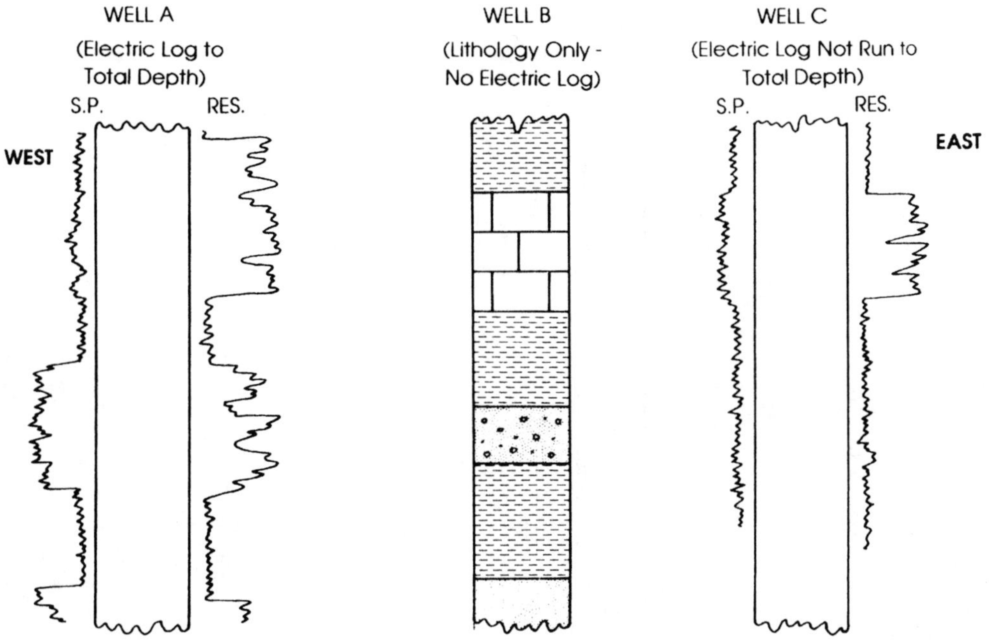

FIGURE 3.23 Electric logs.

Exercise 3-6 SEDIMENTARY STRUCTURES

During the 1930s and early 1940s in the Lakewood and M Street area of Dallas, Texas, the builders designed small brick homes whose design appeal was "enhanced" by using selected artful sandstones from the Ouachita Mountains of southeastern Oklahoma. The selected stones were used on porches and patios, around windows and doors, and on fireplaces. While newer houses have moved to different architectural styles, these old houses provide an excellent "laboratory" for the study of sedimentary structures.

In this problem, one portion of a rock façade on a randomly located house in the M Street area will provide the data. Figure 3.24(a) shows five rocks used in the brick- and rock-work on the house. To help interpret these rocks, the following definitions are provided.

> *Flute cast*: A sole mark consisting of a raised, oblong bulge on the underside of a siltstone or sandstone. The steep or blunt, rounded upcurrent end gradually flattens and merges into the bedding plane in a downcurrent direction.

Load cast: More irregular than a flute cast. It is formed by loading of sand on a soft, unconsolidated underlying mud and shows differential settling. A load cast is characterized by not having any distinction of upcurrent and downcurrent ends.

Tool mark: A current mark produced by the impact against a muddy bottom of a solid object swept along by the current. The mark may also show continuous contact with the bottom as a groove cast. The tools were often sticks, shell fragments, pebbles, fish bones, or seaweed; thus a wide variety of mark shapes could exist.

Determine the current directions and types of sedimentary structures in each of the sandstones in Figures 3.24(a) and 3.24(b). Did the stonemasons set the pieces with consistent orientation?

FIGURE 3.24(a) Oklahoma turbidite sandstones on façade of a Dallas, Texas, house.

FIGURE 3.24(b) Oklahoma turbidite sandstones on façade of a Dallas, Texas, house.

Exercise 3-7 GRAND CANYON

Now that your knowledge of sedimentary environments and correlation methodologies has increased, return to Exercise 2–6, Interpreting the Geologic History of the Grand Canyon, and Figure 2.28.

1. Explain how the Tonto Group (Tapeats, Bright Angel, and Muav) could represent part of a transgression in the Grand Canyon area.

2. What type of environmental and climatic conditions are represented by the Redwall Limestone?

3. Assume that a laterite soil (see Glossary) developed on the upper surface of the Redwall Limestone. What are the environmental implications of this development prior to the deposition of the Supai?

4. Within the overall Paleozoic sequence of geologic units seen within the walls of the Grand Canyon, which strata represent a sequence in which the sea level is falling?

Exercise 3-8 CENTRAL TEXAS UPLIFT

This exercise asks you to draw a stratigraphic column, given field data, and then to interpret the stratigraphy and sedimentary environment.

The following paragraphs give lithologic descriptions of a measured section of rocks that are exposed along the San Saba River in the Central Texas Uplift region west of Austin, Texas. Using these data, construct a measured stratigraphic section at a scale of 1 inch equals 200 feet on the graph paper provided in Figure 3.25. Prepare this section in basically the same manner as that shown on Exercise 3–1. Include on the stratigraphic section such information as lithology, names, and ages of the lithologic units. In addition, leave room to record any distinctive color, minerals, fossils, or sedimentary structures associated with a given unit.

After completing the graphic presentation, use it as the basis for answering the interpretive questions about the geologic history of this portion of Texas during the Cambrian and Ordovician periods.

DATA: Written Stratigraphic Descriptions

Directly overlying the Town Mountain Granite of the Precambrian is the Riley Formation of the late Cambrian. This formation is composed of the following members: the Hickory Sandstone, the Cap Mountain Limestone, and the Lion Mountain Sandstone. The Hickory Sandstone is 360 feet thick and is made up of three parts. The 20-foot basal portion is an arkosic conglomerate containing numerous ventifacts (see Glossary). The middle part (50 feet) is a silty sandstone with a buff-to-light-reddish color. The upper part is dark, well-cemented, well-sorted sandstone, 290 feet in thickness.

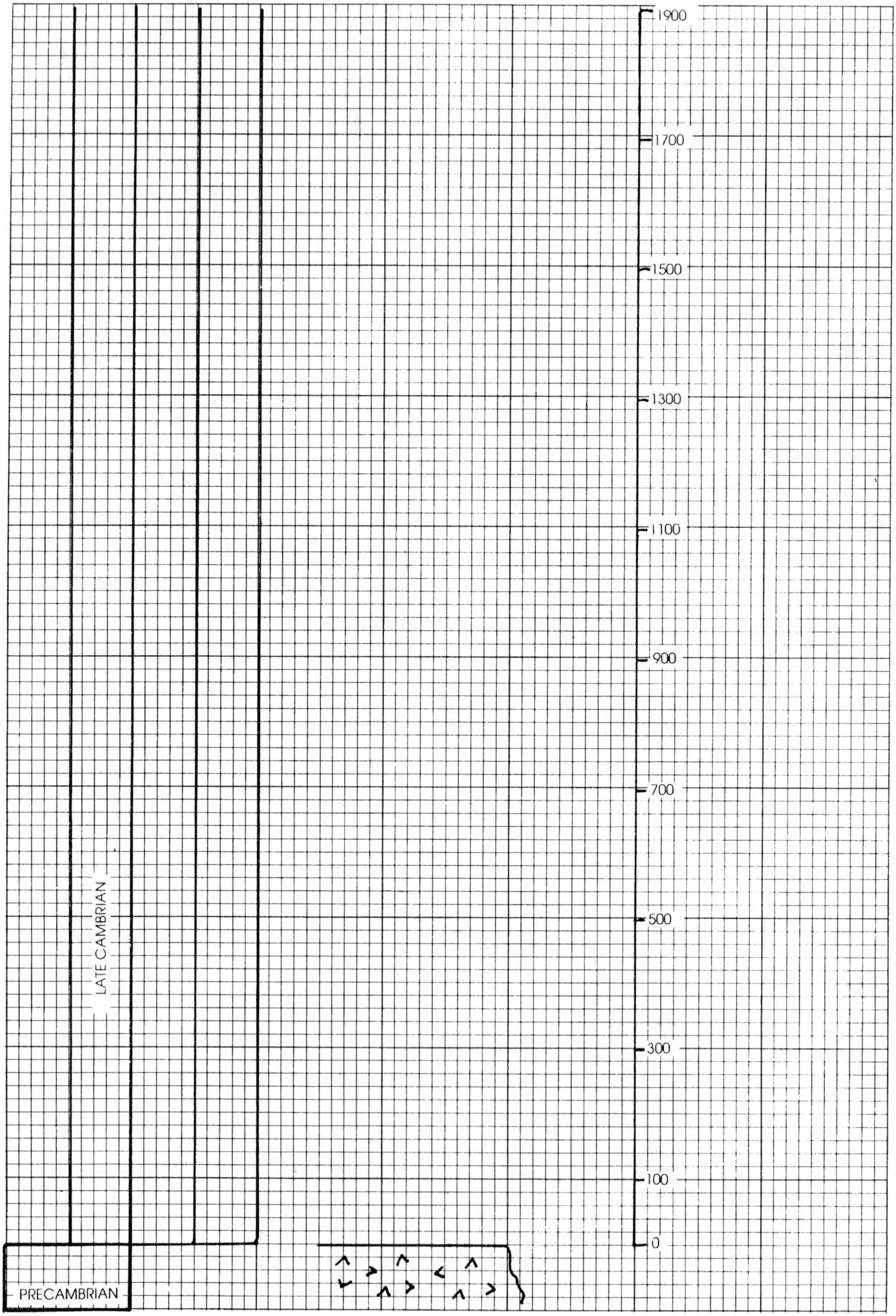

FIGURE 3.25 Graph paper for construction of stratigraphic columns.

Above the Hickory Sandstone is the Cap Mountain Limestone, which is 280 feet thick. The contact with the Hickory Sandstone is gradational. Several thin sandstone layers like those in the upper part of the Hickory are present in the lower 100 feet of the Cap Mountain. The limestone of the Cap Mountain is grey to brown and contains grains of glauconite, trilobite fragments, and other marine fossils that increase in abundance toward the top.

The uppermost unit, the Lion Mountain Sandstone (37 feet thick), is predominantly massive glauconitic sand. It includes 10 feet of fragmental limestone with trilobites at the base.

Overlying the Riley Formation is the Wilberns Formation of Late Cambrian age. It is composed of four members: the Welge Sandstone, the Morgan Creek Limestone, the Point Peak Shale, and the San Saba Limestone.

The Welge Sandstone member is an 18-foot bed of red sandstone without glauconite.

The Morgan Creek Limestone (120 feet) is a medium to coarsely grained glauconitic limestone containing a few small biohermal reefs. It is purple at its base but is predominantly dark grey.

The Point Peak Shale (160 feet) is a green calcareous shale interbedded with a few lenses of limestone that contain a few small biohermal reefs.

The San Saba Limestone (280 feet) is a yellow-to-buff glauconitic limestone with large algal reefs. Marine fossils are abundant, and the bedding is massive.

Overlying the Wilberns Formation is the Ellenburger Group, of Cambrian–Ordovician age. The contact of the Ellenburger Group with the Wilberns Formation appears to be gradational within the Llano region of Texas. Regionally, there is some evidence for an unconformity at this boundary. The lower part of the Ellenburger Group (500 feet) is a light-yellow-to-buff, very fossiliferous dolostone. It is massively bedded and contains markedly less glauconite than the underlying Riley and Wilberns formations. The base of the Ellenburger Group is marked by the presence of the Cambrian fossil *Lytospira gyroceras* and other Cambrian trilobites and brachiopods. The upper part of the Ellenburger Group (100 feet) is similar in lithology but contains fossils of the earliest Ordovician age. The top of the Ellenburger is an unconformable surface marked by a zone of red-stained, rounded chert fragments separating it from an overlying Middle Devonian limestone that is present throughout the area.

a. What kind of rock is the Town Mountain? Was it intruded before or after the overlying Early Paleozoic rocks? How can you tell?

b. What type of unconformity separates the Town Mountain Granite from the base of the Riley Formation? Defend your answer.

c. What does the presence of ventifacts and arkosic fragments indicate about the source area and its probable climate prior to Late Cambrian time?

d. Was the sea transgressing or regressing during the deposition of the Cap Mountain Limestone? Cite all supporting evidence.

e. How was the Riley Formation differentiated from the Wilberns Formation in the field?

f. Describe the water conditions that existed during the deposition of the San Saba Limestone.

g. Do lithostratigraphic and time-stratigraphic units always coincide? Relate your answer to the evidence in this section.

h. Which time-stratigraphic units are missing at the top of the section? (See the Geologic Time Scale on the inside back cover.)

Exercise 3-9 EAST TEXAS OIL PROSPECT

Through an inheritance from an uncle who lived in Tyler, Texas, you have obtained a piece of land in Texas. The uncle's property has never been explored or drilled for oil, but there is oil production on surrounding acreage. A small independent petroleum company, the Crooked Hole Petroleum Company of Kilgore, Texas, has offered to lease your land in order to drill a wildcat or initial oil well. Since you are now the landowner and must give permission to the petroleum company before it can drill, you decide to make an independent evaluation of the prospect before signing the lease. You decide to use some of the college geology learned a while back, and you ask to look at some of Crooked Hole's geological information. The company geologist lets you review some of his data but not his interpretations. Therefore you must analyze the data for yourself.

As part of your investigation, you need first to find out which stratigraphic units and types of oil-bearing structures exist on the land adjacent to your property. Second, you need to evaluate the possibility that some oil-bearing structure and reservoirs may extend from beneath the neighboring oil field into the strata underlying the surface of your land.

Data

The company geologist lets you review data that includes a series of electric logs and rock samples derived from well cuttings and cores obtained from nearby wells. The data at each test point consist of samples of the first rocks that occur just below the regional, nonproducing, Cretaceous shale bed, a very good correlation marker in most of the wells near your land. Since each of the samples was directly below this regional shale horizon, it is assumed that all of the rock types belong to either the Sour Lake Formation or the High Island Formation. Both of these formations were exposed to erosion at the same time during the development of the unconformity. Therefore, by mapping these two formations, you can draw a paleogeographic map of the region as it existed just prior to burial by the overlying widespread Cretaceous shale unit.

Samples

Each of these sample descriptions is keyed to the sample location map in Figure 3.26.

1. Limestone, grey, coarse-grained, fossiliferous (*Ceraurus*—trilobite; *Platystrophia*—brachiopod).
2. Sandy shale.

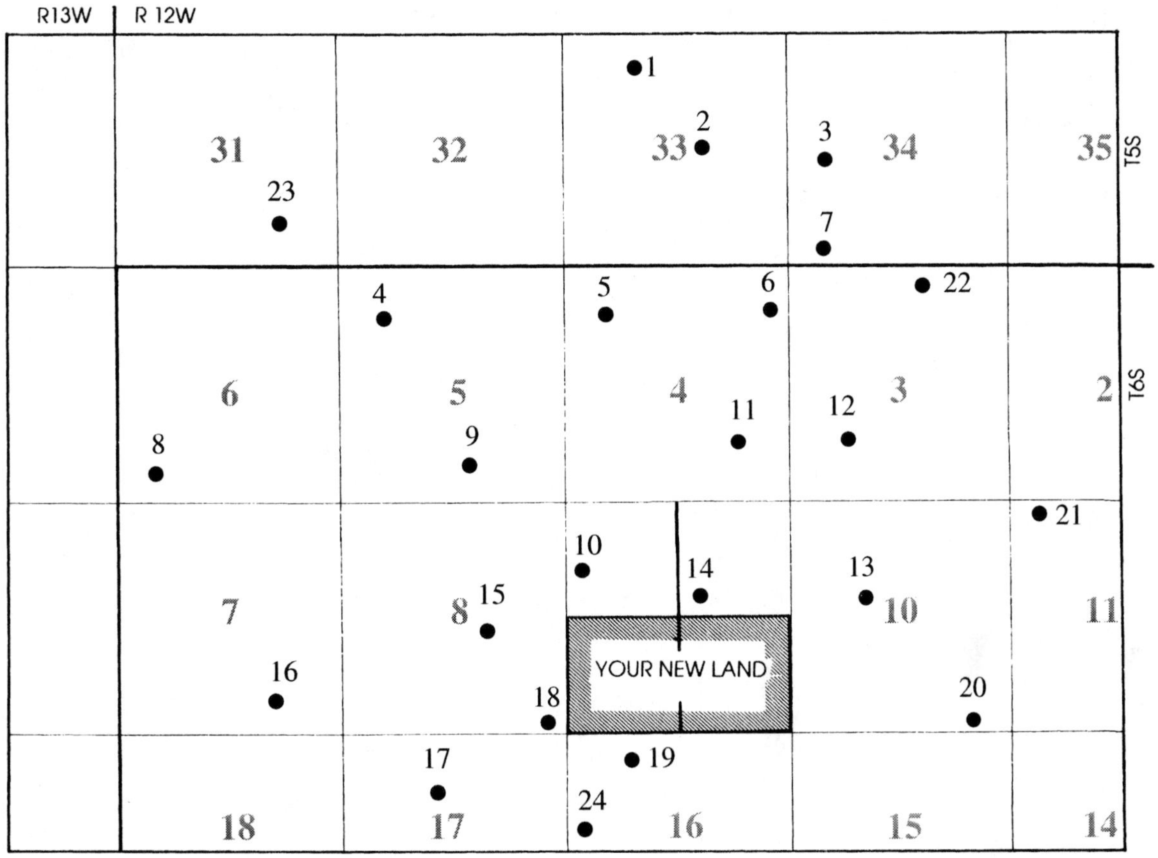

FIGURE 3.26 Sample location map.

3. Sandy shale.
4. Limestone, grey, coarse-grained with some chert, fossiliferous (*Strophomena*—brachiopod; *Isotelus*—trilobite).
5. Shale, grey, both marine and terrestrial fossils.
6. Sandstone, light yellow, well sorted, unfossiliferous.
7. Sandy shale.
8. Limestone, grey, cherty (*Isotelus*—trilobite; *Hebertella*—brachiopod).
9. Shale with some fragments of limestone, sparsely fossiliferous.
10. Shale, dark grey to black, Lagoonal fauna (*Myalina*—bivalve).
11. Sandstone, buff, well sorted, unfossiliferous, contains oil.
12. Sandy shale, dark grey, marine fossils.
13. Sandy shale, fossiliferous (*Mesolobus*—brachiopod) (Penn).
14. Sandstone, light yellow to buff, well sorted, unfossiliferous, some free oil.
15. Shale (much like unit 9) (*Composita*—brachiopod; *Girtyocoelia*—sponge).
16. Shale, dark grey, fossiliferous.
17. Sandy shale, lagoonal fauna.
18. Sandstone, buff, good sorting, some shale, trace of oil.
19. Sandstone.
20. Shale, grey, marine fossils, some sand grains (*Mesolobus*—brachiopod).

21. Shale, grey, marine fossils.
22. Shale.
23. Limestone.
24. Sandy shale, grey, marine.

Questions

1. Using the sample location map (Fig. 3.26) as a base, draw a lithofacies map showing the distribution of limestone, sandstone, sandy shale, and shale in the area (see Fig. 3.7). Each of the sample locations is a drilled wellsite.

2. Determine the age of the three major lithofacies using the ranges of the fossils. *Note*: To answer this question, look forward to a section of Chapter 4, "Use of Fossil Assemblages in Age Determinations" on page 151–152, which illustrates the methodology to be used.

Limestone: ______________________________

Shale: ______________________________

Sandstone: ______________________________

3. Interpret the relationship between these rock units by completing the stratigraphic cross section in Figure 3.27. (Draw a line on the sample location map to indicate the location of the cross section.)

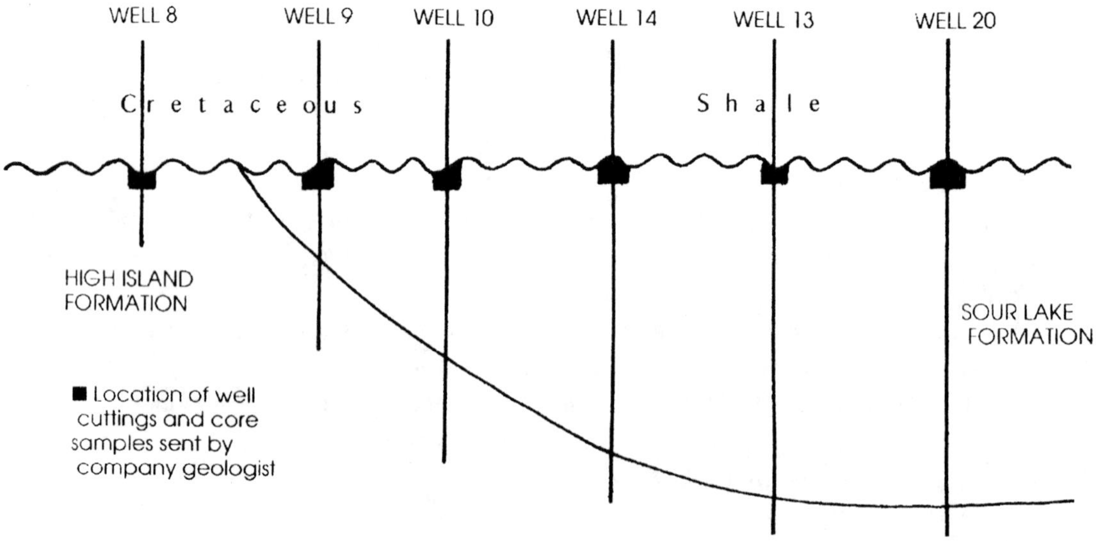

FIGURE 3.27 Stratigraphic cross section.

4. What is the significance of the contact between the limestones and the adjacent shales?

5. How does this contact differ from the contacts between the sandstones, sandy shales, and shales that make up the Sour Lake Formation?

6. Describe the kind of unconformity that separates the overlying Cretaceous shale from the underlying Sour Lake and High Island formations.

7. Would your land be considered a good petroleum prospect? Why?

8. If your answer to Question 7 is yes, give the exact location, using the land grid system, for the best 160-acre prospective drilling site.

Exercise 3-10 GEORGIA COASTAL PLAIN

This problem is designed to demonstrate how regional subsurface data can be used to interpret the geologic history of a specific physiographic province of the United States.

Procedure

1. The location of each of the data points is placed on the generalized geologic map of Georgia (see Fig. 3.28). Connect the points to form the cross section A-A′.
2. The location of each wellsite is plotted on the accompanying page of graph paper (Fig. 3.29). At each site, sketch a preliminary vertical line for each well. Next mark off on this vertical line the total depth and the formation contacts for each well. Remember that the well data are given as depths below land surface and not below sea level! Use a vertical scale of 1 inch = 1,000 feet, and use sea level as a datum plane.
3. Sketch a profile of the ground surface.
4. Connect the surface and subsurface data points to complete the cross section.

Data*

Well #1—Mr. R. L. Laury; 3 miles south of Greensboro; Greene County, Georgia
Elevation: 750 feet; total depth: 152 feet
Formation Tops: 0 to 12 feet—soil
12 to 152 feet—schist: Precambrian?

*The Well-log data were selected and modified from "Petroleum Exploration in Georgia," Information Circular Number 38, 1970, Department of Mines, Mining and Geology, Geological Survey of Georgia.

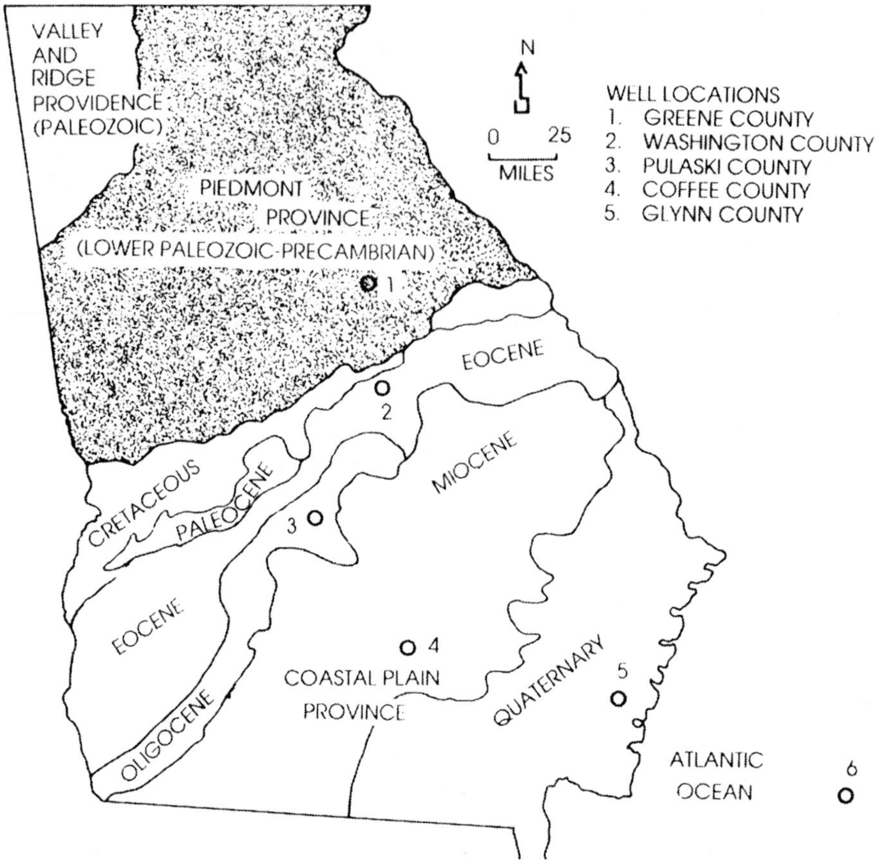

FIGURE 3.28 Generalized geologic map of Georgia.

Well #2—Middle Georgia Oil and Gas Company; 12 miles northwest of Sandersville; Washington County, Georgia

Elevation: 500 feet; total depth: 605 feet

Formation Tops: 0 to 250 feet—soil; Tertiary?
250 to 350 feet—basal Upper Cretaceous
350 to 605 feet—schist: Precambrian?

Well #3—Mr. R.O. Leighton; 5 miles northeast of Hawkinsville; Pulaski County, Georgia

Elevation: 300 feet; total depth: 5,050 feet

Formation Tops: 0 to 200 feet—few samples: Oligocene?
200 to 550 feet—Eocene
550 to 1,900 feet—Upper Cretaceous
1,900 to 2,450 feet—Lower Cretaceous
2,450 to 5,050 feet—schist and gneiss

Well #4—Carpenter Oil Co.; 2 miles southeast of Sapps Still, Coffee County, Georgia

Elevation: 250 feet; total depth: 4,150 feet

Formation Tops: 0 to 450 feet—Miocene

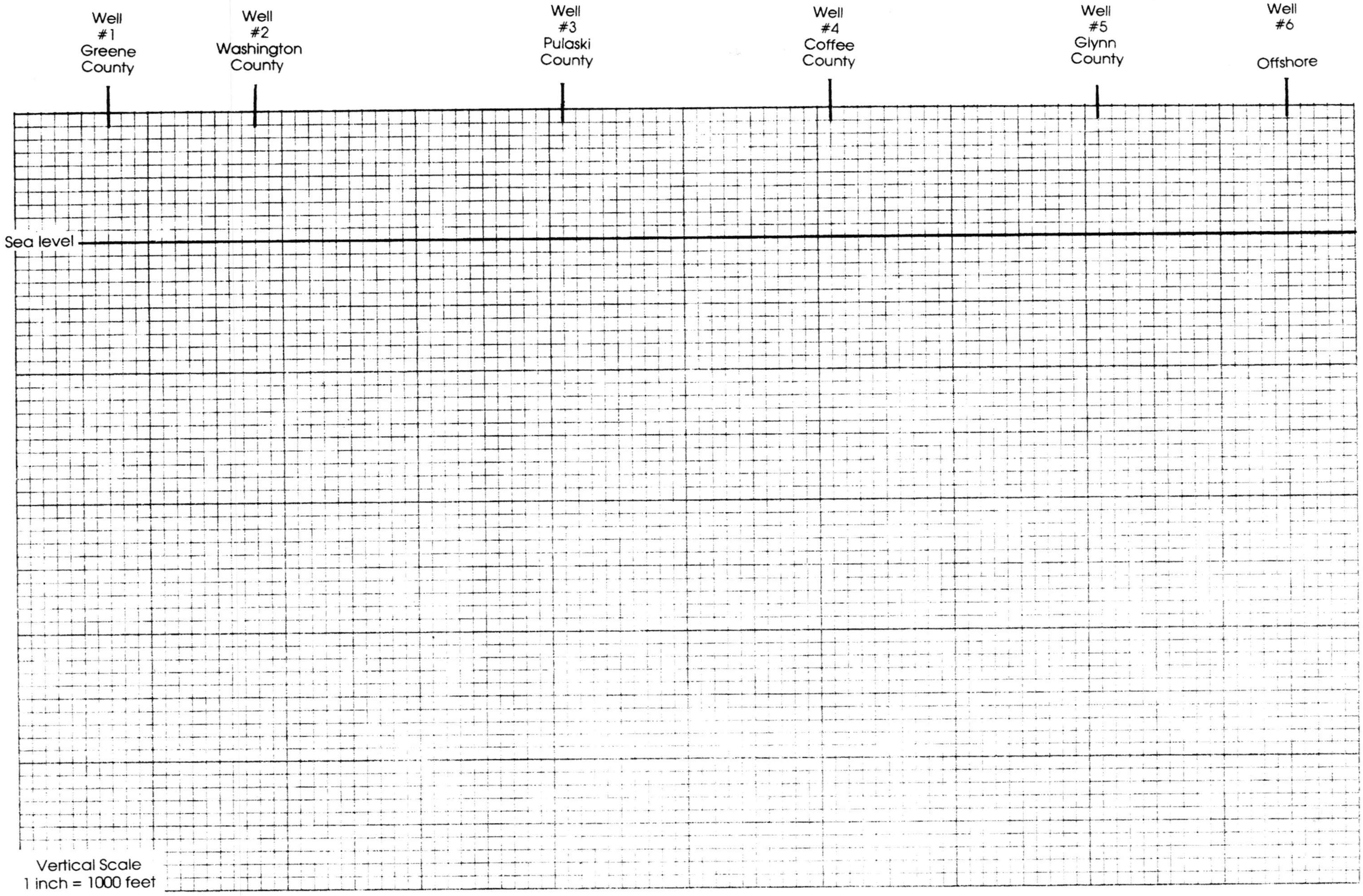

FIGURE 3.29 Graph paper for Exercise 3.10.

450 to 1,050 feet—Oligocene
1,050 to 1,850 feet—Eocene
1,850 to 3,750 feet—Upper Cretaceous
3,750 to 4,100 feet—Lower Cretaceous
4,100 to 4,150 feet—Pre-Cretaceous sediments

Well #5—Humble Oil and Refining Co; 11 miles northwest of Brunswick; Glynn County, Georgia

Elevation: 10 feet; total depth: 4,750 feet

Formation Tops: 0 to 50 feet—soil
50 to 150 feet—Pliocene?
150 to 650 feet—Miocene
650 to 1,450 feet—Oligocene
1,450 to 2,300 feet—Eocene
2,300 to 2,550 feet—Paleocene?
2,550 to 4,100 feet—Upper Cretaceous
4,100 to 4,700 feet—Lower Cretaceous
4,700 to 4,750 feet—Pre-Cretaceous sediments

Well #6—Ocean Producing Co., COST G.E.-1 Well, Atlantic Ocean, 74 miles east of Jacksonville, Florida, on Blake Plateau

Elevation of drilling platform: 98 feet; water depth: 136 feet; total depth: 13,194 feet TVD (true vertical depth)

Formation Tops: 0 to 136 feet—Seawater
136 to 390 feet—Holocene/Pleistocene
390 to 570 feet—Pliocene
570 to 750 feet—Miocene
750 to 1,230 feet—Oligocene
1,230 to 3,480 feet—Eocene
3,480 to 3,750 feet—Paleocene
3,750 to 5,950 feet—Upper Cretaceous
5,950 to total depth—Lower Cretaceous

Questions

1. What kinds of rock lie beneath the Cretaceous and younger strata of the Georgia Coastal Plain?

2. Are any unconformities present in the cross section? If so, indicate their positions with the appropriate symbol on the cross section. If present, what types of unconformities are they?

3. Utilizing information from the cross section, discuss in as much detail as possible the geologic history of Georgia's coastal plain during the following eras.

a. Paleozoic

b. Mesozoic

c. Cenozoic

Exercise 3-11 REMOTE SENSING AND RADAR IMAGERY

The term *remote sensing* refers to any method of gathering information about an object without making actual contact with it. The most common form of remote sensing with which everyone is familiar is sight, the ability to see the world around us. Many times geologists need to preserve visible evidence by taking photographs for future analysis in a laboratory. Thus photography, particularly aerial and satellite photography, is an important remote-sensing tool. Additional methods of remote sensing include gravity and magnetic studies, subsurface radar imaging, and seismic profiling. Aerial photographs and other remote sensing methods are used extensively by earth scientists because they provide a synoptic overview of the earth's surface. Aerial photographs provide a better perspective on the interrelationships among surface features such as towns, rivers, outcrops, topography, and agricultural activity.

Earth scientists may use remote sensing in many different ways. A hydrologist may study ice distribution and annual patterns in the mountains to try to determine whether glaciers are growing or receding. This research can help estimate how much water will be available for people in a valley during the spring and summer. The environmental scientist may use remote sensing to help decide how large a population an area can support, the best place for industrial complexes, where a new highway should be routed, or the best place for refuse disposal. Remote sensing can also be used to identify and monitor sources of pollution. Similarly, a forest worker may use infrared photography to spot diseased trees, and a sociologist may use color photography to study the changes in various neighborhoods of a city. The military was responsible for the initial development of remote-sensing technology, and the military still makes extensive use of many techniques. The earth resource (Landsats) and weather satellites orbit the earth and send back pictures to help earth scientists and others locate natural resources and monitor major weather systems, tropical storms, and the overall sea state of the ocean. A succession of remote-sensing data taken over the years has an additional benefit, because the data can document changes in the same area over an interval of time.

Another form of remote sensing that has found increased application in geologic studies is radar imagery. The most frequently used radar technique is SLAR, or sidelooking airborne radar. Radar images are grainier than photography, have lower resolution (photography can distinguish much smaller objects than radar), are more difficult to obtain because of instrument complexity, and are more expensive than ordinary aerial photographs. Radar does have some advantages, however, since it can penetrate cloud cover (but not rain), which handicaps photo interpretation. Radar works much like a TV camera, except that a TV camera is a *passive* sensor, detecting reflected energy from an independent illuminator (the sun). Radar is an *active* sensor, sending out pulses that illuminate the terrain and are reflected back to the airplane's receiving equipment. Therefore, radar has another advantage because it can be used at night. Some radar systems can penetrate vegetation and "look at" the ground below; this greatly facilitates study of the geology in a jungle area, where photographic interpretation would otherwise be very difficult.

An object on a photograph is light if it reflects sunlight and dark if the object absorbs light (that is, dark in color or shadowed). However, the object in a radar image is bright if a large amount of the radar pulse is reflected back and dark if little or no energy is returned, regardless of the sun's brightness or if it is at night. The radar return depends mostly on the angle between the radar beam and the object and on how "rough" the object is. For instance, water is smooth and thus reflects away all the radar energy at the same angle at which it is received. None is returned to the radar receiver; therefore water is black on radar. When the radar strikes a hillside facing the airplane, most of the energy is reflected directly back, and the hill looks very bright. If the backside of the hill is blocked from the radar beam, it is in shadow and appears dark. A sandy beach is smooth and would appear darker than a "rough" beach covered with cobbles and boulders.

In the questions that follow, four maps are provided of an area around Dardanelle, Arkansas. Each map provides the basic information needed to answer the following questions. (The text and illustrations for this exercise, Figures 3.30, 3.31, 3.32, 3.33, and Table 7.2, have been adapted from a geologic training exercise provided to the author by personnel at Phillips Petroleum Company, Bartlesville, Oklahoma.)

Questions

1. Refer to the topographic map in Figure 3.30.

a. On which side of the Arkansas River is the town of Dardanelle (east or west)?

b. What forms the boundary between Pope and Yell counties?

c. Note the large meander in the Arkansas River southeast of Dardanelle. What is the relationship between the river and Gibson Lake?

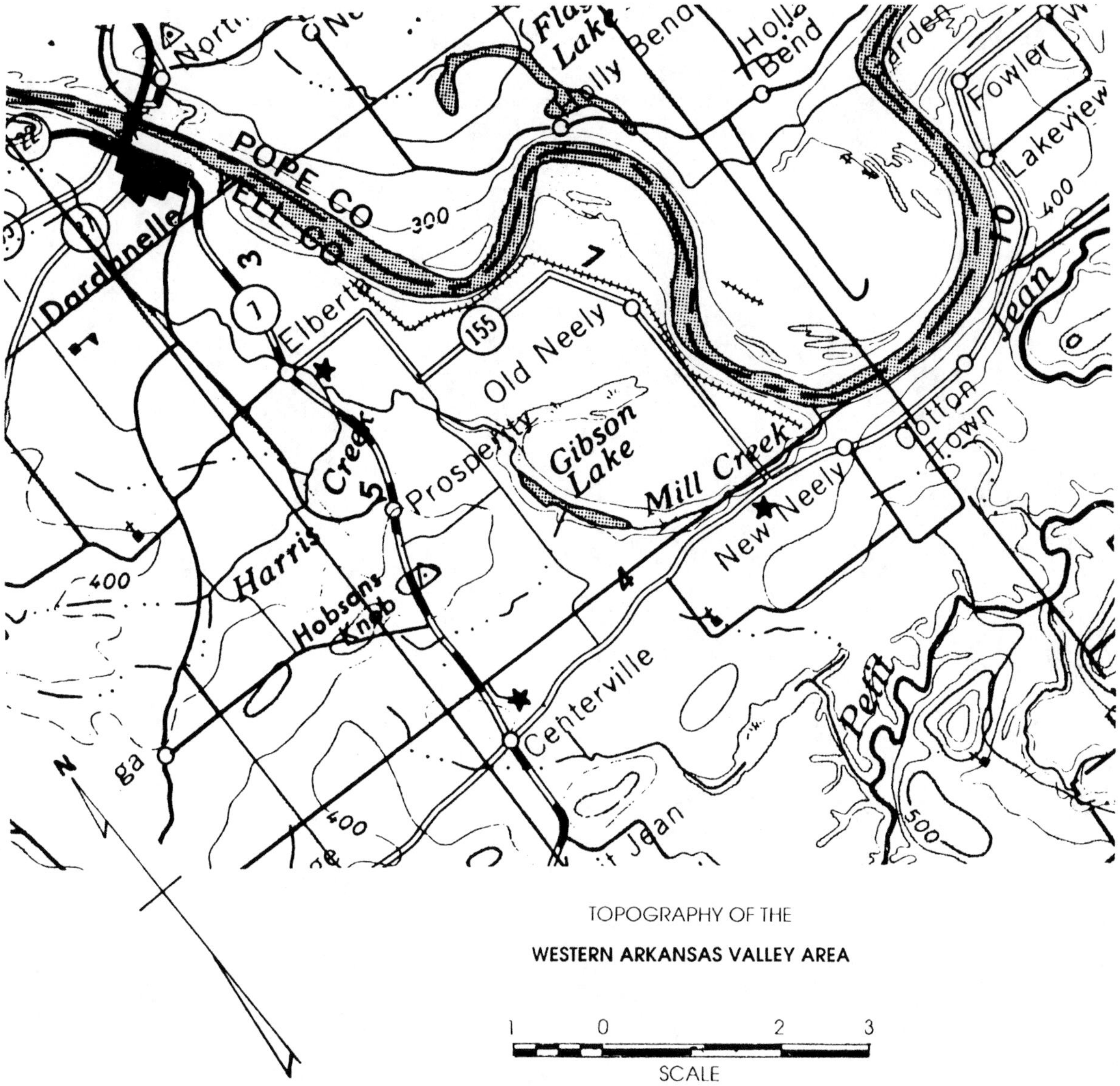

FIGURE 3.30 Topographic map of the western Arkansas Valley area.

2. Refer to the aerial photomosaic of the vicinity of Dardanelle, Arkansas (Fig. 3.31).

   a. Find Dardanelle, Gibson Lake, and Petit Jean Creek on the photograph.

   b. Note while looking at the photomosaic that some of the photographs in the mosaic have different tones. This tonal difference is due to photoprocessing. Also, note that the northern end of Gibson Lake is slightly offset, due to minor flight-path differences.

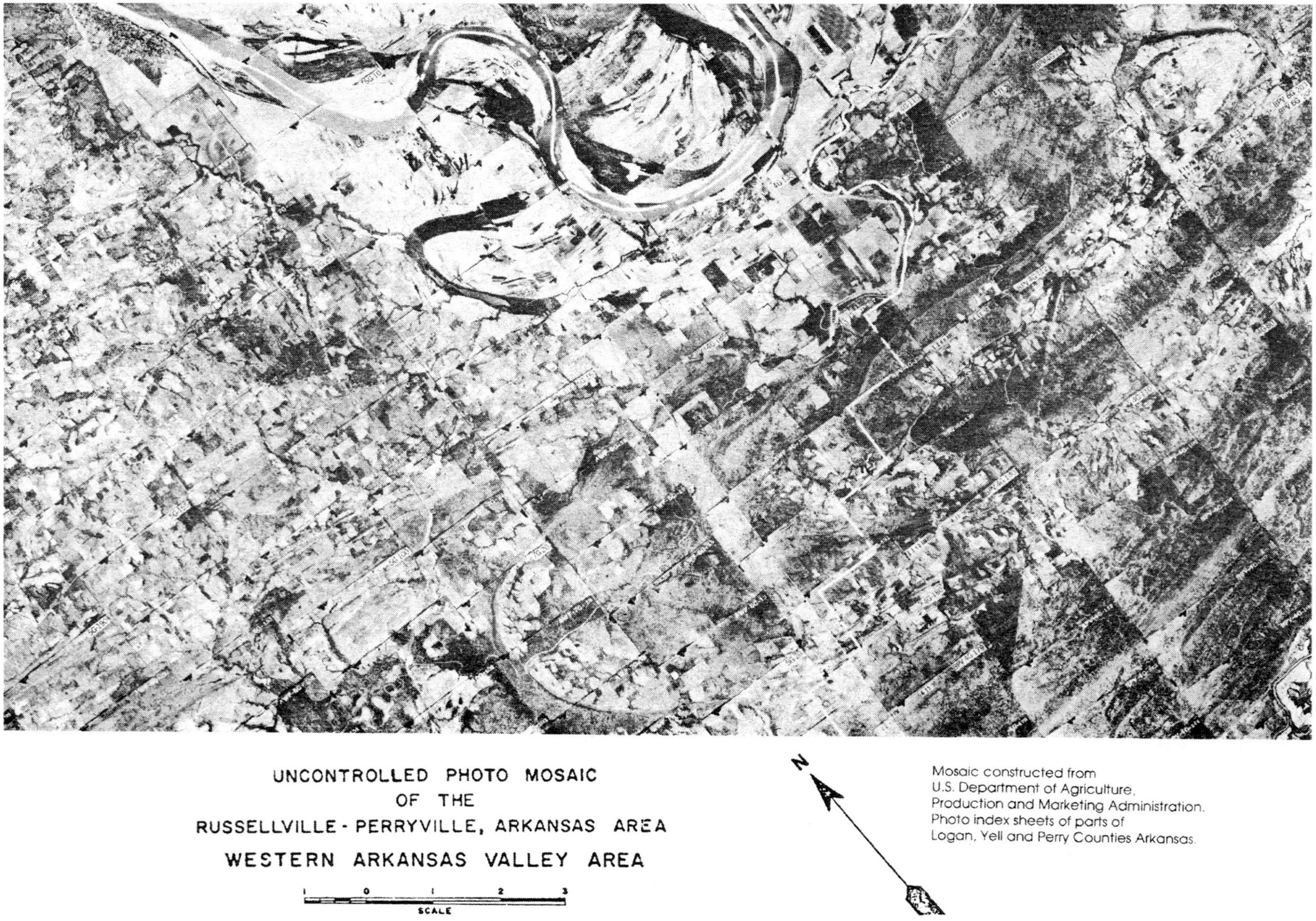

FIGURE 3.31 Photomosaic of the vicinity of Dardanelle, Arkansas.

c. Study the pattern of photo strips on the mosaic. Was the direction of the flight path north–south or east–west?

d. Is there evidence of the direction from which the sunlight was coming?

e. Does this mean that the sun was at a high angle (overhead) or a low angle (near the horizon)?

f. Would it be possible to emphasize the trend of mountains or topography in a photo by choosing a high sun angle or low sun angle? Which sun angle would be best, low or high?

3. Refer to the SLAR imagery of Dardanelle, Arkansas (Fig. 3.32). This mosaic has fewer images covering larger areas, and the flight path is from northwest to southeast.

a. In what general direction was the radar "looking" (northeast or southwest)?

b. What is your evidence for the answer to the previous question?

c. Would low-sun-angle photographs give the same effect? Why or why not?

d. Which mosaic gives the clearest detail, aerial photograph or SLAR? Why?

e. Why is the river light grey on the photo mosaic but black on the radar image?

f. Which mosaic provides a better initial evaluation of overall topographic structure?

g. Note that forest areas add roughness and show brighter and more apparent relief than cultivated or grassy areas on the SLAR mosaic. Thus, evaluating the distribution of vegetation types would depend on textural and tonal variations. For detailed agricultural studies, which mosaic would be preferred, the radar or the photo?

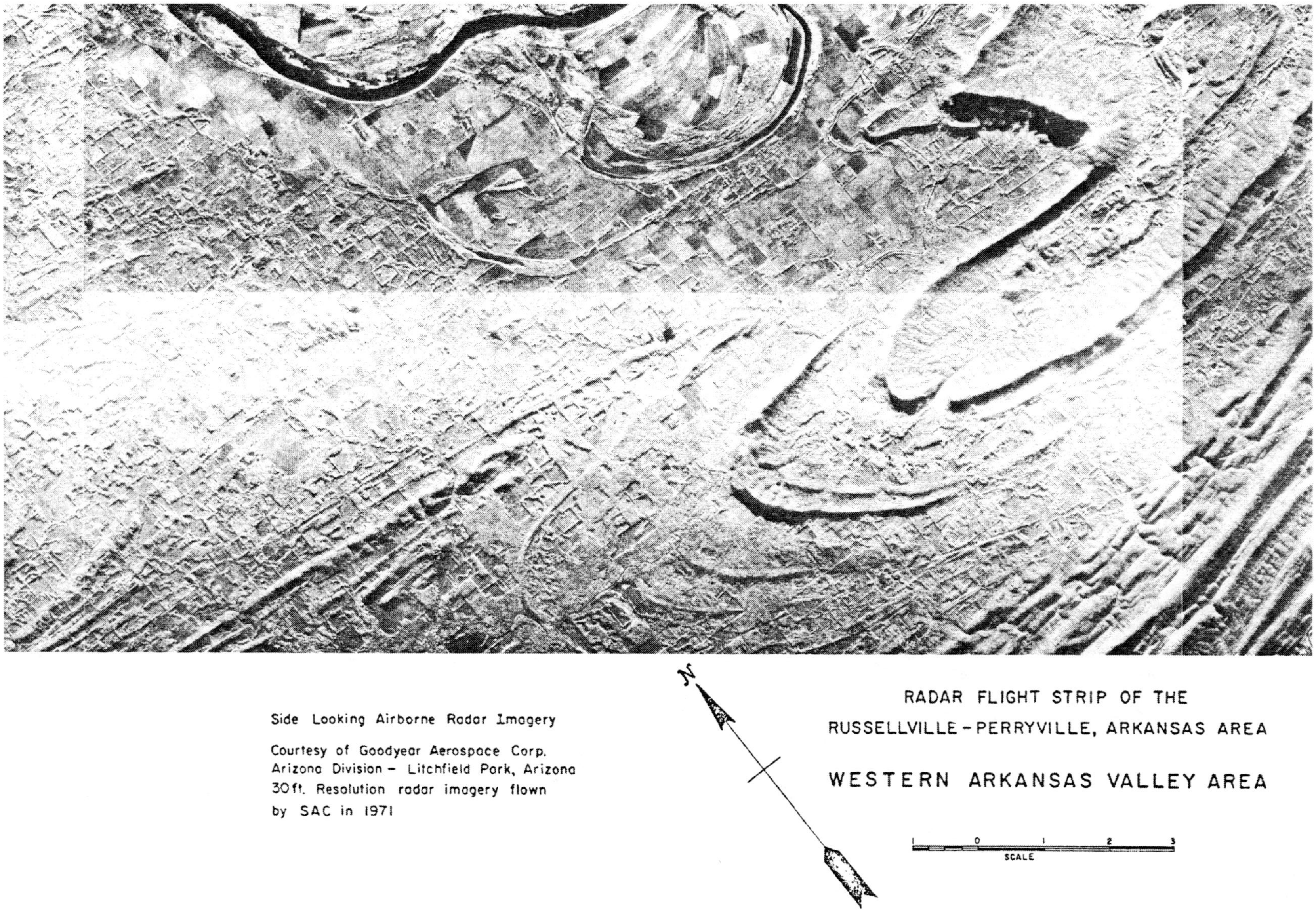

Side Looking Airborne Radar Imagery
Courtesy of Goodyear Aerospace Corp.
Arizona Division - Litchfield Park, Arizona
30 ft. Resolution radar imagery flown
by SAC in 1971
N
RADAR FLIGHT STRIP OF THE
RUSSELLVILLE - PERRYVILLE, ARKANSAS AREA
WESTERN ARKANSAS VALLEY AREA
1 0 1 2 3
SCALE

FIGURE 3.32 SLAR image of Dardanelle, Arkansas.

h. For land use planning and mapping (categorizing land for purposes of agriculture, industry, business, residency, and the like), which mosaic would be better?

i. What are the bright "blips," or "spots," across the Arkansas River at Dardanelle?

j. Look again carefully at the large meander southeast of Dardanelle. What is different about it when comparing the two images?

k. How can you determine which mosaic (photo or radar) is newer?

l. Notice that the Arkansas River has meandered frequently and has left many meander scars on the floodplain. Which mosaic (photo or radar) is more effective in showing these features? Why?

m. Examine the prominent U-shaped structural feature five miles south-southwest of the large meander scar. Do the rocks appear to be layered or crystalline? (*Hint:* These rocks form the frontal part of the Ouachita Mountains extending across Arkansas and eastern Oklahoma.)

n. What evidence indicates the general direction of the feature's structural axis, and what direction or azimuth is indicated?

o. Cite evidence to show that the structure is one of the following: a syncline, an anticline, a monocline, a dome.

p. In which direction does the structure plunge, and what evidence is available to indicate this direction?

q. Where would the youngest rocks be located in this structure? (Describe the area on the SLAR mosaic.)

4. See Figure 3.33, the geologic map of Dardanelle, Arkansas and vicinity, and Table 3.2, the map's legend.

a. On the large structure south of the large bend in the Arkansas River (marked Poteau syncline), color in red pencil the McAlester Formation (Pm) and lightly shade the Upper Atoka Formation (Pau) with a lead pencil. Examine the colored geologic map and discuss what type of structure is present. How does this compare with your answer to the SLAR image in part **p** of Question 3?

b. Why are there several small patches of Pm on the top of the mountain?

TABLE 3.2 Map Legend

Sedimentary Rocks

Period	Symbol	Formation		
QUATERNARY	Qal	Surficial deposits		
PENNSYLVANIAN	IPb	Boggy fm		
	IPs	Savanna fm		
	IPm	Mc Alester fm		
	IPh	Hartshorne ss		
	IPau	Upper Atoka fm	IPa	Atoka, undivided
	IPam	Middle Atoka	IPa	Atoka, undivided
	IPal	Lower Atoka	IPa	Atoka, undivided
	IPjv	Johns Valley Shale		
CARBONIFEROUS	Cs-j	Stanely-Jackfork fm undivided		

Map Symbols

Paved road

Improved road

Unimproved road

308 + Photo center

Photo Dip Symbols:

- Dip component
- Less than 1°
- 1-3°
- 3-7°
- 7-15°
- 15-30°
- Over 30°
- Vertical
- Overturned beds

Field measured dip

Tie bar, connecting outcrops of same fm

Formation contact, dashed where inferred

Fault dashed where inferred

Thrust Fault teeth on upthrown side

Contorted bedding

Lineation

Anticline

Syncline

Overturned Anticline

Overturned Syncline

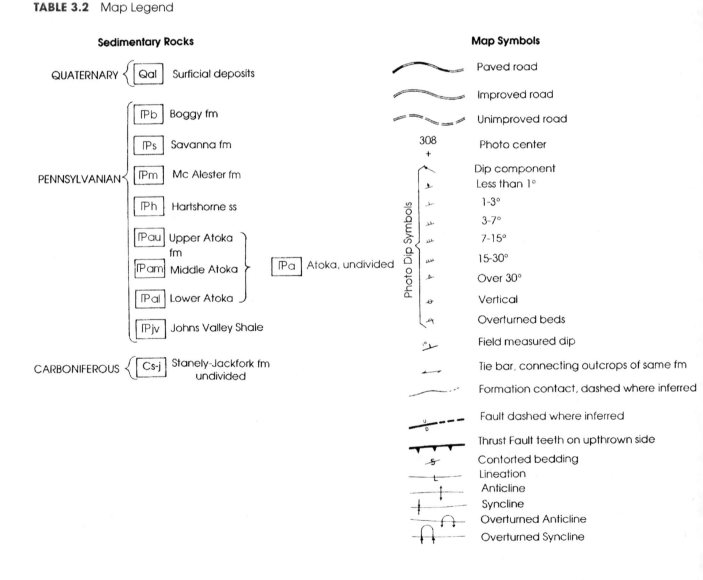

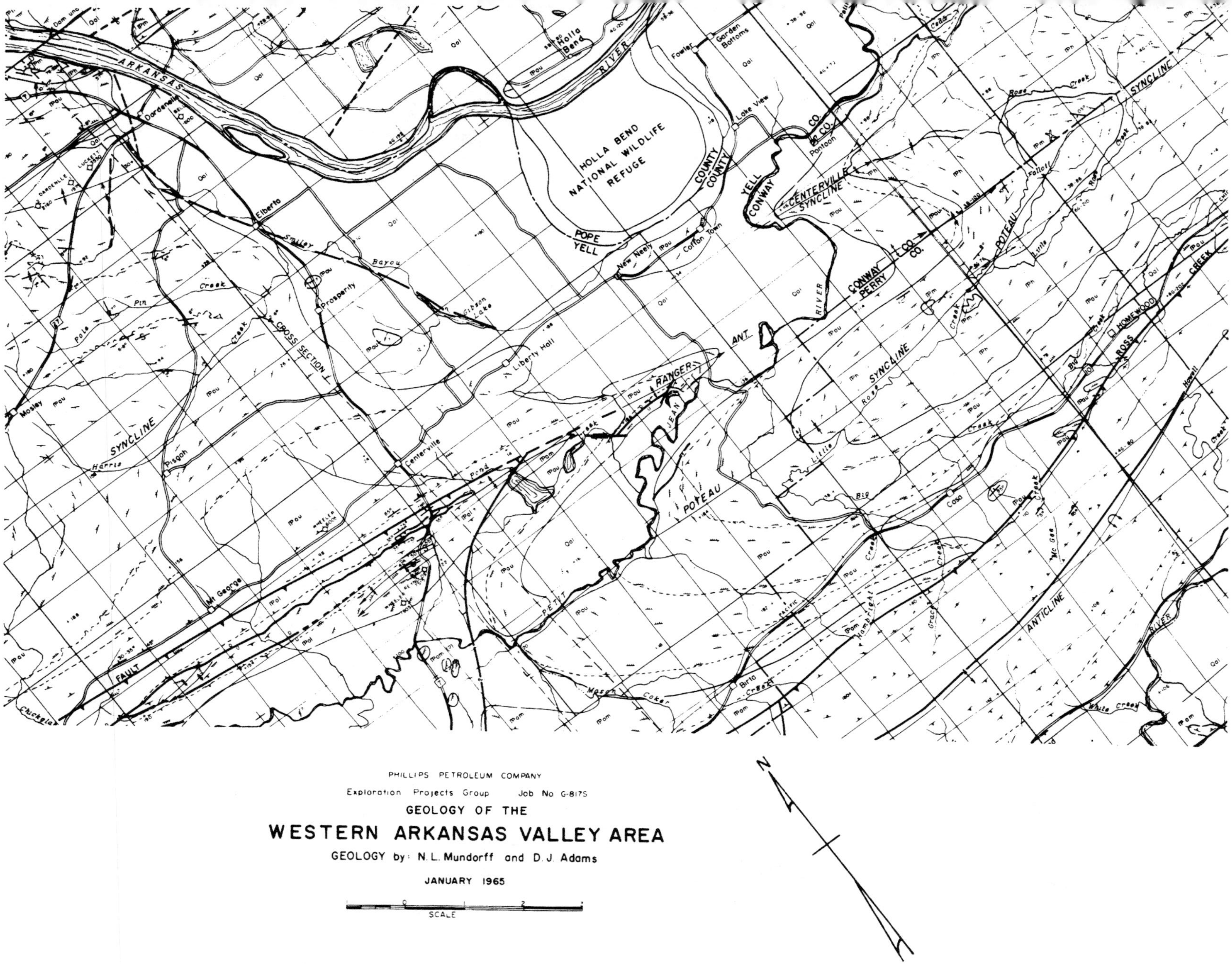

FIGURE 3.33 Geologic map of Dardanelle, Arkansas.

CHAPTER

4

Paleontology

In the previous chapters you studied the lateral and vertical relationships of sedimentary rock strata and the concept of relative time. Now we will examine these rocks for their fossil content. Fossils are evidence or remains of formerly living organisms, most often preserved in sediment in one of the many depositional environments you have studied. By using the principle of uniformitarianism and an understanding of modern organisms—their habitats and relationships to other organisms within an ecosystem—you can expand your ability to interpret ancient sedimentary rocks, paleoenvironments, and paleogeography.

DEFINITION OF A FOSSIL

The word *fossil* comes from the Latin *fossilis*, meaning "dug up." In the early 1800s, when geology was becoming a distinct science, the term *fossil* was applied to almost any object that was found or "dug up." Today, geologists restrict the definition of a fossil to include only the naturally preserved remains or traces of animals or plants that lived in the geologic past. By this definition, an Egyptian mummy cannot be considered a fossil because it is not a naturally preserved organism. Similarly, an arrowhead cannot be considered a fossil because it was fashioned by a person and thus would be called an artifact. Raindrop impressions could not be classified as fossils since they were not formed by any living organism.

Perhaps one of the more confusing aspects of the definition of a fossil concerns the geologic past. Many geologists define the geologic past as being any time prior to recorded human history, whereas others set the dividing line between past and present at the beginning of the Holocene epoch, or about 11,000 years ago. Such a restriction appears to be unwarranted for the following reasons: (1) a measured age cannot be assigned to the rocks containing the fossils under investigation, (2) there is nothing in the fossilization process relating directly to the passage of time, and (3) there are variations in the animal's hard parts and composition as well as the chemical parameters of depositional environments. Cases exist where Miocene fossils are better preserved than some modern seashells.

Some people refer to coal or oil as "fossil" fuels. These people are using the term *fossil* to imply old age. Specifically, the individual plants that make up coal are fossils; coal itself is a sedimentary rock.

The vast majority of fossils are found in sedimentary rocks. The concentration and distribution of fossils vary markedly, both vertically and horizontally, within sedimentary units. In some cases an entire sedimentary unit may be composed of fossils; a good example is coquina limestone.

The Study of Fossils

Paleontology is the area of geological study that deals with the examination and interpretation of fossils. The paleontologist uses fossils in several different ways. Three of the most important ways are: (1) field identification of fossils as an aid in geologic mapping,

(2) fossils as indicators of past environments, and (3) study of a fossil's taxonomic classification and evolutionary development to gain a better understanding of its role as an indicator of geologic time.

NATURE OF THE FOSSIL RECORD

Preservation of Fossils

The potential preservation of plants and animals as fossils is greatly enhanced by two favorable conditions. The first is the organism's possession of durable internal or external hard parts. The second is the organism's relatively rapid burial within a sedimentary environment. In most cases, the soft parts of an organism are destroyed and little or no evidence of them is preserved in the geologic record. In very rare instances, the entire animal is preserved; some examples are insects trapped in resin that can turn to amber and mammoths frozen in ice. Another frequently cited case of near-perfect preservation involves a brown residue of the soft parts of plants and animals. This carbon film is what remains of plants and animals after they undergo *carbonization* or *distillation* processes by which the volatile components (oxygen, hydrogen, and nitrogen) of organic substances are "burned off" during decay and preservation. The carbon films of soft-bodied animals that were discovered at the turn of the century by C.D. Walcott in the Middle Cambrian Burgess Shale of western Canada represent some of the most famous examples of this kind of animal preservation. Another similar process, which preserves animals almost in their entirety, is *desiccation*, or water evaporation from the tissues. Bodies of Native Americans who lived in North and South America approximately 10,000 years ago have been preserved through desiccation.

Most of the fossils used to interpret earth history and to reconstruct ancient environments are organisms whose hard parts have been preserved. Even these hard parts commonly undergo some form of alteration after an organism's death. Since the hard parts of most invertebrate organisms are composed of calcium carbonate, silica, or chitin, and since the bones of most vertebrates are composed primarily of calcium phosphate, alteration during their transportation and burial should be expected. Modes of preservation can be subdivided into the following categories: (1) preservation without alteration, (2) preservation with alteration, (3) molds and casts, and (4) trace fossils. The following section details these methods of preservation.*

Modes of Fossil Preservation: Preservation Without Alteration

Soft parts (rare)—Freezing of organisms, such as the mammoths of Siberia. Mummification of remains in dry climates. Entrapment of organisms in resin (amber) or in oil seeps.

Hard parts—Unaltered shells, bones, teeth, most commonly of calcium carbonate, calcium phosphate, silicia, and chitin; cellulose and wood.

Modes of Fossil Preservation: Preservation with Alteration

Leaching—Chemical dissolution of the most soluble portions of the remains, commonly resulting in bleached and pitted shell and bone.

---

*Source: After J. D. Cooper et al. 1986, *A Trip Through Time*, 2nd ed., Table 3.1, p. 74. New York: Macmillan Publishing Co.

Carbonization—Change by chemical action of original plant or animal material to a thin film of carbon that outlines the shape of part or all of the organism.

Permineralization—Deposition in pore spaces of buried remains by underground solutions of mineral material, most commonly calcium carbonate, silica, pyrite, or dolomite, which are unlike original shell or bone composition. Examples are petrification of wood and bone.

Recrystallization—Conversion into a more stable form (such as the calcite form of calcium carbonate) of less stable compounds (such as the aragonite form of calcium carbonate of some clams and snails).

Replacement—Complete dissolution and replacement (or nearly so) by new mineral material of original material (such as shells or bones); common replacement minerals are calcite, dolomite, quartz, and pyrite.

Modes of Fossil Preservation: Molds and Casts

Molds—Removal by dissolution of organic material buried in sediment; void left in the rock is a mold (for example, an imprint); molds can be internal (expressing the shape of the inside of a shell or other feature) or external (expressing the shape of the outside of the object) [Fig. 4.1(a)].

Casts—Filling of a mold (void) with sediment or mineral material, thus preserving the shape (internal or external) of the organic feature [see Fig.4.1(b)].

Modes of Fossil Preservation: Trace Fossils

Tracks and trails—Footprints of animals, including indications of movements by invertebrates (Figs. 1.5, 1.8, and 4.2).

Burrows—Excavations made by worms and other animals, such as clams, crabs, shrimp, or fish, as they tunnel into sediments (Fig. 1.19).

Borings—Holes bored through shells by predator snails or other organisms; holes bored into rock by rock-boring organisms such as clams, worms, and certain crustaceans.

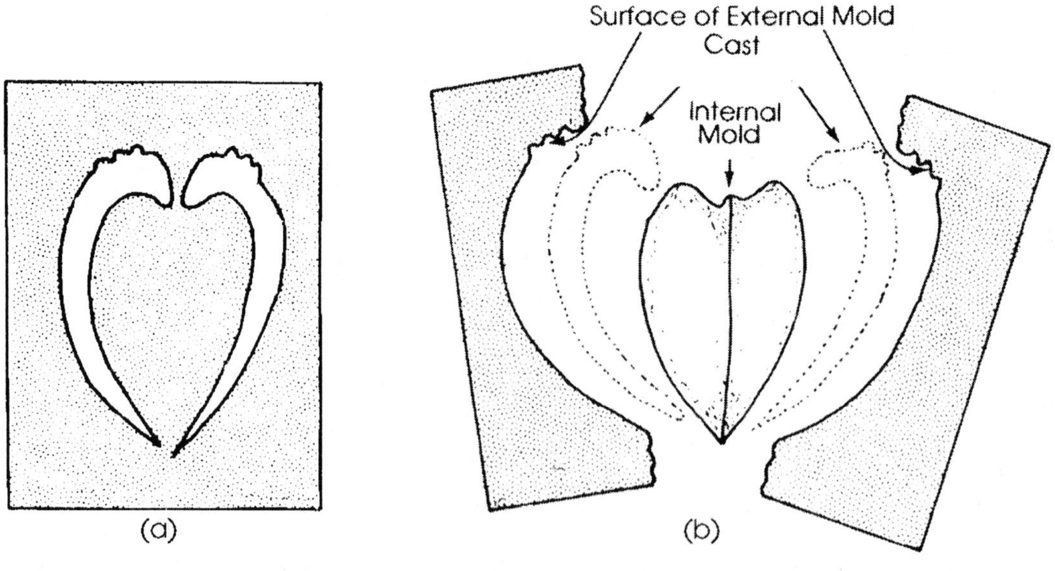

FIGURE 4.1 Molds and casts: (a) Cavity left in a rock after the fossil has been dissolved (mold). (b) Rock is broken apart: An internal mold will be the matrix material that fills inside. If the mold were filled with a secondary material, it would form a casting of the original shell.

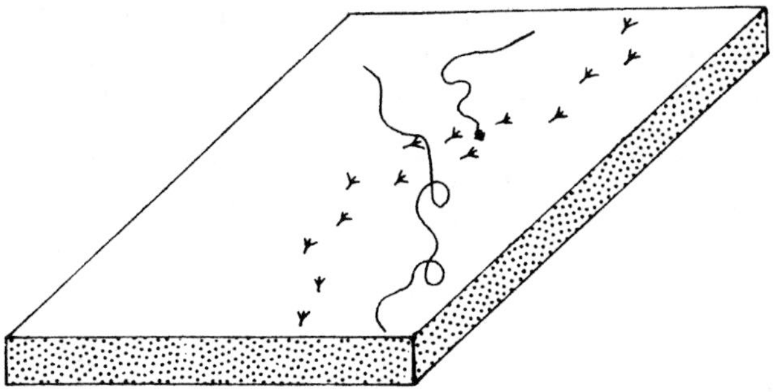

FIGURE 4.2 Bird tracks on soft sediment.

Coprolites—Fossilized animal excrement; may give evidence of diet, animal size, and habitat.

Chemical fossils—Trace of organic acids such as are found in Precambrian sediments.

Bias of the Fossil Record

The fossilization process favors the preservation of marine animals with hard parts, so the fossil record gives a somewhat distorted view of life at any given time. Table 4.1 lists the known animal species living today and gives the estimated percentage of the overall total population that each species constitutes. The table refers to numbers of species and not to numbers of individuals within a species. Consider the fact that both the whooping crane and the anopheles mosquito are each a species, but there are fewer than 200 individual whooping cranes living today, whereas there are uncountable individual anopheles mosquitoes. Insects are the dominant living animal group, yet the percentage of fossil insects is only about 1%. No doubt this reflects the fact that insects have chitinous exoskeletons that chemically break down after death, thus greatly diminishing their preservation potential.

Even animals with hard parts are subject to postmortem destruction, which adds even greater bias to the fossil record. Shells accumulating on a beach may be broken and destroyed by wave action; thus the abrasion of these and other animal parts may be so great as to render them unrecognizable as fossils. The effects of the mechanical destruction may be duplicated in the laboratory by using a rotational tumbler. In one experiment that began initially with a known mixture of bivalves, gastropods, limpets, bryozoans, starfish, echinoids, and calcareous algae, all of the bryozoans, starfish, echinoids, and calcareous algae were destroyed after 183 hours of tumbling. As a result of this destruction, the relative abundance of the bivalves, gastropods, and limpets was greatly enhanced. Thus, in the experiment, the animals with strong shells became dominant in the final assemblage owing to the destruction of those without such shells. Because of this

TABLE 4.1 Subdivision of Living Animal Species (percent)

Protozoans	2.5%	Arthropods	82.0% (75% are insects)
Porifera	<0.1	Brachiopods	<0.1
Cnidarians	0.8	Echinoderms	<0.1
Mollusks	7.0	Vertebrates	3.0
Bryozoans	<0.1	All others	5.0

**Source*: After J. D. Cooper et al. 1986, *A Trip Through Time*, 2nd ed., Table 3.1, p. 74. New York: Macmillan Publishing Co.

destruction, the final assemblage was highly biased. Vertebrate carcasses left in open terrestrial environments are subjected to scavenging, which often causes the bones to become disarticulated and dispersed. Chemical weathering in the open can dissolve and weaken bone, literally turning it to dust.

Mechanical, chemical, and biological alterations within the depositional environment can change the composition of any original fossil assemblage, so the fossil record must always be viewed as biased and often atypical of ecological diversity at any given time.

Diversity

Diversity of an animal, plant, or other group can be measured by the number of families, genera, or species within the group. Throughout geologic time the diversity of groups has fluctuated greatly, reaching peaks at some times and decreasing significantly at others. Many groups seem to have undergone a long period of decline before their extinction, whereas others seem to have become extinct rather abruptly.

The fossil record may also be biased by the chemical and biological destruction of shell material. Most shells are made of calcium carbonate, which may dissolve in acidic waters. Shells may be completely dissolved over a relatively short span of time or over millions of years. Destruction or disarticulation of shells may also be caused by the biological action of scavengers and bacteria.

The fossil diversity of a group through time may be a function of environmental or climatic conditions. An example of the diversity patterns through time for one animal group can be illustrated by reviewing the diversity diagram in Figure 4.28, which shows the expansion and contraction of an arthropod class called the trilobites. Much information can be read from a diagram of this type. Trilobites first appeared in the Cambrian period and subsequently underwent rapidly increasing diversity until the Ordovician, when the group experienced an evolutionary explosion. This time period represents the phase of maximum development for the trilobites. After the Ordovician their diversity slowly declined until extinction occurred during the Permian period.

Fossils as Environmental Indicators

Fossils provide important geologic clues to the environmental ecological conditions that existed when the rocks containing them were formed. Heavy-shelled organisms are often associated with more turbulent or higher-energy environments, whereas delicately shelled organisms are indicators of relatively calm or quiet-water conditions.

Since soft parts are rarely preserved, inferences concerning the original environmental behavior of fossil animals and plants can be obtained only from detailed studies of their modern analogs. Research studies have also established the relationship between a modern organism and its fossil counterpart. Study of the modern chambered nautilus can shed some light on the anatomy and habits of ancient ammonites. These studies also establish the relationships between organisms and the environmental conditions needed to support life. Once geologists have come to understand the life habits of modern animals, application of the principle of uniformitarianism enables them to relate their environmental knowledge to the fossil faunal and floral record. By establishing the various environmental tolerances of modern animals and plants, geologists gain valuable insight into environmental parameters such as paleosalinity, paleotemperature, and paleoclimate.

One major problem in using fossils as environmental indicators is that an organism can be transported and reburied in sediment of a different depositional environment. Often, a fossil that has been transported will show signs of abrasion or wear, but one fossil does not provide sufficient basis for predicting paleoenvironmental conditions. Therefore, geologists usually try to collect a *suite*, or *assemblage*, of fossils from a formation or area to produce more meaningful interpretive results.

MAJOR CONCEPTS OF ECOLOGY AND PALEOECOLOGY

Ecology is the study of the relationships of living organisms to the chemical and physical conditions in their environment and to other organisms. Paleoecology attempts to do the same for ancient organisms. As previously mentioned, a thorough knowledge of modern organisms' life habits is essential in paleoecologic interpretations.

The three primary spheres of habitation for living organisms are the atmosphere, the hydrosphere, and the lithosphere. No species spends its entire life cycle in the air, but many species spend various amounts of time in the atmosphere. The most common habitats are in the hydrosphere (including both salt water and freshwater) and on land (terrestrial habitats).

Life Habits Within an Ecosystem

Organisms within any given ecosystem can pursue a variety of different life habits. In terrestrial ecosystems, creatures can be mobile, moving around within a variety of habitats or even flying in the air. Some organisms, such as trees, are rooted in the soil. Many animals burrow in the soil, and microbes of various types live in the soil as well.

Within all the various habitats that comprise a marine ecosystem, organisms fall into three categories: planktonic, nektonic, and benthic. Each of these categories of marine organisms can be described and illustrated as follows.

1. *Planktonic* organisms float in water currents or swim only weakly. Plankton consist of both phytoplankton (plantlike producers) and zooplankton (consumers). Figure 4.3 shows some common planktonic forms.

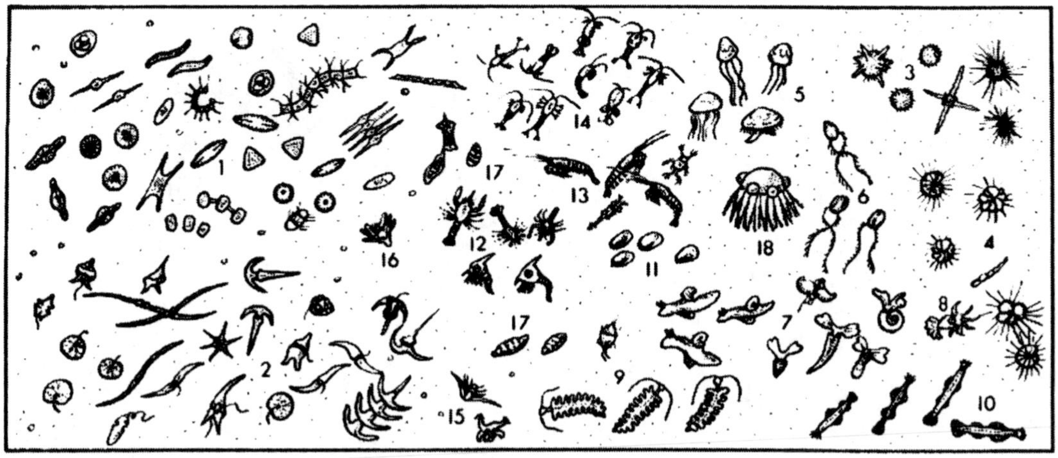

FIGURE 4.3 Planktonic marine organisms: 1. Algae (diatoms) 2. Algae (flagellates) 3. Protozoa (radiolarians) 4. Protozoa (foraminifera) 5. Cnidaria (medusae) 6. Ctenophora (comb-jelly) 7. Mollusca (snails) 8. Mollusca (nautiloid) 9. "Worms" (annelids) 10. "Worms" (arrow-worms) 11. Arthropoda (ostracods) 12. Arthropoda (crab larvae) 13. Arthropoda (krill) 14. Arthropoda (copepods) 15. Echinodermata (starfish larvae) 16. Echinodermata (sea urchin larvae) 17. Chordata (tunicates) 18. Hemichordata (graptolite). Drawn by Terry Chase.

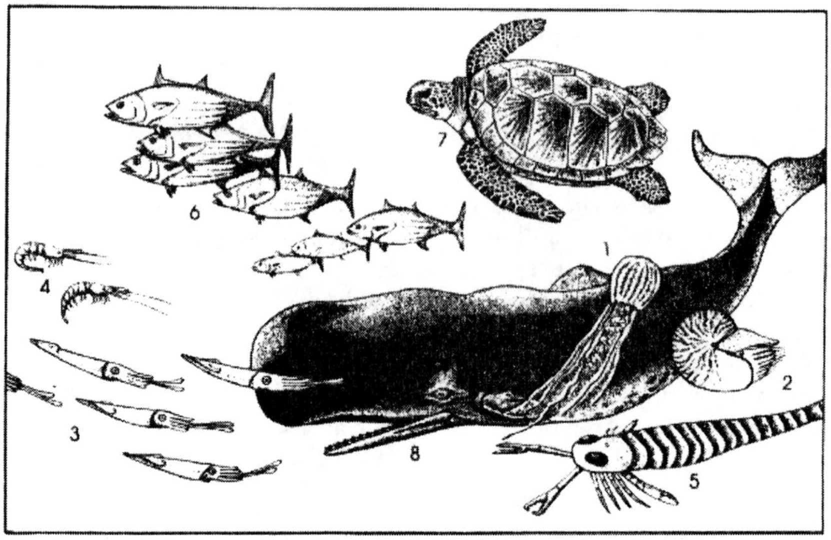

FIGURE 4.4 Nektonic marine organisms: 1. Cnidaria (jellyfish) 2. Mollusca (nautiloid) 3. Mollusca (squids) 4. Arthropoda (shrimp) 5. Arthropoda (eurypterid) 6. Chordata (fish) 7. Chordata (reptile) 8. Chordata (mammal). Drawn by Terry Chase.

2. *Nektonic* marine organisms swim strongly and are not confined by wave or current forces. Figure 4.4 depicts common nektonic animals.
3. *Benthic* organisms are bottom dwellers. Benthic organisms pursue a variety of lifestyles. *Vagrant* (mobile) benthic creatures move about freely on the bottom (Fig. 4.5). *Sessile* benthic organisms are stationary. Some live above, but are attached directly to, the substrate (*epifauna*: Fig. 4.6), while others live within the substrate in boreholes or burrows (*infauna*: Fig. 4.7).

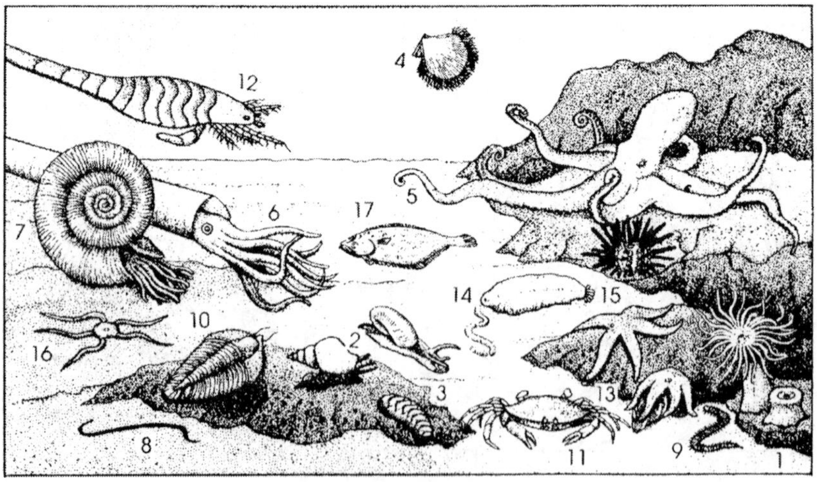

FIGURE 4.5 Marine vagrant benthic organisms: 1. Cnidaria (anemones) 2. Mollusca (snails) 3. Mollusca (chiton) 4. Mollusca (scallop) 5. Mollusca (octopus) 6. Mollusca (nautiloid) 7. Mollusca (ammonite) 8. "Worms" (ribbon worm) 9. "Worms" (annelid) 10. Arthropoda (trilobite) 11. Arthropoda (crab) 12. Arthropoda (eurypterid) 13. Echinodermata (starfish) 14. Echinodermata (sea cucumber) 15. Echinodermata (sea urchin) 16. Echinodermata (brittle star) 17. Chordata (fish). Drawn by Terry Chase.

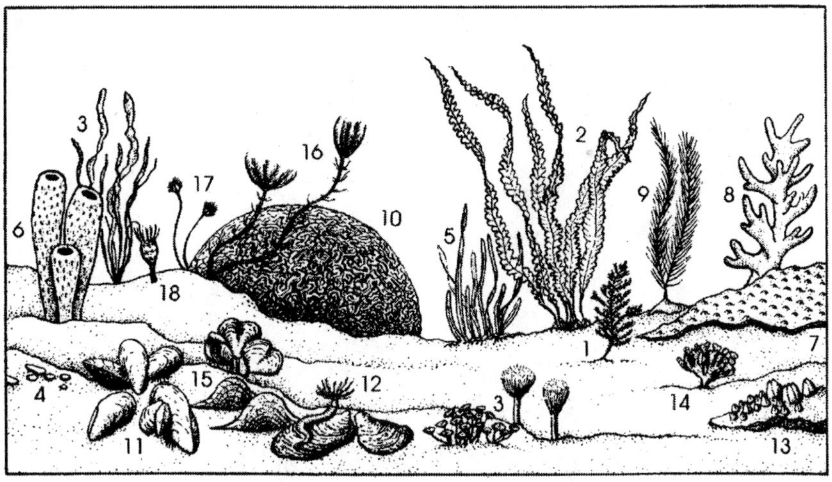

FIGURE 4.6 Marine sessile benthic organisms (epifauna): 1. Algae (red) 2. Algae (brown) 3. Algae (green) 4. Protozoa (foraminifera) 5. Plantae (angiosperm) 6. Porifera (sponge) 7. Porifera (stromatoporoid) 8. Cnidaria (hydrozoan coral) 9. Cnidaria (sea whip) 10. Cnidaria (coral) 11. Mollusca (mussels) 12. "Worms" (annelid) 13. Arthropoda (barnacles) 14. Ectoprocta (bryozoan) 15. Brachiopoda (lampshells) 16. Echinodermata (crinoids) 17. Echinodermata (blastoids) 18. Echinodermata (cystoid). Drawn by Terry Chase.

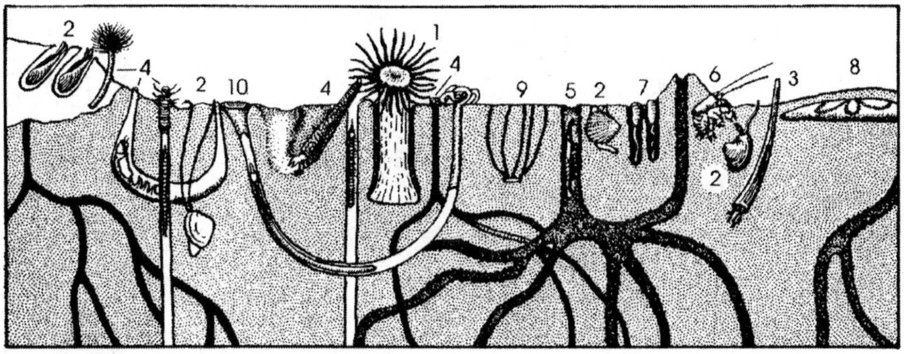

FIGURE 4.7 Marine sessile benthic organisms (infauna): 1. Cnidaria (burrowing anemone) 2. Mollusca (clams) 3. Mollusca (tusk shell) 4. "Worms" (annelids) 5. Arthropoda (ghost shrimp) 6. Arthropoda (mole crab) 7. Brachiopoda (inarticulates) 8. Echinodermata (sand dollar) 9. Echinodermata (brittle star) 10. Chordata (acorn worm). Drawn by Terry Chase.

Cycle of Materials and Energy Within an Ecosystem

Wherever the ecosystem, it contains certain characteristic components such as nutrients, producers, consumers, and decomposers. *Nutrients* consist of organic and inorganic molecules that act as fertilizer for the producers (plants or other photosynthetic organisms). *Producers* use these nutrients as well as solar energy and chlorophyll to create carbon-hydrogen-oxygen compounds (food) by the chemical process of photosynthesis. *Consumers* (animals or other nonphotosynthetic organisms) must ingest producers in order to carry on their life functions. There are several types of consumers in an ecosystem: *herbivores*, which eat the producers directly; *carnivores*, which eat the herbivores or other carnivores; *omnivores*, which eat both flesh and vegetation; *scavengers*, which eat dead flesh; and *parasites*, which gain nourishment from a living host. *Decomposers* (bacteria and fungi) transform the dead plant and animal material back to nutrients, which, in turn, start the cycle over again as fertilizer for producers.

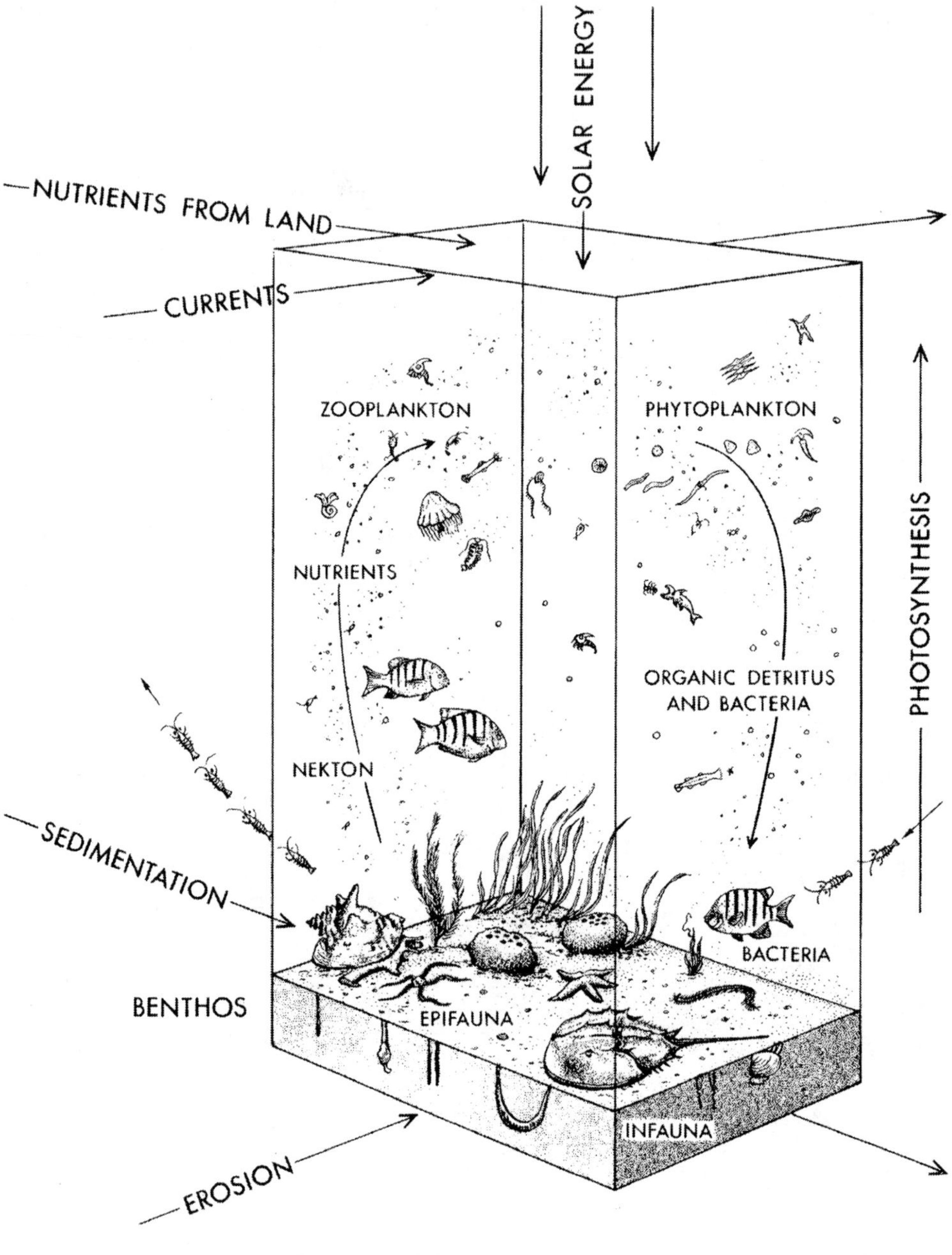

FIGURE 4.8 Interrelationships of life habits within the marine ecosystem. Drawn by Terry Chase.

In general, the interrelationships of life habits within the marine ecosystem are as depicted in Figure 4.8.

BIOSTRATIGRAPHY AND BIOZONES

Fossil animals and, less commonly, fossil plants are used to determine the relative geologic age of a sediment, the nature of paleoenvironments, and the paleogeography that characterized a particular area. Fossils can also aid in correlation. There are a number of ways to use fossils and their ranges (total time of existence) to help correlate rock units from one area to another. Two major types of biozones are the range zone and the concurrent range zone. The *range zone* uses the range in time of one species or genus. Certain fossils (particularly floating or swimming species) are especially useful for age (or time) determinations. These fossils are known as *index fossils*, and to be so classified they must fulfill the following three important conditions:

- The organism should have lived during a relatively short period of geologic time (usually an epoch or less) and should have been quite abundant.
- The organism should have had a wide geographical distribution that was not strongly influenced by facies changes.
- The organism should have had a distinctive morphology or appearance to be easily recognized. Fossils that are considered for use in geologic dating are usually studied in great detail by specialists in paleontology.

Another type of zone, the *concurrent range zone*, relies on plotting the overlapping ranges of several species. First and last occurrences of fossils are used to separate one zone from another as older species become extinct and newer ones enter the geologic record. Often certain species have distinct preferences for certain environmental conditions, such as salinity and substrate. The use of a variety of different species to define a zone can eliminate the problem and aid in correlation across facies changes.

USE OF FOSSIL ASSEMBLAGES IN AGE DETERMINATIONS

One of the more important tasks in historical geology is the determination of the age of rocks on the earth's surface. Since a high percentage of these surface rocks are sedimentary and fossiliferous, the relative age of such rock strata can be determined by the following technique, once a series of index fossils has been identified. See page 156 for a discussion of taxonomic categories, such as phylum, genus, and species.

Step 1. The phylum level of each fossil must first be identified (use the key on pages 157–159). Next, the name for the individual fossil must be determined. (For detailed work, the species names for most fossils will have to be located in paleontological reference books.)

Step 2. Once the genus or species is known, the geologic range for that fossil may be plotted on a geologic time diagram such as Figure 4.9. Each fossil should be plotted separately.

Step 3. This step is the most important in age determination. Three fundamental assumptions must be made: (1) that all of the fossils have been collected from the same formation or horizon, which represents one period of deposition; (2) that all of the fossil organisms lived together and are thus representative of life during one interval of geologic time; and (3) that the complete geologic range of each fossil is known. Therefore, from the sample diagram, an interval of geologic time can be determined in which all of the fossils could have mutually coexisted. In this example, that time is the Permian.

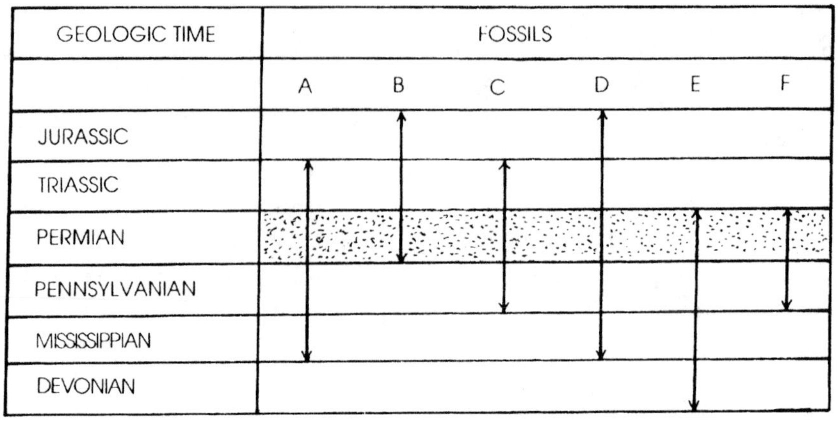

FOSSIL ASSEMBLAGE FOR ABOVE CHART

Fossil A (Mississippian-Triassic)
Fossil B (Permian-Jurassic)
Fossil C (Pennsylvanian-Triassic)
Fossil D (Mississippian-Jurassic)
Fossil E (Devonian-Permian)
Fossil F (Pennsylvanian-Permian)

FIGURE 4.9 Example of overlapping geologic fossil ranges for age determination.

EXERCISES

Exercise 4-1 DETERMINING ROCK AGE FROM FOSSIL ASSEMBLAGE, SOUTHWEST MISSOURI

Suppose that a geologist had collected the following suite of fossils from a quarry in southwestern Missouri. Given the ranges of each fossil, determine the age of the rock in the quarry. Document how you reached your answer (see Figure 4.9).

Quat	
Tert	
Cret	
Juras	
Trias	
Perm	
Penn	
Miss	
Dev	
Sil	
Ord	
Camb	

Specimen 1—Ordovician–Permian
Specimen 2—Silurian–Mississippian
Specimen 3—Cambrian–Triassic
Specimen 4—Devonian–Eocene
Specimen 5—Ordovician–Jurassic
Specimen 6—Mississippian–Recent

a. What system of rocks was sampled? (Remember, *system* refers to rocks deposited during a geologic period.)

____

b. Which is the best index fossil?

____

c. Which is the poorest index fossil?

____

d. If you had only two fossils, which two would give you the age of the rock without your needing any others?

____

Exercise 4–2 DETERMINING AGE OF LIMESTONE FROM FOSSIL ASSEMBLAGES

The following list of fossils represents an assemblage collected from a limestone. Using these fossils and their geologic ranges in the manual, determine to which geologic system the formation belongs.

The geologic system is ______________________

Quat	
Tert	
Cret	
Juras	
Trias	
Perm	
Penn	
Miss	
Dev	
Sil	
Ord	
Camb	

Archimedes (Bry., Fig. 4.18)
Myalina (Biv., Fig. 4.22)
Bellerophon (Gast., Fig. 4.24)
Aulopora (Cnid., Fig. 4.17)
Loxonema (Gast., Fig. 4.24)
Sigillaria (Plant., Fig. 4.35)
Fenestrellina (Bry., Fig. 4.18)
Dielasma (Brach., Fig. 4.20)
Microcyclus (Cnid., Fig. 4.17)

Exercise 4-3 CRETACEOUS FOSSIL

While fishing from the banks of the Missouri River in eastern Montana, you notice some shiny fragments of fossils lying in the grey shaly mud of the banks. The nearby outcrops are known to be composed of Cretaceous strata. Upon closer inspection, you discover that the fragments were from a coiled "critter" that had very ornate "curved" lines across the very shiny pearly or lustrous outer layer. What is this fossil "critter"?

Exercise 4-4 DETERMINING AGE OF LIMESTONE, NORTHEAST OKLAHOMA

A geology teacher and his students went to a limestone quarry in northeastern Oklahoma and collected a group of fossils. After the students identified the following group of fossils, they concluded that the rocks belonged to the ______________________________ system.

Quat	
Tert	
Cret	
Juras	
Trias	
Perm	
Penn	
Miss	
Dev	
Sil	
Ord	
Camb	

Coral	Brachiopod	Bryozoan	Gastropod
Favosites	*Composita*	*Fenestrellina*	*Trepospira*
Lophophyllidium	*Enteletes*		
	Juresania		

Exercise 4-5 DETERMINING AGE OF LIMESTONE

Determine the geologic system of the major limestone unit in the center of the two columns in Figure 4.10. For all of the animals found in the same stratigraphic unit, there should be one common interval of geologic time. This period of geologic time is ______________________________.

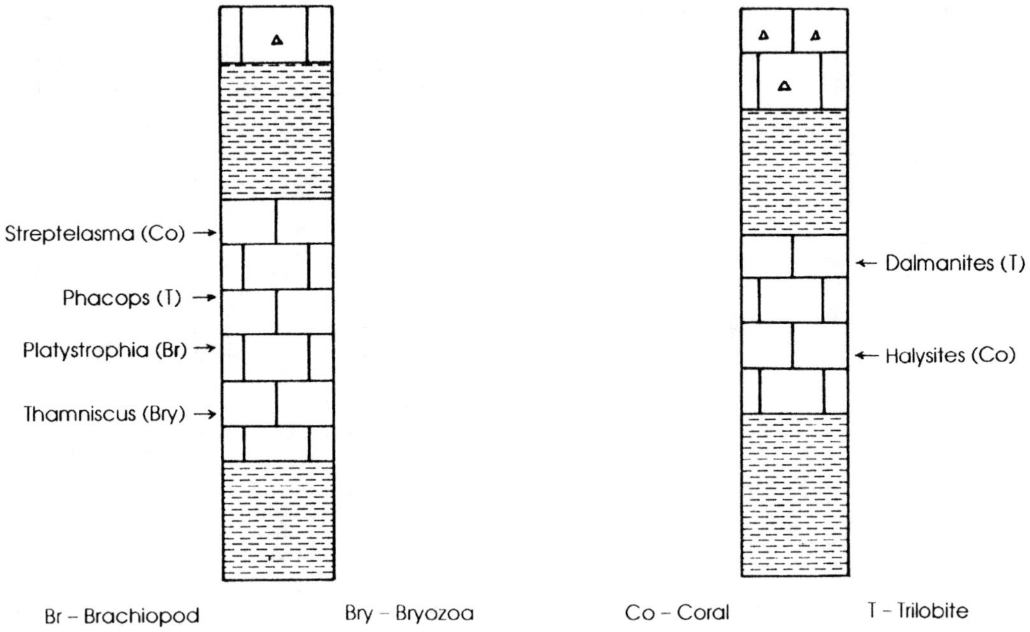

FIGURE 4.10 Stratigraphic columns for Exercise 4-5.

Exercise 4-6 USING TRILOBITES TO DETERMINE ROCK AGE

All of the fossils named in the columns in Figure 4.11 are trilobites (see Fig. 4.28). During what geologic system was the cherty limestone deposited?

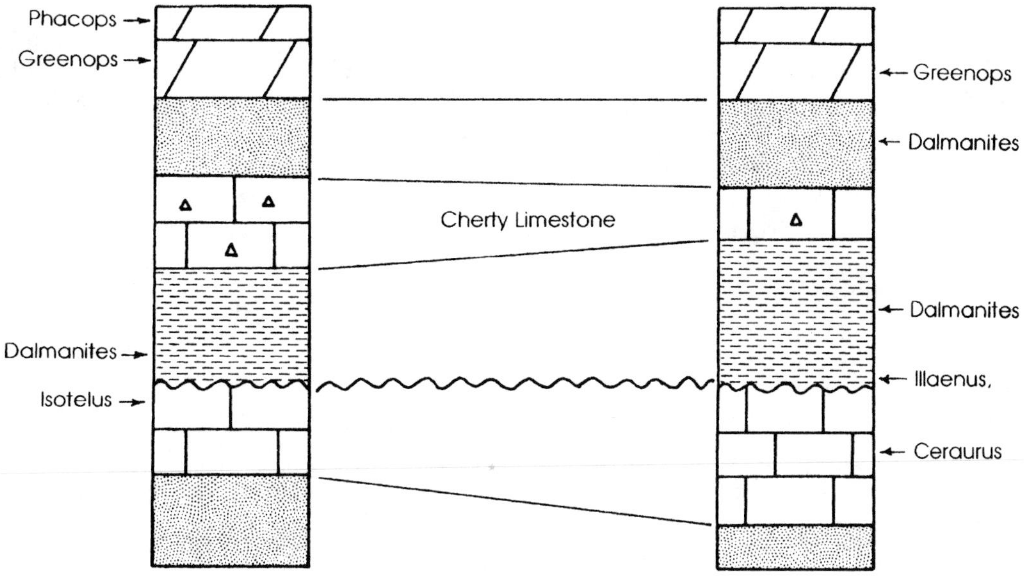

FIGURE 4.11 Stratigraphic columns for Exercise 4-6.

Exercise 4-7 AGE DETERMINATION AND CORRELATION

Correlate the stratigraphic columns in Figure 4.12 on the basis of fossils, and determine the approximate missing interval (periods) of geologic time represented by the unconformity. This interval is ____________________________.

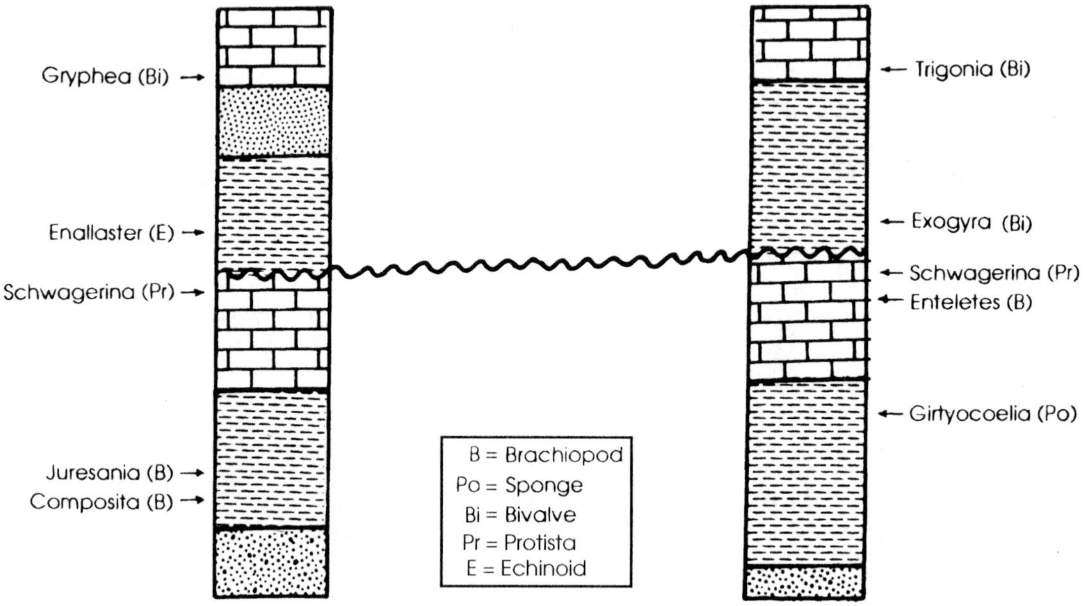

FIGURE 4.12 Stratigraphic columns for Exercise 4-7.

CLASSIFICATION OF ANIMALS AND PLANTS

A Swedish naturalist, Carl Linnaeus (1707–1778), in his publication *Species Plantarum* (1753), proposed a binomial classification scheme for naming plants, and in his *Systema Naturae* (1758), a scheme for animals. His classification system is still used today. The *species* is the basic unit of the Linnaean classification system. In geology, the term *species* is commonly defined as a restricted group of plants or animals whose individual differences are quite limited. In turn, the major characteristics of these restricted groups (the species) are distinct enough to differentiate them from other groups of similar organisms. A biologist would further require in the definition that members of a species be able to interbreed. The geologist, working only with the preserved fossil record, must assume that interbreeding took place.

The following is an example of the Linnaean classification for a vertebrate and an invertebrate animal:

Kingdom	Animalia	Animalia
Phylum	Chordata	Brachiopoda (Cambrian–Holocene)
Class	Mammalia	Articulata
Order	Carnivora	Orthida (Cambrian–Permian)
Suborder		Orthidina (L. Camb.–L. Dev.)
Family	Canidae	Plectorthidae (L. Ord.–U. Sil.)
Genus	*Canis*	*Hebertella (M. Ord.–U. Ord.)*
Species	*Canis familiaris*	*Hebertella sinuata (U. Ord.)*

Note that generic and specific names are always set in italics. Also keep in mind that the plural of *genus* is *genera* but that the plural of *species* is *species*. The total life span of a fossil group becomes more restricted as the species level of identification is approached, and most index fossils are designated at the specific level of classification.

To establish changes in animals and plant groups through geologic time or to use fossils to date a specific geologic event, a geologist must be able to identify specific fossils. Therefore, when each fossil is collected at an outcrop, it is cleaned, identified, and cataloged. To give a fossil a name the geologist must compare it with many others that have been collected worldwide. The detailed identification and classification of fossils is a complex branch of geology known as *paleontology*.

Only a limited number of fossils are presented in this manual. Owing to continuing research, some of these fossils have been renamed or their geologic ranges modified. A major reference for more detailed study of most fossils is the *Treatise of Invertebrate Paleontology* by R. C. Moore, University of Kansas Press (1964). The fossils presented in this section are primarily teaching tools rather than detailed reference materials. Generalized diagrams that indicate when various phyla flourished with greater dominance during geologic time are also presented. Because new species of fossils are being discovered constantly by paleontologists, fossil animal species now outnumber living forms. Consequently, the classification of these animals and the interpretation of their relationships to the general evolutionary development of life forms have changed in the last couple of decades. The following classification scheme is merely a capsule summary of more detailed animal and plant classifications. There is no single taxonomic scheme that is agreed upon by all biologists and paleontologists.

KEY TO IDENTIFYING MAJOR INVERTEBRATE FOSSIL PHYLA

The use of morphologic characteristics to classify animals or plants is termed *use of a dichotomous key*. To follow the keying process is to unlock the identity of the plant or animal.

To use this key, start with the first statement. If the fossil in question is colonial, go to the pages indicated. If the fossil is not colonial, determine the symmetry and go to the pages indicated. Follow the directions given in the subsequent statements, which contain characteristics of the group being investigated. Eventually the key will lead to the proper name for the phylum to which the fossil belongs. For your convenience, definitions and sample illustrations* are provided.

1. Colonial—Animals that live in close association with others and usually cannot live as separate individuals (many grow as an attached group). Go to:

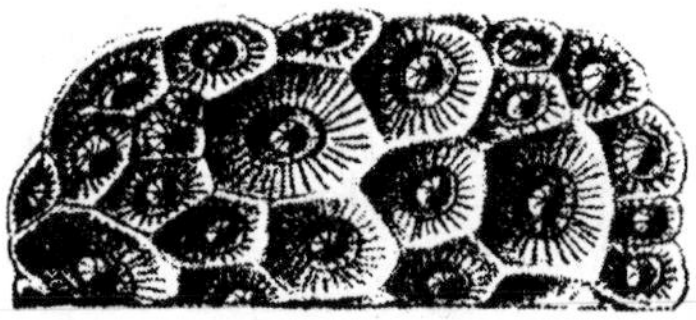

COLONIAL

Phylum Porifera, Fig. 4.16

Phylum Bryozoa, Fig. 4.18

Phylum Cnidaria, Fig. 4.17

Phylum Hemichordata, Fig. 4.31

*Source: The Illinois Geological Survey's "Guide to Beginning Fossil Hunters," by C.W. Collinson (1956).

2. Noncolonial (solitary)—Usually refers to an organism that lives as an individual and not as part of a colony.

a. Pentameral symmetry—Arrangement of the organs or shape to give the organism a five-rayed appearance (like a starfish). Go to:

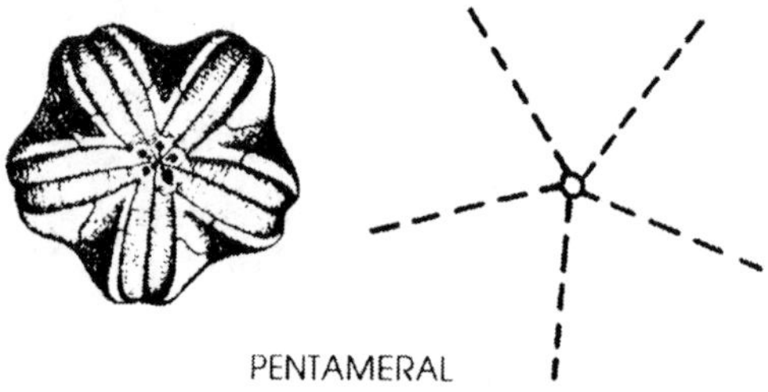

Echinodermata, attached, Fig. 4.29 and Fig. 4.30

Echinodermata, unattached, Fig. 4.30

b. Bilateral symmetry—Arrangement of the organs or shape of an organism such that, when divided in half, the opposite sides share equal or mirror images.

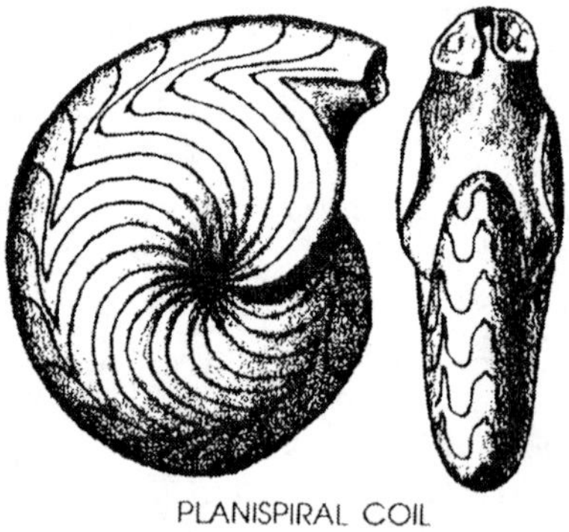

Coiled in planispiral manner with sutures. Go to Phylum Mollusca, Class Cephalopoda, Figs. 4.25, 4.26

Cigar-like shape. Go to Phylum Mollusca, Class Cephalopoda, Fig. 4.25

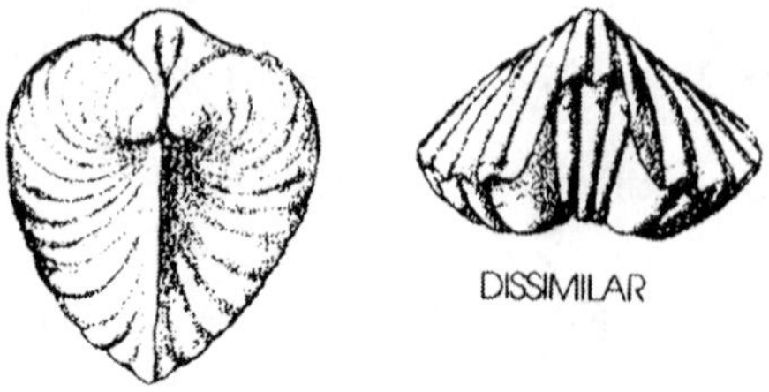

Bivalved—Two valves similar and mirror images. Go to Phylum Mollusca, Class Bivalvia, Figs. 4.21, 4.22

Bivalved—Two valves dissimilar, symmetry through shells. Go to Phylum Brachiopoda, Figs. 4.19, 4.20

Trilobed. Go to Phylum Arthropoda, Class Trilobita, Fig. 4.28

c. Radial symmetry—Circular, symmetrical about a central point. Cup or horn shape. Go to Phylum Cnidaria solitary corals, Fig. 4.17

d. Coiled in a nonchambered (hollow) cone-shaped (conispiral) or flat (planispiral) coil. Go to Phylum Mollusca, Class Gastropoda, Figs. 4.23, 4.24

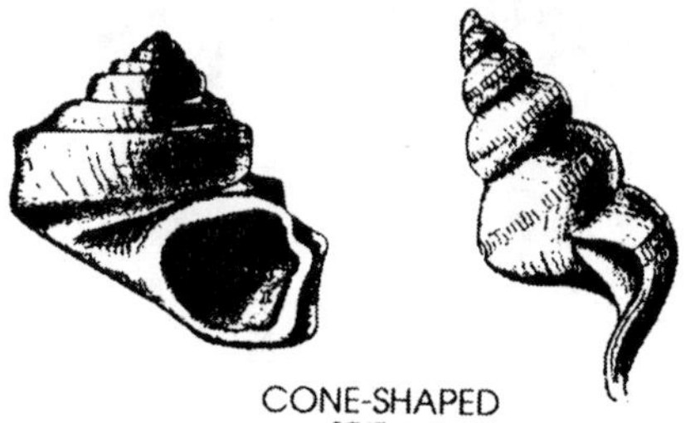

PALEONTOLOGY

CLASSIFICATION AND DESCRIPTIONS OF SELECTED KINGDOMS, PHYLA, AND CLASSES

KINGDOMS: Monera, Protista, Animalia, Plantae

I. KINGDOM MONERA

Kingdom Monera consists of modern species of bacteria and cyanobacteria (photosynthetic bacteria also known as blue-green algae). All living members of Kingdom Monera consist of single, prokaryotic cells (very small cells having no nuclear wall or organelles). Ancient members of Kingdom Monera were the earliest fossil life forms and have been found in Archean rocks as old as 3.5 billion years. Mats of cyanobacteria form structures called *stromatolites* in modern intertidal tropical waters, and cyanobacteria also have been discovered in ancient stromatolite structures found in Archean and Proterozoic rocks.

Modern stromatolites grow in the shallow tidal waters of Shark Bay on the northwest coast of Australia. Figure 4.13(a) shows a Precambrian stromatolite, *Collenia*, which can be found worldwide. Individual 2–3 foot wide stromatolites undergo a growth sequence, as shown in Figures 4.13 (b)–(d). At night the mat of

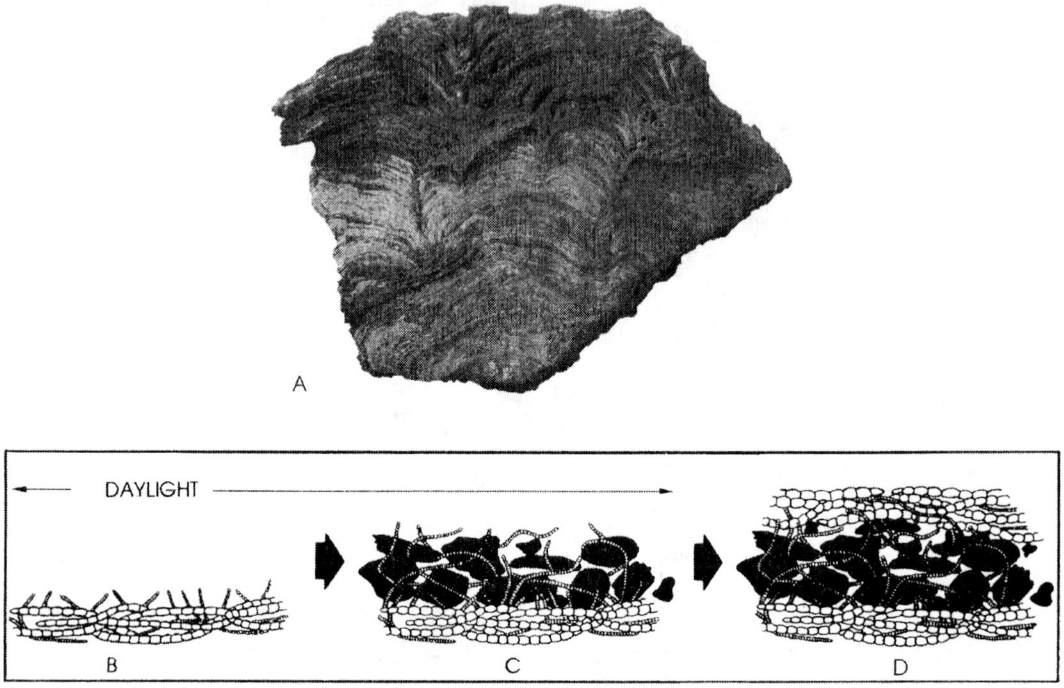

FIGURE 4.13 (a) Collenia (Precambrian-Cambrian) DK. (b), (c), (d) Stromatolite growth sequence. Drawn by Terry Chase.

algae is dormant [Fig. 4.13 (b)]. As sunlight increases, the algae begin to grow. As the wind and surf energy in the environment increase during the day, sediment is moved in the shallow water, with some being trapped in the growing filaments of algae [Fig. 4.13 (c)]. During the late afternoon, the wind, surf, and sediment movement decrease, but the algae continue to grow until darkness. In this phase the algae growth binds the trapped sediment [Fig. 4.13 (d)].

II. KINGDOM PROTISTA (PROTOCTISTA)

Members of Kingdom Protista consist of single eukaryotic cells (cells having a nuclear wall and organelles). This is a varied group and probably should be classified into several kingdoms. Some members are photosynthetic and others (protozoa) must consume other protists. Four major types are discussed here: foraminifera, radiolaria, diatoms, and coccoliths (Fig. 4.14).

A. Foraminifera (Alveolate group)

Foraminifera are a marine protista group in which some members are planktonic (zooplankton) and others are benthic. Foraminifera are protozoans that secrete tiny shells, called *tests*, of calcite or create tests of cemented silt grains (called *agglutinated*). The structure of the test is a single chamber or a series of chambers, and the size is about that of a grain of sand ($^1/_2$ to 2 mm). Some fossil forms, such as *Nummulites*, were considerably larger (15–25 mm). The protoplasm of the living cell extends out from the main opening, or aperture, and also from pores in the test (see Fig. 4.14). This external net of pseudopodia traps food particles for digestion inside the cell. Foraminifera reproduce by alternation of sexual and asexual phases.

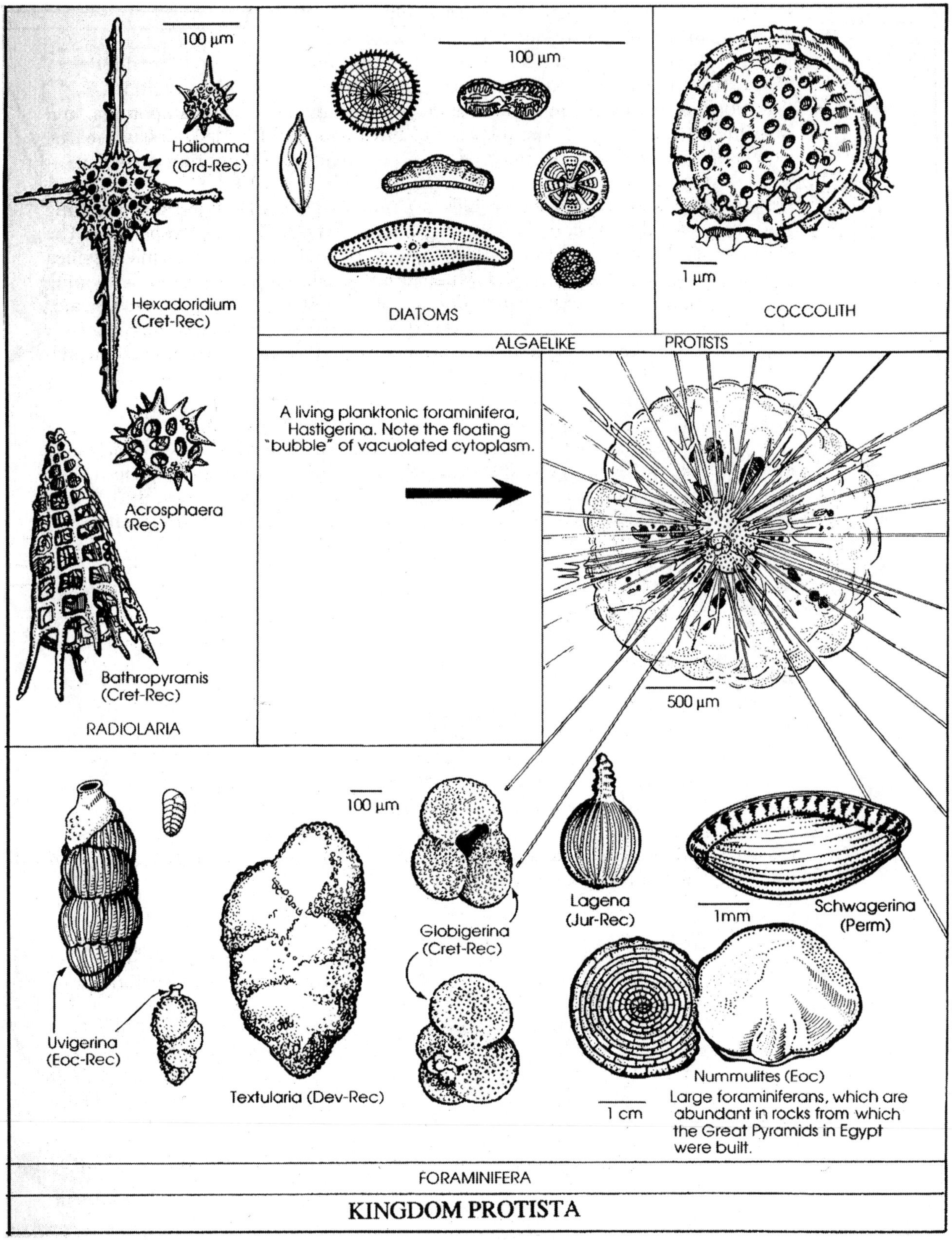
100 µm
Haliomma
(Ord-Rec)
Hexadoridium
(Cret-Rec)
Acrosphaera
(Rec)
Bathropyramis
(Cret-Rec)
RADIOLARIA
100 µm
DIATOMS
1 µm
COCCOLITH
ALGAELIKE PROTISTS
A living planktonic foraminifera, Hastigerina. Note the floating "bubble" of vacuolated cytoplasm.
500 µm
100 µm
Lagena
(Jur-Rec)
1mm
Schwagerina
(Perm)
Globigerina
(Cret-Rec)
Uvigerina
(Eoc-Rec)
Textularia (Dev-Rec)
Nummulites (Eoc)
1 cm
Large foraminiferans, which are abundant in rocks from which the Great Pyramids in Egypt were built.
FORAMINIFERA
KINGDOM PROTISTA

FIGURE 4.14 Kingdom Protista. Drawn by Lucy Mauger.

In the Paleozoic era, agglutinated foraminifera were most common. Wheat-shaped sand-sized calcareous forms with a complex chamber structure, the fusulinids, are good index fossils for the Late Paleozoic. Foraminifera with calcite tests were more common in the Mesozoic and Cenozoic. Foraminifera are extremely useful in biostratigraphy and can be used to subdivide geologic time into finer intervals called *zones*.

B. Radiolaria

Radiolaria are marine zooplankton that secrete a test of opaline silica in spherical, helmet-shaped, and spiny forms, commonly with open pores. Radiolaria became abundant in the Mesozoic era. As in the foraminifera, the cell is protozoan, and pseudopodia extend from openings in the lattice of the test to trap food particles. Some radiolaria contain algae within their tissue, which supply them with oxygen. Radiolaria are smaller than foraminifera, silt to fine sand in size, and are abundant in modern seas.

C. Diatoms (Chromista group)

Diatoms are a form of algae and are therefore photosynthetic organisms. They first appeared in the Early Mesozoic and became abundant later in that era. These algae secrete minute silt-sized tests of opaline silica that are usually round or oval-shaped. The two valves of the test fit together like a box and its lid. They are found in fresh water as well as marine waters.

Diatoms and radiolaria are the primary components of deep-sea siliceous oozes, and after lithification these deposits form one variety of the sedimentary rock chert.

D. Coccoliths

Coccoliths are extremely small calcareous platelets secreted by single cells of photosynthetic yellow-green algae that are abundant today as phytoplankton in the sea. These algae first appeared in the Triassic. Their shell fragments, along with foraminifera, are an abundant component of pelagic calcareous oozes. Coccolith platelets are the primary constituent of the sedimentary rock chalk, such as those that compose the cliffs near Dover on the English Channel.

III. KINGDOM ANIMALIA

Ediacaran biota

This is an interesting group of organisms, thought to be mostly animals but including many enigmatic specimens. These organisms were first discovered in the Ediacara Hills of Australia. The known time range of Ediacaran organisms is 600–544 million years (Upper Precambrian, Vendian period), and the distribution is worldwide: *Australia*: Flinders Ranges, Officer Basin, Amadeus Basin, Mt. Skinner, Georgina Basin; *Africa*: Namibia, Algeria; *Europe*: Russia White Sea, Ukraine, Northern Norway, Urals, England; *Asia*: Siberia, China, Iran; *North America*: Newfoundland, British Columbia, North Carolina, northwestern Canada.

The kinds of organisms found in Ediacaran faunas are quite varied. Trace fossils are generally horizontal burrows made by some kind of mobile creature with a head area, possibly with bilateral symmetry. A second group includes discs [*Cyclomedusa, Mawsonites* (Fig. 4.15)], which are thought to be benthic, possibly an anemone base or holdfast. Another interpretation holds that these discs are giant

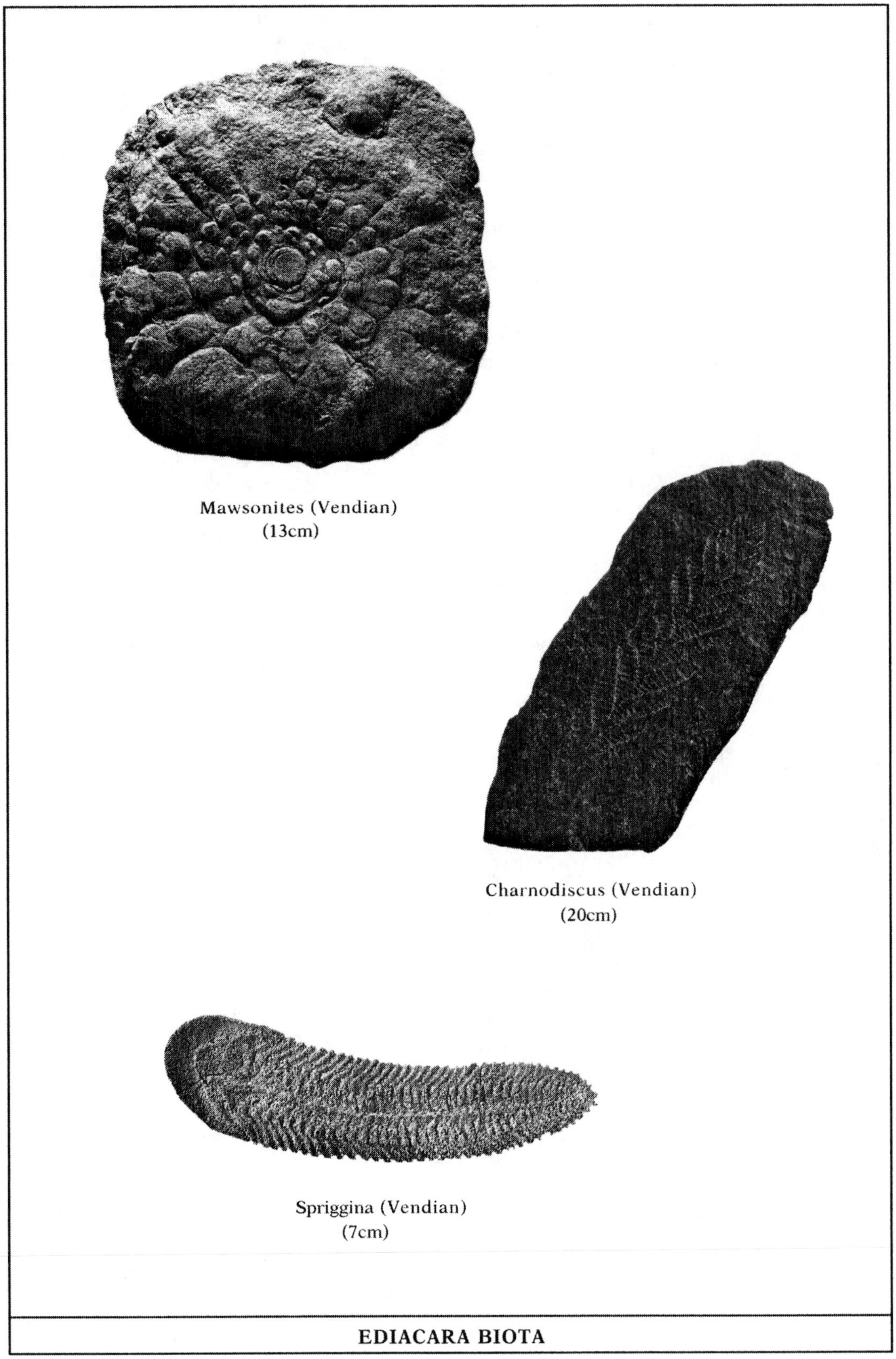

FIGURE 4.15 Ediacara biota. DK.

protists (xenophyophores), in which case they would be classified in Kingdom Protista. Seilacher (1982) has classified some as "Vendobionts," with a quilted structure and parallel linked tubular segments such as the frond fossil *Charnodiscus* (Fig. 4.15). Some think *Dickinsonia* may be a flat vendobiont, but others classify it as a segmented worm. Another enigmatic fossil is *Spriggina* (Fig. 4.15), which has a crescent-shaped structure at one end that some have interpreted as a head, including it with trilobites. Efforts to identify eyes or other structures have failed, and others classify *Spriggina* with the frond fossils. Vendobionts may be a separate, extinct kingdom, an extinct group of Cnidarians, or an extinct animal phylum. McMenamin and Seilacher suggest passive absorption of nutrients directly into cells as a feeding strategy, possibly with symbiotic algae. Retallack (1994) suggests that Ediacarans are a type of lichens (symbiotic fungi and algae). For examples, see McMenamin, 1998, *The Garden of Ediacara*, chapter 2, "The Sand Menagerie." Also visit the Vendian exhibit at the University of California's Museum of Paleontology at Berkeley. Exercise 4–9 directs students to this Web site for an examination of these enigmatic fossils.

Ecologically the Ediacarans seem to be benthic, found in original life position. Most appear to have lived in shallow water in the photic zone. Since they have not been disrupted by wave action, it seems they were held intact by microbial mats. Some are found in deep-water turbidite deposits. Large predators seem to have been absent, but there is evidence that something, perhaps sponges, bored into their coverings. Feeding strategies of Ediacarans include burrowing, passive absorption, or filter feeding. Frond fossils were elevated above the seafloor. A few Ediacarans appear to continue into earliest Cambrian (?), depending on where the boundary is drawn.

Phylum Porifera (Cambrian to Recent)
The phylum Porifera (see Fig. 4.16) consists of sponges, stromatoporoids, and possibly archaeocyathids. Members of this phylum are multicelled but have no true organs or tissues. There are 1500 genera of modern sponges, of which 80% are marine and 20% are freshwater. They are common in the deep sea, especially siliceous glass sponges.

Sponges consist of two layers of cells separated by jellylike material containing amoeboid cells that carry on bodily functions and secrete skeletal components. Most sponges have spicules of silica or calcite, a combination of spicules and spongin, or some have an encrusting calcareous skeleton that engulfs the spicular skeleton. The living tissue of the sponge surrounds a central cavity. The sponge draws in water through the pores in the outer layer. Food is trapped by collar cells that possess a flagellum (a whiplike structure) and cilia (fringing hairlike structures); the food is digested and passed to the amoeboid cells, which distribute it to other body cells. Sponges are important as an element in reef fabrics. Hexactinellids (glass sponges) are generally confined to deeper off-reef waters; others prefer shallower waters. In the Ordovician of central North America, they are important in stromatoporoid-bryozoan-algal reefs. Other important occurrences are in the Silurian of Tennessee, the Devonian–Mississippian of New York and Pennsylvania, and the Texas Permian reef complex.

Archaeocyathids (*Incertae sedis*) (Cambrian) are an extinct group of Early Paleozoic organisms that sometimes are classified with sponges and sometimes as a separate phylum. Their skeleton is composed of two cones, one inside the other, separated by vertical partitions called *septae*. The central cavity is open, as in modern sponges.

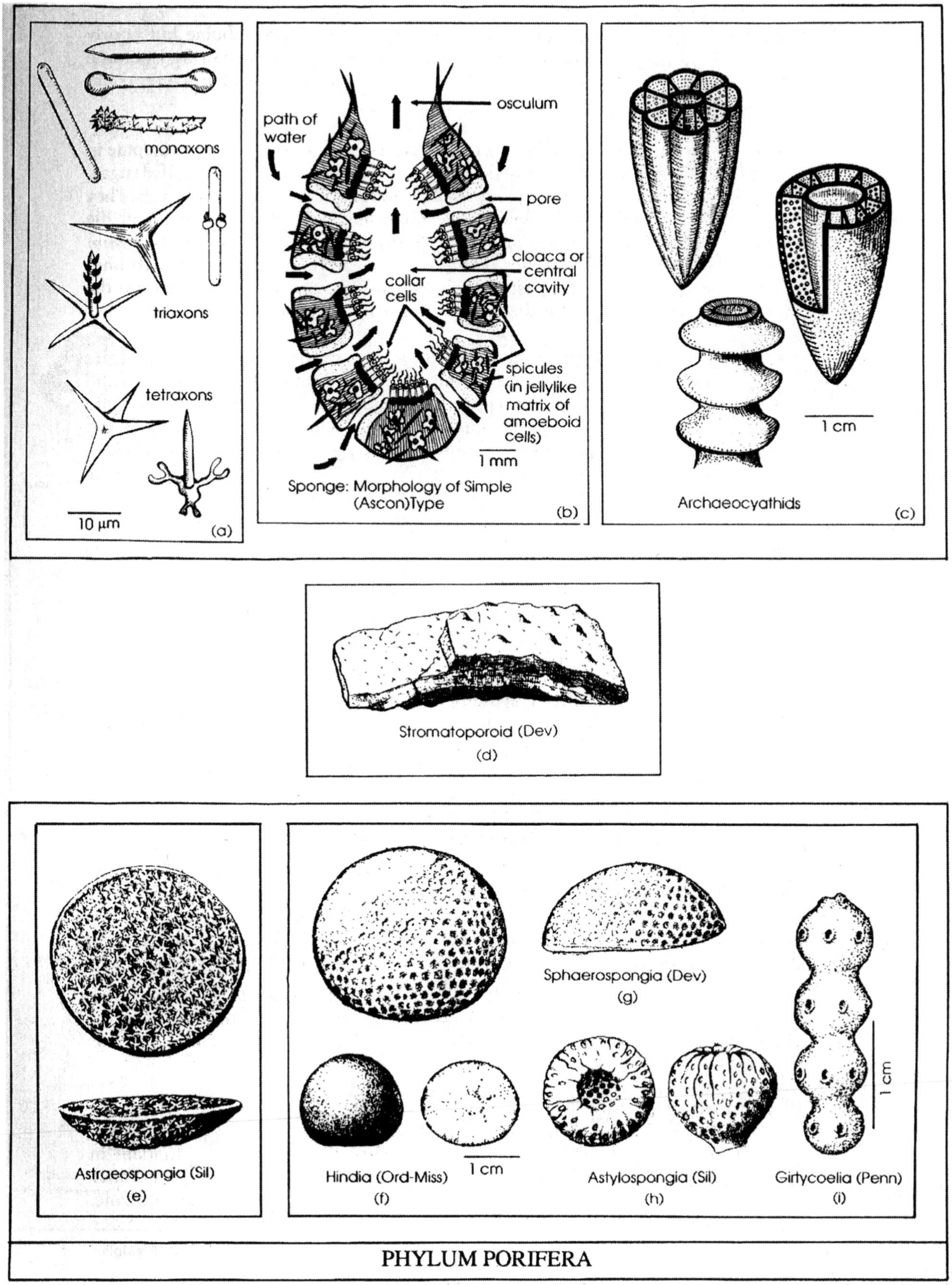

FIGURE 4.16 Phylum Porifera: (a–c) Lucy Mauger (d) Terry Chase (e–i) The Illinois Geological Survey's "Guide to Beginning Fossil Hunters," by C. W. Collinson (1956).

Archaeocyathids are distributed on all continents, and in the Lower Cambrian of Russia they can be used in stratigraphic correlations. They were most abundant in the Lower Cambrian and waned toward the Middle Cambrian; only one species is found in the Upper Cambrian. Because of their limited stratigraphic range, they make excellent index fossils [see Fig. 4.16(c)].

Stromatoporoids (Cambrian to Oligocene) are an extinct group of reef-building sponges of particular abundance in the Silurian and Devonian. They are characterized by star-shaped grooves on their growth surface (astrorhizal canals). Their skeletons are preserved as calcite, but they may have originally been composed of aragonite [see Fig. 4.16(d)].

Phylum Cnidaria (formerly Coelenterata) (Precambrian to Recent)
Members of this phylum include the corals, sea anemones, and jellyfish. They exist as polyps or medusae, or they alternate stages. The polyp stands on a base, with mouth and tentacles extended upward. The medusa floats with mouth and tentacles extending downward. See Figure 4.17 for examples.

Classes: Hydrozoa, Scyphozoa (hydra and jellyfish)
These classes are rare in fossil form and are not discussed here.

Class Anthozoa (Precambrian to Recent)
Class Anthozoa includes the corals. Corals have a polyp stage only, with no medusa stage in the life cycle. Though some corals are soft and have no calcareous skeletons, hard corals secrete aragonite skeletons. The skeleton is tube-shaped and has walls that extend upward as the polyp grows; the tube is called a *corallite*. As the polyp grows, it lifts its base and secretes a support plate beneath it. A flat plate is called a *tabula*, and small plates along the edge of the corallite are called *dissepiments*. In addition to the tabula and dissepiments, corals secrete radial plates that stand vertically between the folds of tissue at the base of the polyp. These vertical walls, or *septae*, look a little like the section dividers in a grapefruit. Hard corals are either solitary or colonial. *Solitary* corals have corallites that are not attached to any other corallite. In *colonial* corals, the corallites are attached to one another, forming colonies of various sizes and shapes.

Three important groups of hard corals in the fossil record are the Tabulates, Rugosans, and Scleractinians.

Tabulate corals (Lower Ordovician to Permian) are all colonial. They possess well-developed tabulae but poorly developed septae. Late Devonian extinction affected both tabulates and rugosans adversely. Extinction of many stromatoporoids caused problems for tabulates that associated with them in reefs. Examples are *Favosites* [Fig. 4.17(c)] and *Halysites* [Fig. 4.17(e)].

Rugose corals (Middle Ordovician to Permian) can be either solitary or colonial. They have well-developed septae in sets of four. On the outside of the corallite are coarse ridges called *rugae*. The solitary rugosans have a cup or cone shape and are often called *horn corals*. Solitary corals had no real form of anchorage to the seafloor. They probably colonized soft bottom environments, where they could sink part way into the sediment. Some appear to have lain sideways on the seafloor.

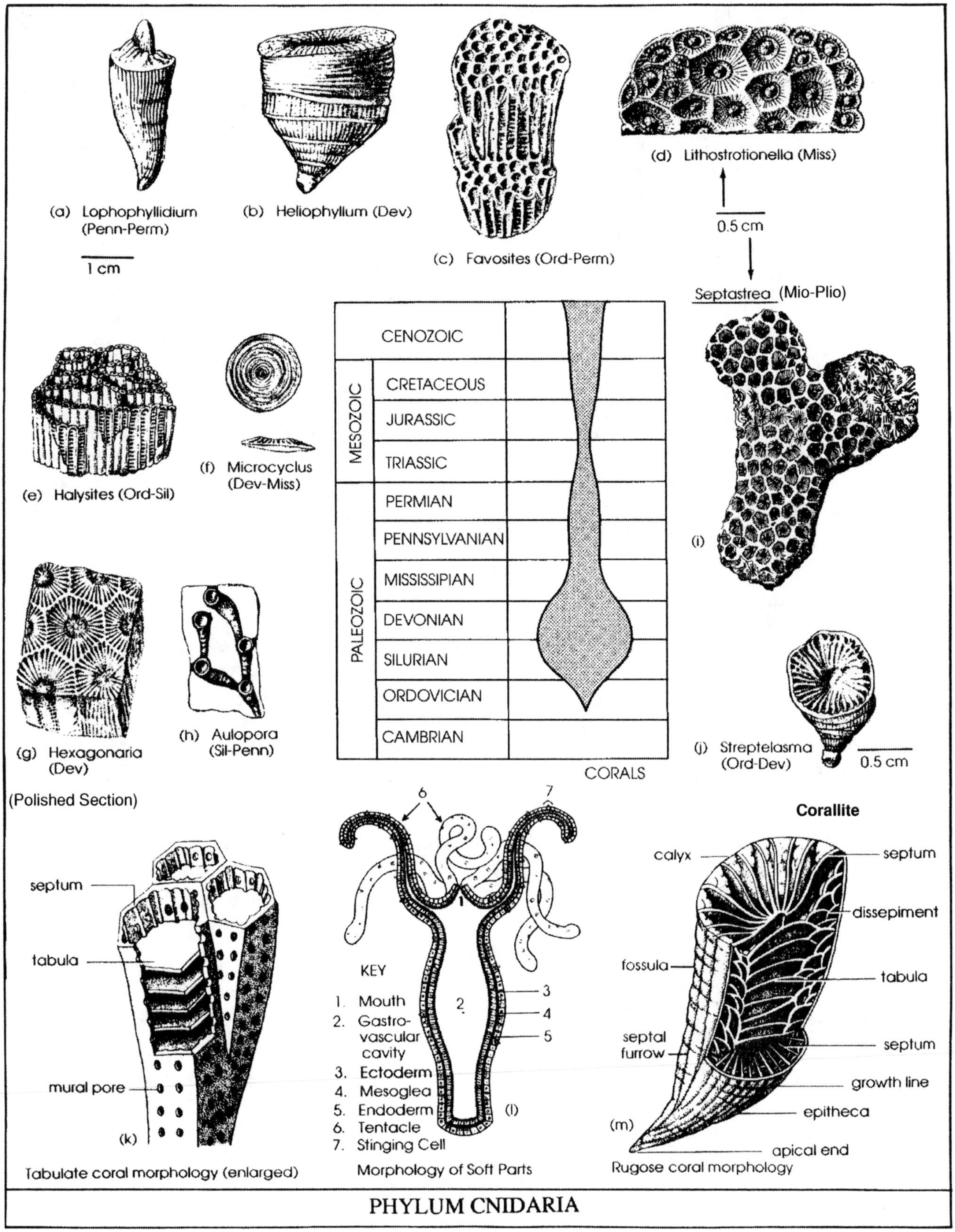

FIGURE 4.17 Phylum Cnidaria: (a–h, j) The Illinois Geological Survey's "Guide to Beginning Fossil Hunters," by C. W. Collinson (1956) (i, l) Lucy Mauger (k, m) Terry Chase.

Colonial rugosans probably did not anchor to a hard substrate but immersed in soft sediment. The weight of the colony would lend stability. Various species were found in fore-reef, reef, and back-reef areas, also in small patch reef areas called *bioherms*. Examples are *Lophophyllidium* [Fig. 4.17(a)] and *Heliophyllum* [Fig. 4.17(b)]. Interestingly, some of these corals can act as geochronometers. The surface of some corals show fine growth lines, sometimes 200 per centimeter (daily growth) grouped into bands (monthly) and broader groupings (yearly). Studies of these growth bands have established the fact that there were more days in a year in the geologic past (Devonian = 400 days per year), showing the slowing of earth's rotation over time.

Scleractinian corals (Middle Triassic to Recent) are reef-building corals in modern seas. They are mostly colonial and possess septa in sets of six. *Septastrea* [Fig. 4.17(i)] is an example from the Pliocene Yorktown Formation of North Carolina. Hermatypic (reef-building) scleractinians have a symbiotic relationship with dinoflagellate algae (zooxanthellae), which live within the coral animal's soft tissue. Benefits of this symbiosis include: carbon produced by algae is used as food by the polyp; oxygen produced by algae is used also; growth rates of the aragonite skeleton are up to three times more rapid than those without symbionts. Reef-building corals are almost all symbiotic with algae, preferring water temperatures between 25 and 29° C and normal salinity. Non-reef-building corals, without zooxanthellae, can colonize deep water; some are colonial, some solitary.

Phylum Bryozoa (Ordovician to Recent)

Bryozoans (Fig. 4.18) were prolific mid-Paleozoic reef-builders. All are colonial, most are marine, but one class is found in freshwater. Unable to tolerate harsh wave action, they inhabit various ocean depths, from shallow shelf to deep water. They need a hard substrate for attachment, but some can root in deepwater muds. Some encrust on hard materials (shells of dead organisms, seaweed); others form standing colonies. Marine species secrete a calcite skeleton and are mostly colonial. Freshwater species do not secrete a calcareous skeleton.

Individual bryozoans reside in a chamber called a *zooecium*. They are rather small, usually less than 1 mm in diameter. The initial member of a colony grows from larva, and the rest of the colony is budded from it. Bryozoa have a U-shaped digestive tract. Ciliated tentacles (*lophophore*) surround the mouth and filter seawater for food particles and oxygen. Retractor muscles pull the tentacles into the zooecium when the animal is disturbed, and water pressure forces the tentacles back out again. Some zooecia are modified to serve as a cleaning mechanism for the colony surface; others serve as brood pouches for fertilized embryos that are later released as larvae.

Paleozoic bryozoans. A group called Trepostomes formed branching colonies or encrusting colonies. *Hallopora* is an example of a branching colony [Fig. 4.18(f)]. Another group, the Fenestrate bryozoans, had free-standing sheetlike colonies with pores through which water could circulate. *Archimedes* had perforated sheets wound around a solid center that twisted like a corkscrew. Often only the center is preserved [Fig. 4.18(g), (j)].

Mesozoic and Cenozoic bryozoans. Bryozoans of the Cheilostomate group have a trapdoor to cover the opening to the zooecium. Most form encrusting colonies that cover shells or reef crevices. These bryozoans are common but inconspicuous in their modern habitats [Fig. 4.18(d)].

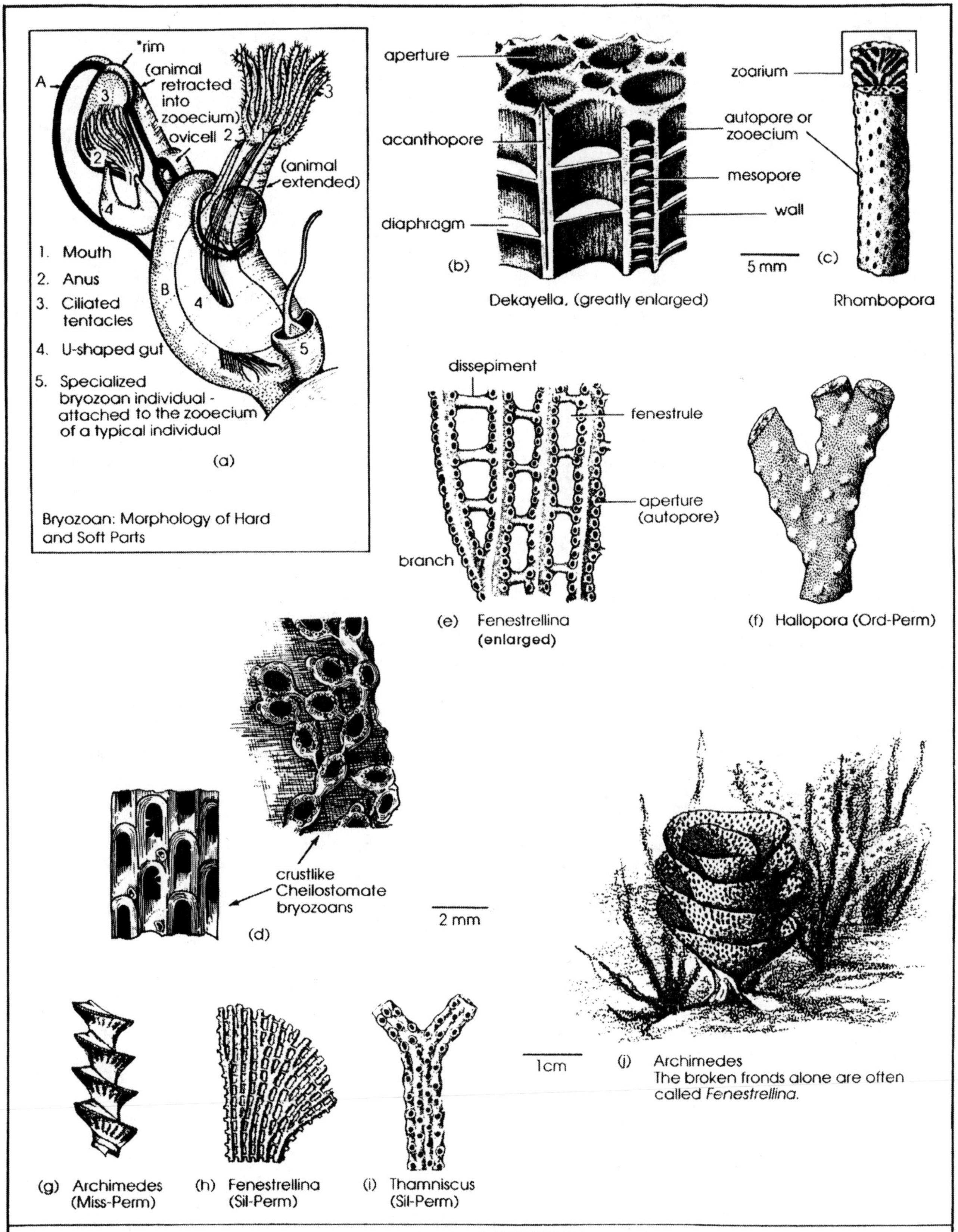

PHYLUM BRYOZOA

FIGURE 4.18 Phylum Bryozoa: (a, d) Lucy Mauger (b–c, e–f) Terry Chase (g–j) The Illinois Geological Survey's "Guide to Beginning Fossil Hunters," by C. W. Collinson (1956).

Phylum Brachiopoda (Cambrian to Recent)
Marine articulate brachiopods are first found in the Early Cambrian, are abundant in Ordovician, Silurian, and Devonian marine strata, but are reduced in numbers after the Devonian. The phylum sustained major extinctions at the end of the Paleozoic. Only two orders survive to the Recent (Figs. 4.19 and 4.20).

Brachiopods are bivalved, with the plane of symmetry running through the center of each valve. The pedicle valve is the larger and often has an opening for the pedicle, which attaches the shell to the substrate or to another shell. The pedicle may be a single stalk or separated into several threads. Adjustor muscles can orient the shell in any direction with the pedicle attaching to the substrate. The brachial valve is smaller and contains an internal support for the lophophore. The lophophore is a coiled organ with comblike filaments that are covered with cilia. The cilia keep a current of seawater coming into the body of the brachiopod when the valves are open. The lophophore traps microorganisms, passes them to the mouth, and obtains oxygen from the water. The mantle is the internal organ that secrets calcite for the shell in articulate brachiopods [Fig. 4.19(a)]. Life habits of brachiopoda generally fall into one of three groups: (1) those that are sessile and fixed in place, attaching by the pedicle or encrusting onto a hard substrate (the pedicle can also attach to a soft substrate by splitting into threads; these brachiopods can survive in areas with strong currents); (2) brachiopods that are sessile but not attached, lying on the seafloor; (3) those that are buried or partially buried in sediment.

A new classification has been drawn up for the brachiopods: Craniiformea, Linguliformea, and Rhynchonelliformea, which is somewhat equivalent to Lingulata, Inarticulata, and Articulata.

Craniiformea (Lower Cambrian to Recent) are a small group with calcite shells, poorly developed teeth and sockets, and no pedicle.

Linguliformea (Lower Cambrian to Recent) includes *Lingula*, the "living fossil," a phosphatic shelled inarticulate brachiopod that has persisted since the Cambrian with little change. It inhabits a brackish to intertidal environment in fine sand, burrows vertically, and anchors with a large pedicle [Fig. 4.19(b)].

Rhynchonelliformea (Lower Cambrian to Recent) dominated brachiopod faunas by the Ordovician. Members of this subphylum have calcareous bivalved shells with teeth and sockets for articulation at the hinge line and a well-developed pedicle. Examples from Figure 4.20 are *Strophomena* (m), *Mucrospirifer* (k), and *Atrypa* (l). In the Late Devonian a crisis developed in existing orders, with the extinction of some groups. After the Devonian the productids arose, curious forms with spines on their valve exteriors. An example is *Juresania* [Fig. 4.20(d)]. Many brachiopod orders went extinct through the Paleozoic and especially at the end of the Permian. Of 18 brachiopod orders, 6 have surviving members. Bivalves seem to have come out ahead after the Permian extinction, colonizing the seafloor to the exclusion of the brachiopods. Also, sea stars arose in the Mesozoic and may have preyed on epifaunal brachiopods.

Phylum Mollusca (Cambrian to Recent)
Members of Phylum Mollusca exhibit bilateral symmetry. A shell of calcium carbonate encloses the soft parts of many mollusks. The shell is secreted by the mantle, an organ common to all mollusks. The visceral mass of mollusks includes digestive, excretory, circulatory, nervous, and muscular systems. Many groups have

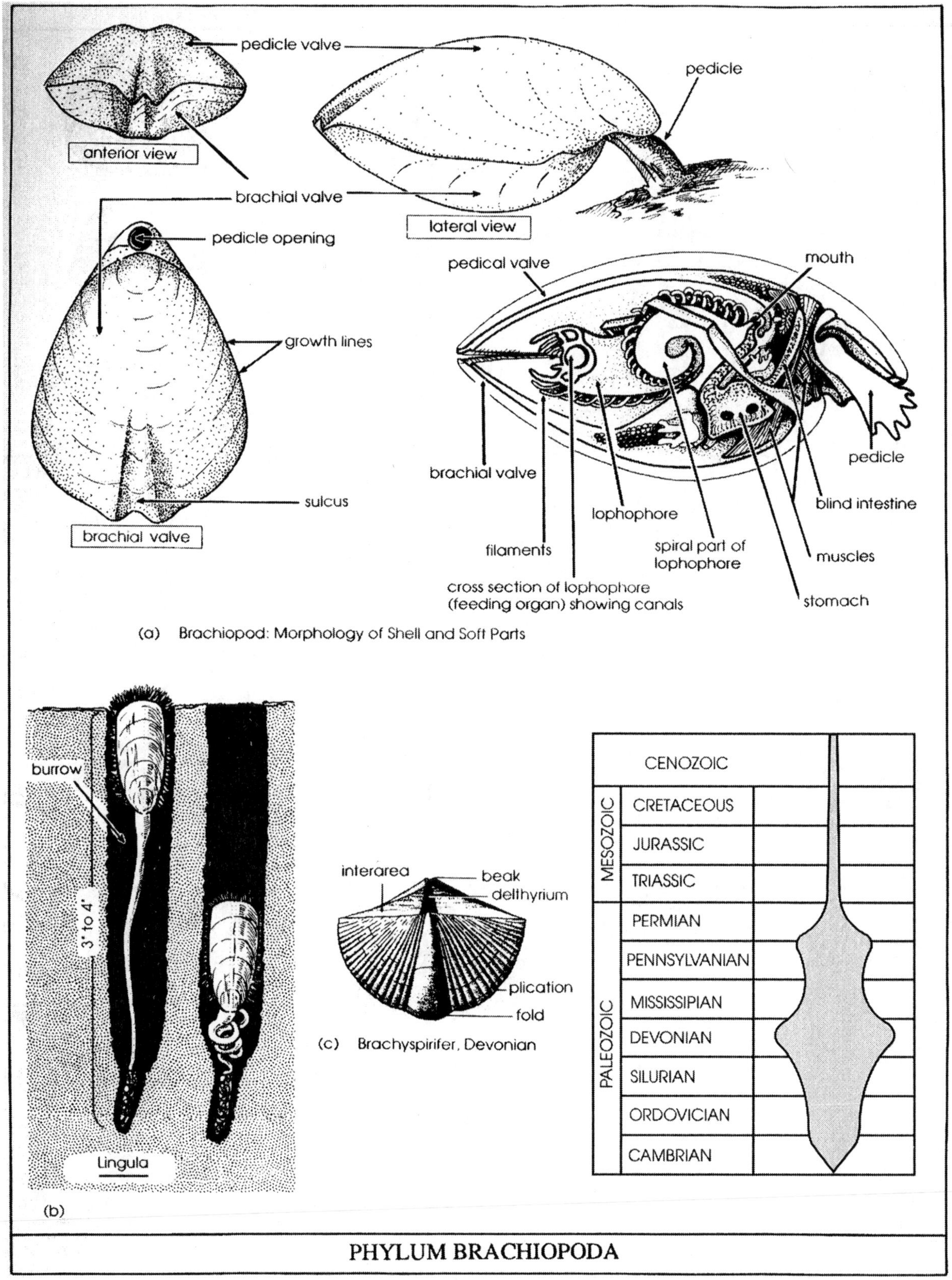

FIGURE 4.19 Phylum Brachiopoda: (a, b) Lucy Mauger (c) Terry Chase.

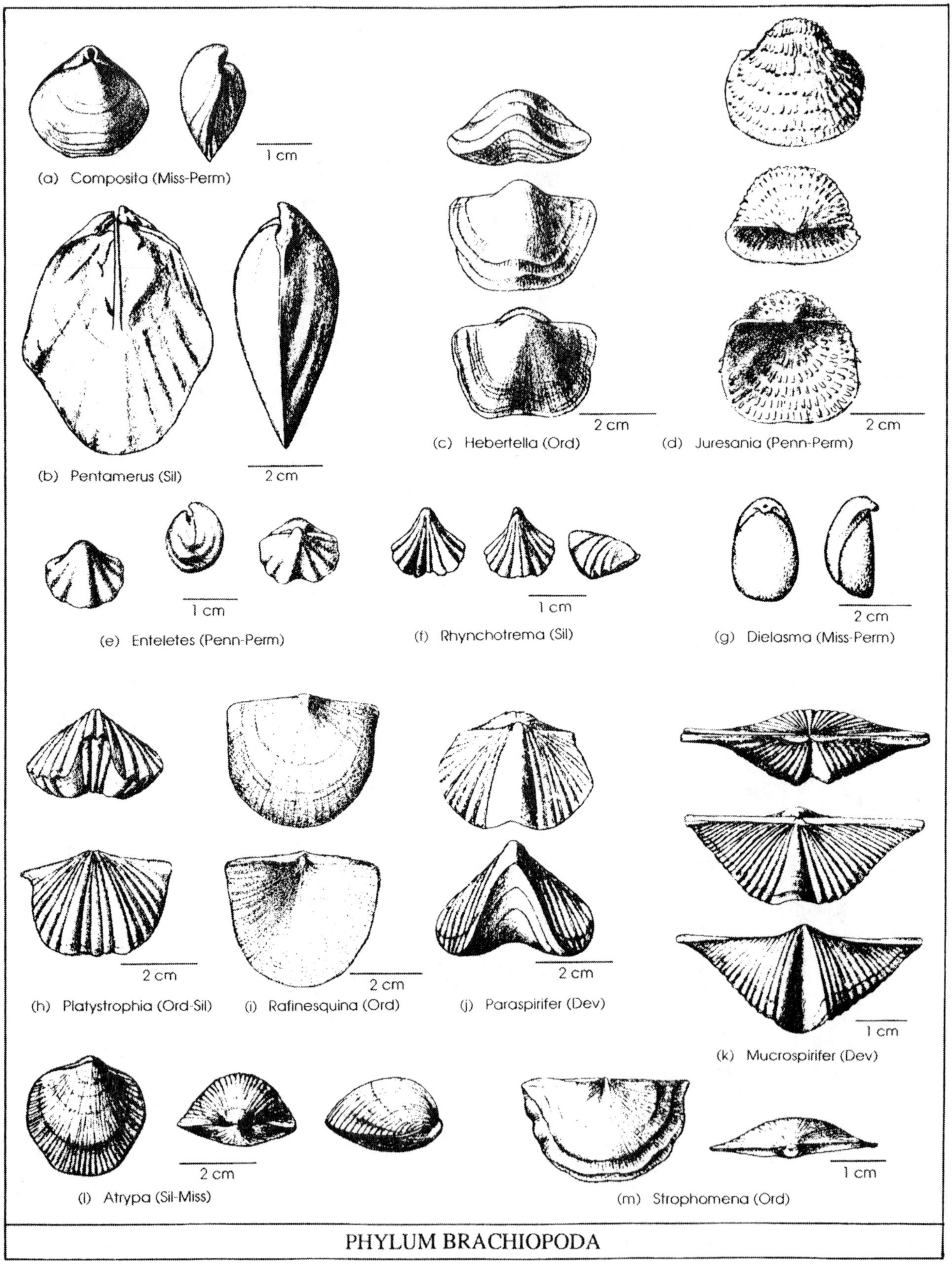

FIGURE 4.20 Phylum Brachiopoda: (a–m) The Illinois Geological Survey's "Guide to Beginning Fossil Hunters," by C. W. Collinson (1956).

a foot adapted for burrowing, movement on the substrate, or food grasping. Three classes important in the fossil record are Bivalvia, Gastropoda, and Cephalopoda.

Class Bivalvia (Cambrian to Recent)
This class (Figs. 4.21 and 4.22) includes clams, oysters, scallops, and mussels. The shells have two valves and are composed of layered calcite and aragonite. In many bivalves, the two valves are mirror images, unlike brachiopod valves, which are different from each other. Muscles hold the valves together, and some bivalves have a flexible ligament that acts as a spring and opens the shells when the muscles relax. One or two muscles are present, and muscle scars are visible on the interior of the shell. An organ called the *mantle* secretes the shell by adding calcite along its edge. The pallial line marks the mantle's attachment to the shell. If retractable siphons are present, the interior of the shell shows an indentation called a *pallial sinus*. This indentation commonly indicates a burrowing habit for the bivalve. The incurrent siphon takes in seawater; the gills remove food and oxygen and move the food to the mouth. Wastes move out of the body through the excurrent siphon. The shells hinge together in the beak area. Teeth and sockets on either valve fit the shells together [Fig. 4.21(a), (b)].

The first bivalves appeared in the Early Cambrian. They became more abundant by the Middle Ordovician. They probably lived in shallow water on top of the sediment or buried at shallow depths, filter feeding. After the Permian, many families of brachiopods became extinct and were replaced ecologically by bivalves.

During the Mesozoic and Cenozoic, burrowers were more abundant, and scallops made their appearance. Following is a discussion of some of the life habits of members of the class Bivalvia.

Shallow burrowers are generally siphonate and often have pallial sinuses. They burrow near the surface–water interface and extend short siphons for feeding and excretion. Examples are Ensis [Fig. 4.22(c)], Venus [Fig. 4.22(h)], and Mercenaria [Fig. 4.21(a)].

Deep burrowers are clams with long siphons to reach the sediment–water interface from the deep burrow. The valves gape to accommodate the long siphons, and the siphons cannot be retracted into the shell.

Attached bivalves such as mussels range from Late Paleozoic to Recent. They attach by byssus fibers to sediment or rock. Rudists [Fig. 4.21(c)] were unique attached bivalves that ranged from Late Jurassic to Cretaceous. The lower valve is conical (like a horn coral), and the upper valve caps the top of cone. They were reef-builders during the Late Mesozoic. The largest rudists were 6 feet in length.

Swimming scallops [Fig. 4.22(i)] range from Carboniferous to Recent. They lie on one valve and can escape predators by snapping the valves together with a single muscle and jetting water from the mantle. Scallops possess light-sensing "eyes" around the outer edge of the mantle.

Cemented oysters [Fig. 4.21(e)] range from Late Paleozoic to Recent. The lower valve grows to conform to the shape of the object to which it is cemented, and oysters can form banks or reefs by cementing to each other.

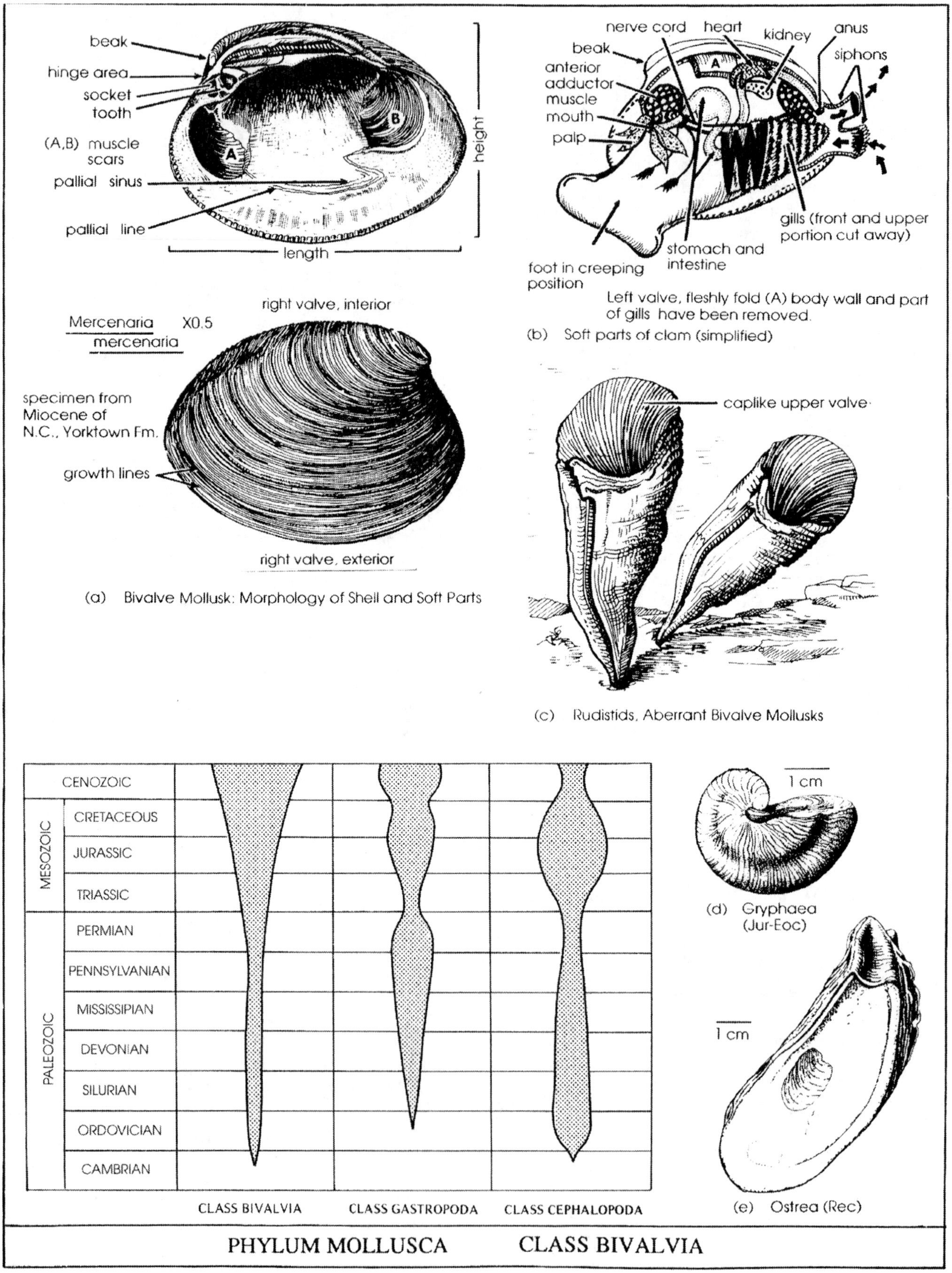
beak
hinge area
socket
tooth
(A,B) muscle scars
pallial sinus
pallial line
height
length
A
B
nerve cord
heart
kidney
anus
beak
siphons
anterior adductor muscle
mouth
palp
gills (front and upper portion cut away)
stomach and intestine
foot in creeping position
Left valve, fleshly fold (A) body wall and part of gills have been removed.
(b) Soft parts of clam (simplified)
right valve, interior
Mercenaria mercenaria X0.5
specimen from Miocene of N.C., Yorktown Fm.
growth lines
right valve, exterior
(a) Bivalve Mollusk: Morphology of Shell and Soft Parts
caplike upper valve
(c) Rudistids, Aberrant Bivalve Mollusks
CENOZOIC
MESOZOIC
CRETACEOUS
JURASSIC
TRIASSIC
PALEOZOIC
PERMIAN
PENNSYLVANIAN
MISSISSIPIAN
DEVONIAN
SILURIAN
ORDOVICIAN
CAMBRIAN
CLASS BIVALVIA
CLASS GASTROPODA
CLASS CEPHALOPODA
1 cm
(d) Gryphaea (Jur-Eoc)
1 cm
(e) Ostrea (Rec)
PHYLUM MOLLUSCA
CLASS BIVALVIA

FIGURE 4.21 Phylum Mollusca, class Bivalvia: (a–c) Lucy Mauger (d–e) Terry Chase.

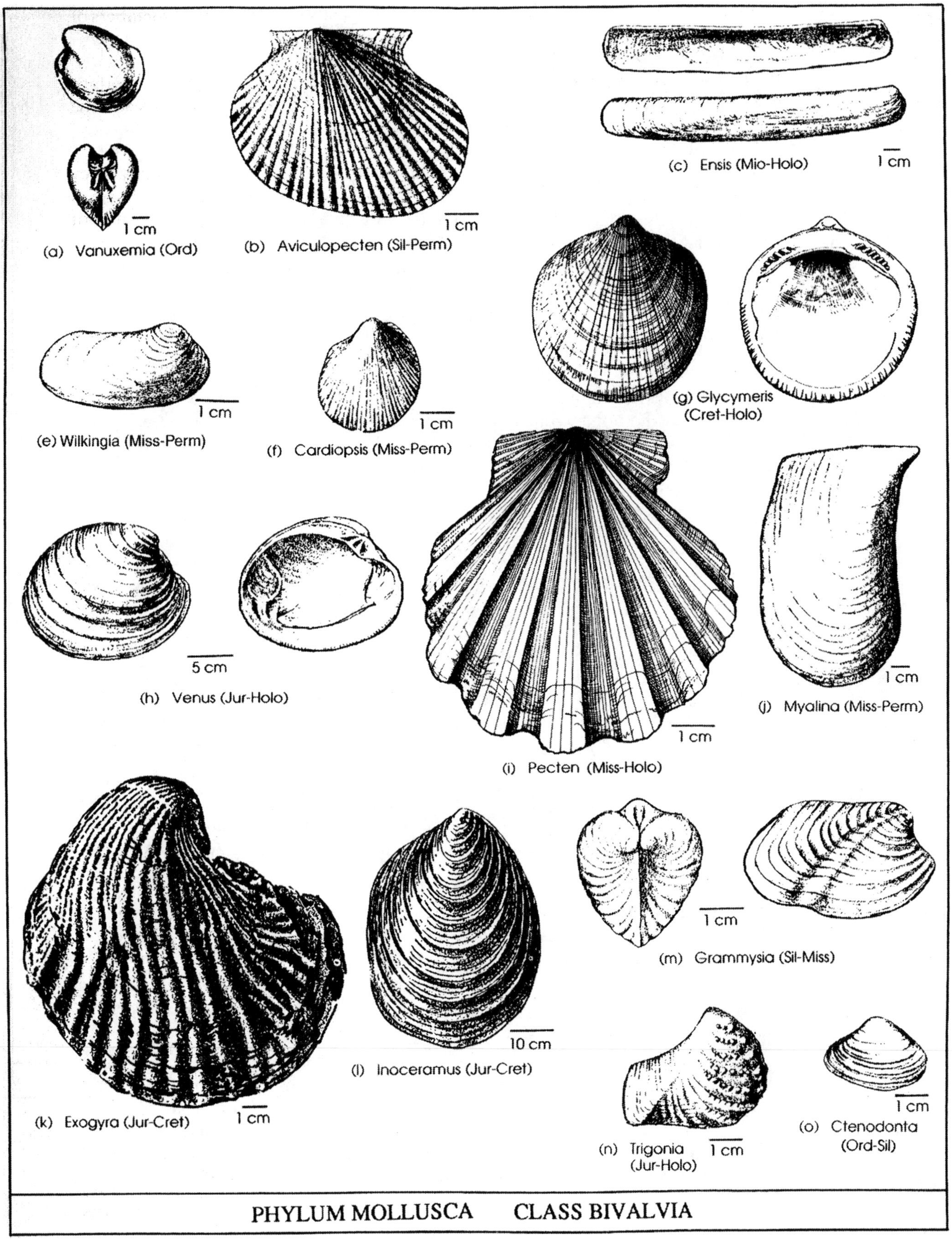

FIGURE 4.22 Phylum Mollusca, class Bivalvia: (a–f, h, j–o) The Illinois Geological Survey's "Guide to Beginning Fossil Hunters," by C. W. Collinson (1956) (g, i) Lucy Mauger.

Borers are clams with rough front edges on their shells for boring into rock, wood, or other shells. An example is the shipworm, *Teredo*. To help in boring into hard substances, some of these bivalves secrete corrosive chemicals.

Class Gastropoda (Cambrian to Recent)
This class (Figs. 4.23 and 4.24) includes snails and slugs. Snails have left the best fossil record. Snails are found in marine and fresh waters and on land. They are the most diverse and abundant class of mollusks. They secrete a single, spirally coiled shell, and their anatomy includes a head, eyes, and sensory tentacles. Snails move along on a foot and feed with a toothlike organ called a *radula*. When disturbed, snails retract into their shells, and many have a cover (*operculum*) that can be positioned over the shell opening (*aperture*). Snails have only one gill, and land-living snails use the mantle cavity as a lung for obtaining oxygen. Snails feed by grazing algae off the sea bottom, deposit feeding, or filter feeding. Some snails are predatory, scraping a hole in the shell of the prey and eating the soft tissue. One type of snail (the pteropod) has a thin shell and is planktonic.

Snails first appear in the fossil record in earliest Cambrian strata (the Tommotian stage), but they did not become abundant until the Ordovician. They were affected by the great Permian extinction, but survived and diversified in the Mesozoic and Cenozoic. Terrestrial and freshwater snails first appeared in the Devonian.

Class Cephalopoda (Upper Cambrian to Recent)
This class (Figs. 4.25 and 4.26) includes the nautilus, octopus, squid, and cuttlefish. The shelled cephalopods are distantly related to the modern nautilus and have a single shell that coils in one plane. Septa divide the shell into chambers. A tube called a *siphuncle* connects the last-formed chamber with previous ones. The body of the animal occupies the last chamber; the rest are filled with gas, which controls the animal's buoyancy in the water. The foot is modified into a set of tentacles for grasping prey. Cephalopods have a head and brain area and a well-developed eye similar to that of vertebrates. A funnel for jet propulsion allows the cephalopod to swim in any direction [Fig. 4.25(a), (c)]. Modern nautiloids live in relatively deep water, 150–300 m, migrating upward at night in search of food.

Early Paleozoic cephalopods were mostly straight-shelled and attained a length of up to 27 feet. Later cephalopods developed a coiled shell. Four major suture patterns developed in cephalopods, reflecting the nature of the interface between the septa and the chamber wall.

1. *Nautiloid*. This straight suture pattern is exhibited in cephalopods from Paleozoic to Recent times.
2. *Goniatite*. This pattern consists of curved sutures and is found only on Paleozoic cephalopods.
3. *Ceratite*. This curved and crenulated pattern is found on cephalopods from the Late Paleozoic to the Early Mesozoic.
4. *Ammonite*. This highly crenulated pattern is found only on Mesozoic cephalopods. Ammonites were extremely abundant in the Mesozoic but became extinct with many other groups of organisms at the end of the Mesozoic. Why the ammonites succumbed and the nautiloids survived is unclear, but it was probably related to differences in their ecological adaptations. Ammonites are extremely valuable for stratigraphic zonation of Mesozoic sedimentary sequences.

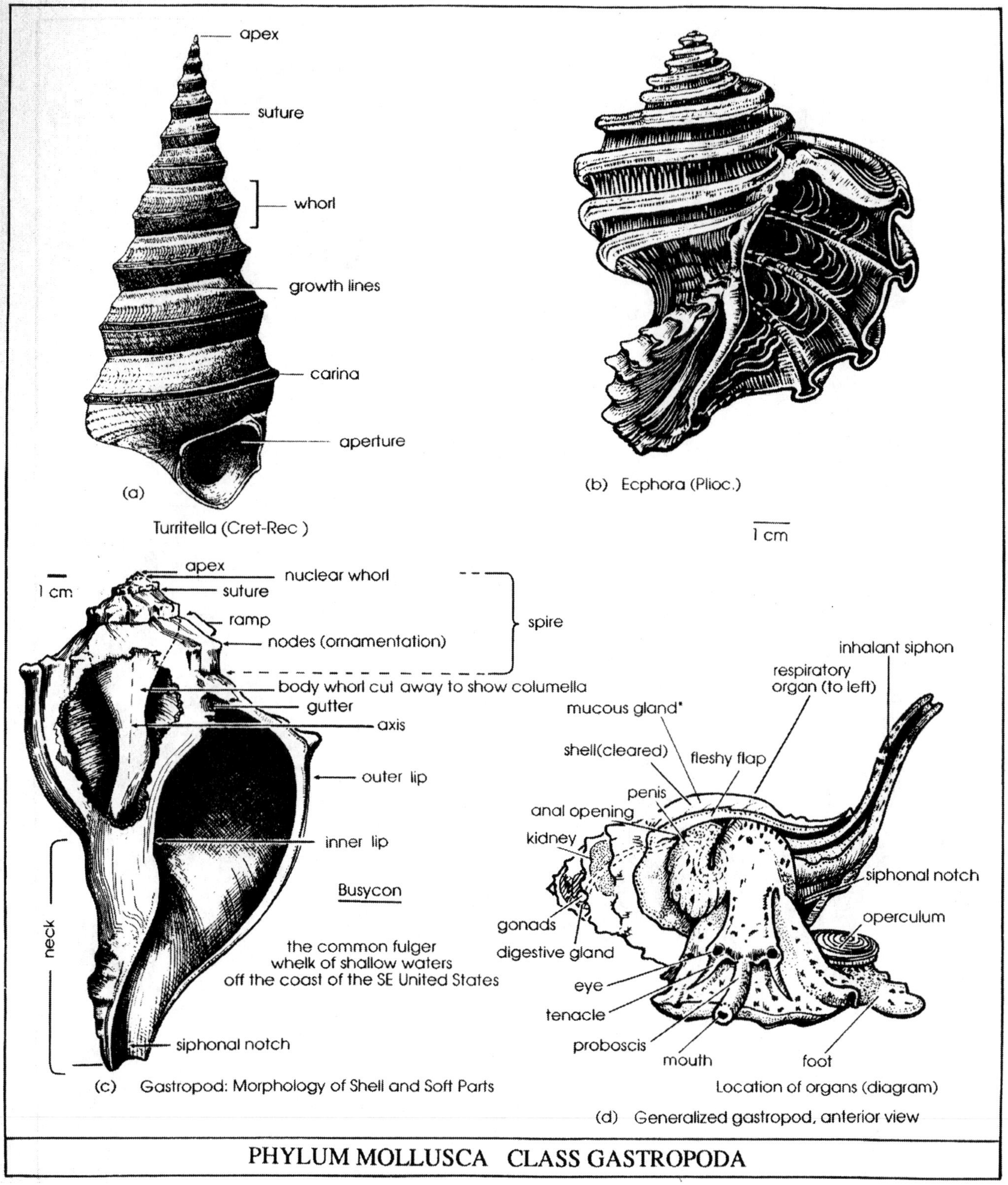

FIGURE 4.23 Phylum Mollusca, class Gastropoda: (a) Terry Chase (b–d) Lucy Mauger.

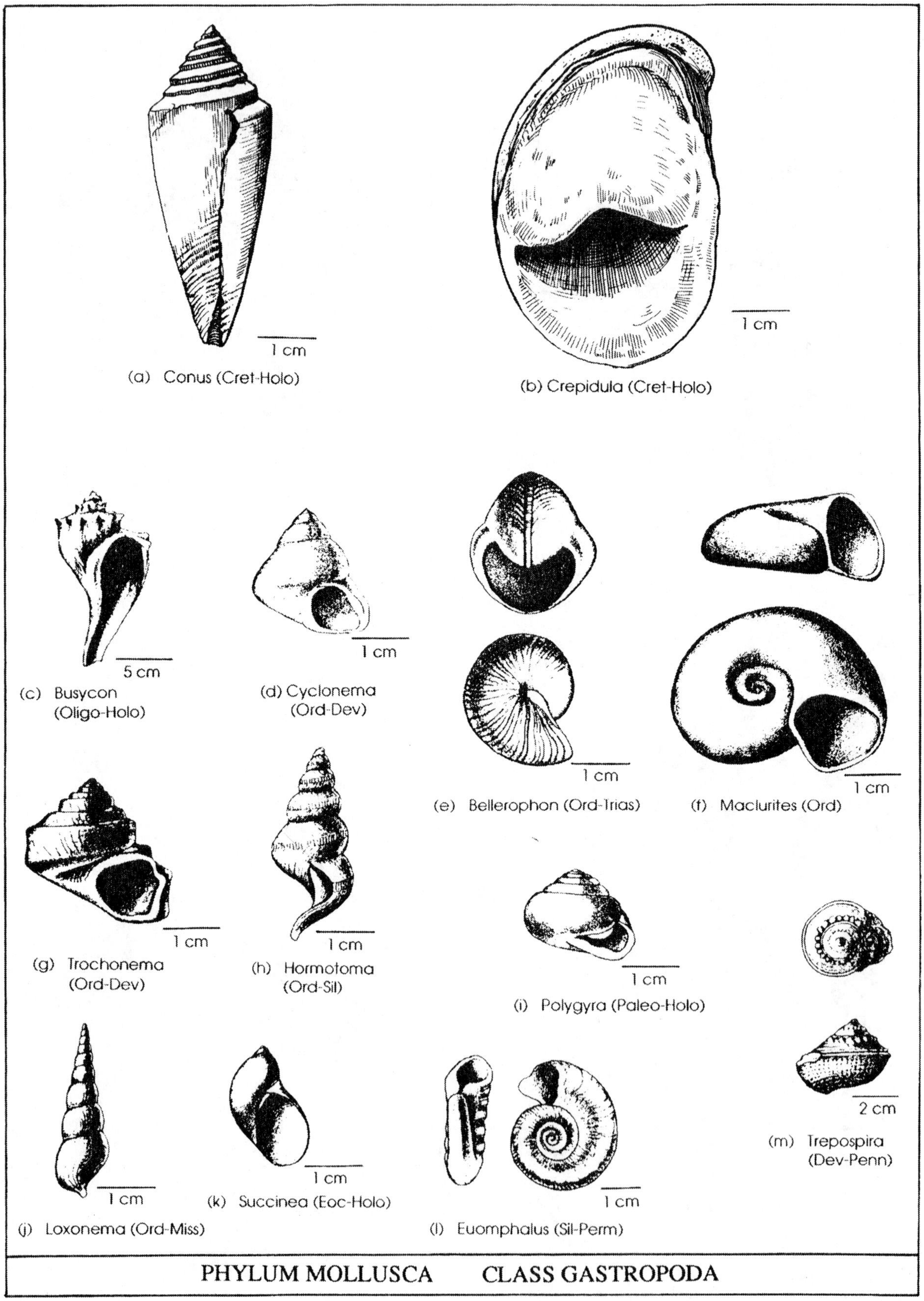

FIGURE 4.24 Phylum Mollusca, class Gastropoda: (a–b) Lucy Mauger (c–m) The Illinois Geological Survey's "Guide to Beginning Fossil Hunters," by C. W. Collinson (1956).

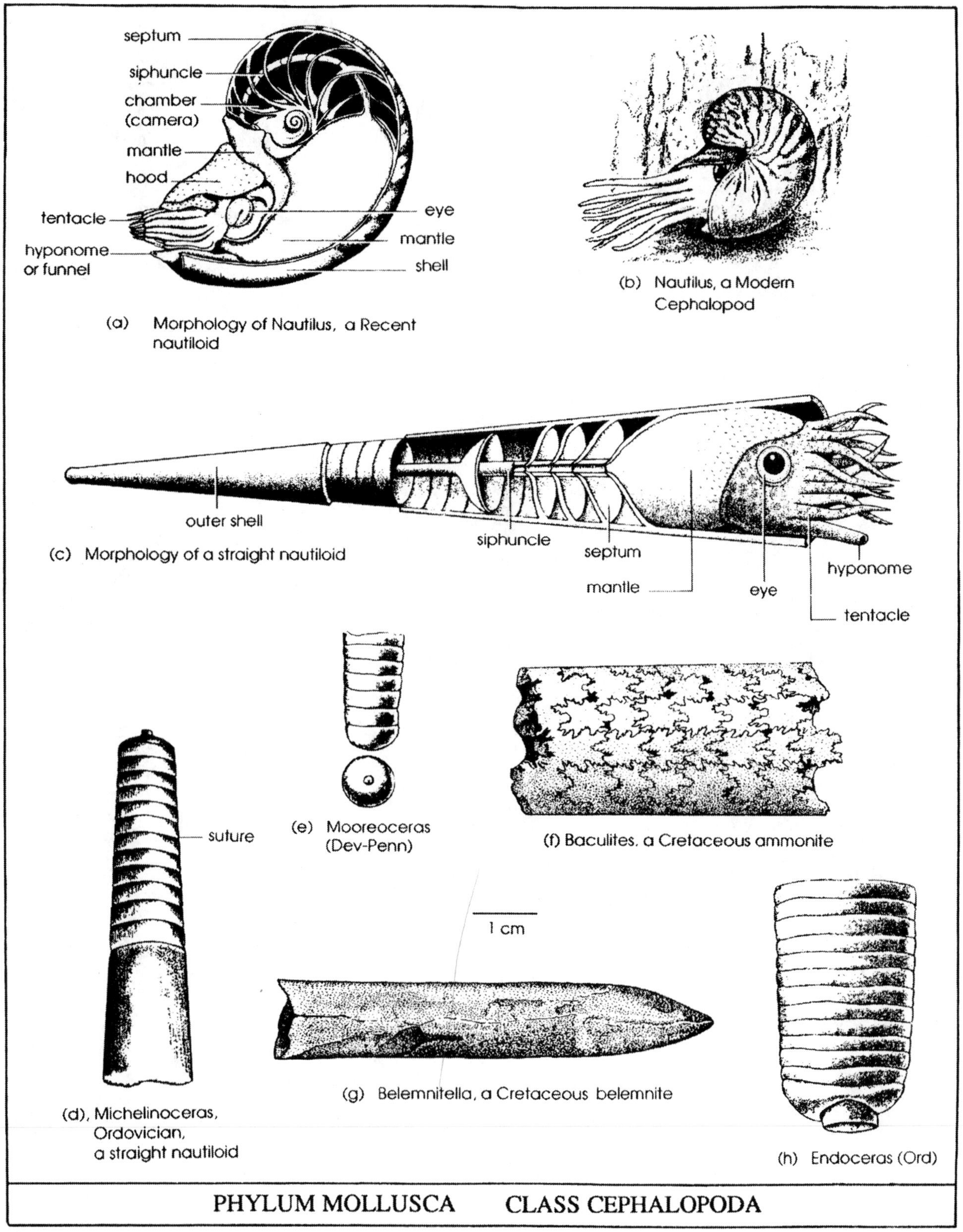

FIGURE 4.25 Phylum Mollusca, class Cephalopoda: (a, c, d, f, g) Terry Chase (b, e, h) The Illinois Geological Survey's "Guide to Beginning Fossil Hunters," by C. W. Collinson (1956).

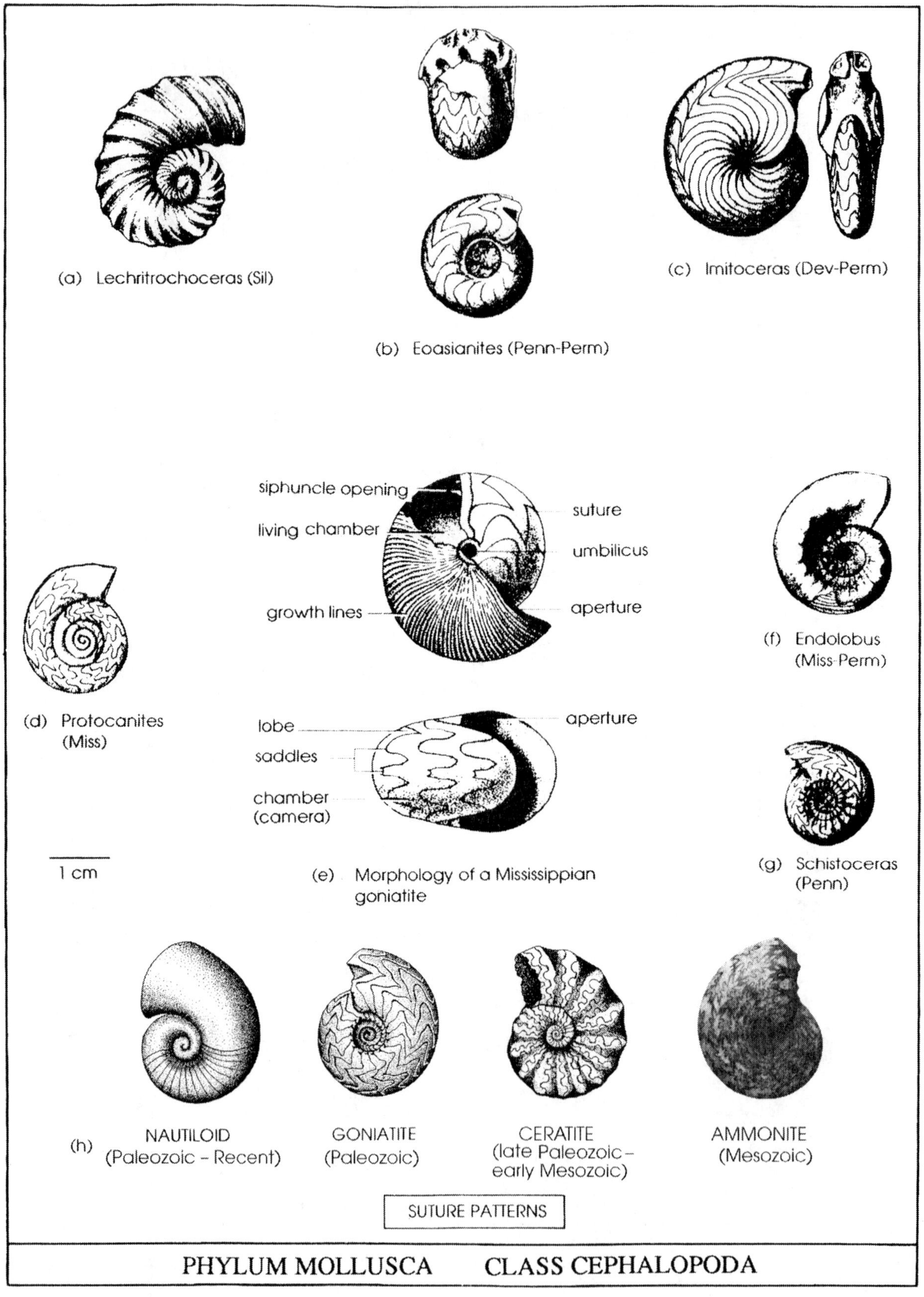

FIGURE 4.26 Phylum Mollusca, class Cephalopoda: (a–d, f–g) The Illinois Geological Survey's "Guide to Beginning Fossil Hunters," by C. W. Collinson (1956) (e) Terry Chase (h) Terry Chase, except Ammonite: DK.

Belemnites [Fig. 4.25(g)] were relatives of the modern squid and possessed cigar-shaped internal calcite supports that are found as fossils in late Mesozoic and early Cenozoic strata.

Phylum Arthropoda (Cambrian to Recent)
Arthropoda (Figs. 4.27 and 4.28) possess an exoskeleton or carapace made of protein. In marine arthropods the carapace is reinforced with calcium carbonate or phosphate. In the generalized arthropod model, the segmented body is divided into three major sections: head, thorax, and tail. Each segment usually has a pair of jointed appendages, specialized for feeding, sensory, or locomotive functions. In marine arthropods, respiration occurs by gills; in terrestrial arthropods, air enters by pores leading to internal tubes. In order to grow, arthropods must shed their exoskeleton and secrete larger ones (molting). Arthropods have a well-developed nervous system and an open circulatory system with a heart, but they have no extensive blood-vessel system. The four main classes of arthropods are Uniramia, Chelicerata, Crustacea, and Trilobita. An alternate classification divides the arthropods into three phyla (Uniramia, Crustacea, and Chelicerata), with the taxonomy of Trilobita somewhat uncertain.

Class Uniramia
This class consists of onychophorans, centipedes, and insects. Most of the Uniramia have a poor fossil record, but occasional insect specimens are found preserved in amber, shales, or fine limestones like the Solnhofen of Germany.

Class Chelicerata
Horseshoe crabs, eurypterids, scorpions, and spiders are chelicerates (Fig. 4.27). Eurypterids (sea scorpions) lived in marine, brackish, and freshwater environments and can be found in rocks of Lower Ordovician to Permian age. They are most common as fossils in the Silurian. The largest eurypterids attained a length of nearly 9 feet. Other members of Chelicerata are poorly preserved as fossils.

Class Crustacea
This class includes ostracods, barnacles, crabs, shrimp, and lobsters (Fig. 4.27). Only the ostracods are common as fossils. Ostracods are bivalved arthropods with shells of chitin and calcite. Their jointed legs can extend between the valves for feeding, swimming, and crawling. Individuals are very small, about 1 mm in size. They live in marine and freshwater environments, most commonly in shallow water. Ostracods range in time from Lower Cambrian to Recent. They are abundant in Paleozoic carbonate rocks, and some species are useful as index fossils.

Class Trilobita (Cambrian to Permian)
Trilobites (Fig. 4.28) are abundant fossils in Lower Paleozoic rocks but became extinct at the end of the Paleozoic. The name *trilobite* comes from the division of the body into three longitudinal lobes—the central, or axial, lobe and two pleural lobes—and three transverse lobes—the cephalon (head), thorax, and pygidium (tail). Since trilobites molted their skeletons in order to grow, many head, thorax, and tail sections are found disarticulated in rocks. To each thoracic segment was attached a pair of legs, which are seldom preserved. The cephalon and pygidium were often composed of fused segments.

Trilobites possessed compound eyes on either side of the glabella. The mouth is on the ventral side below the glabella and directly behind a bulge called the *hypostome*. The appendages were *biramous* (two-branched), with one branch consisting of a

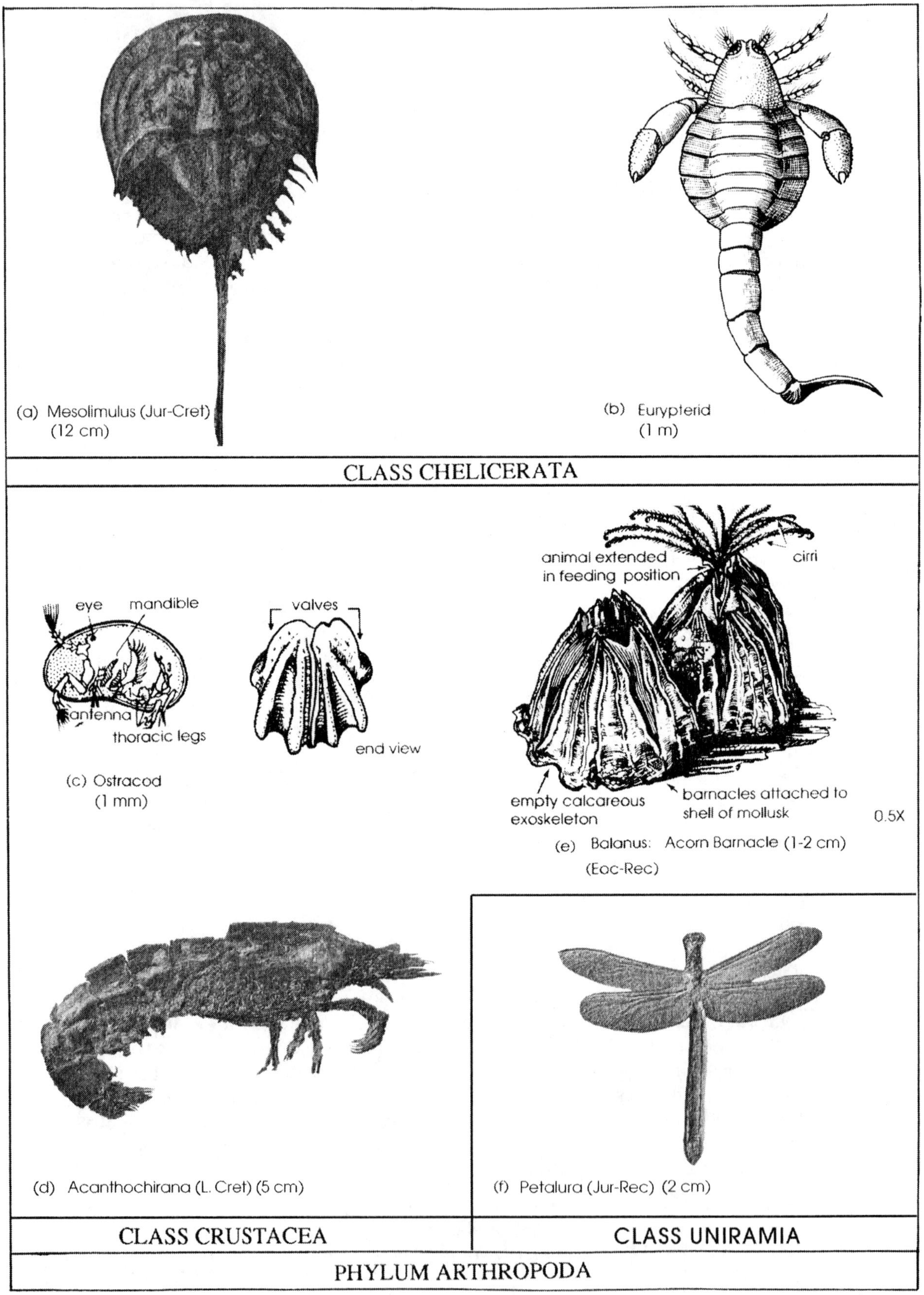

FIGURE 4.27 Phylum Arthropoda, classes Crustacea and Chelicerata: (a, d, f) DK (b, e) Lucy Mauger (c) The Illinois Geological Survey's "Guide to Beginning Fossil Hunters," by C. W. Collinson (1956).

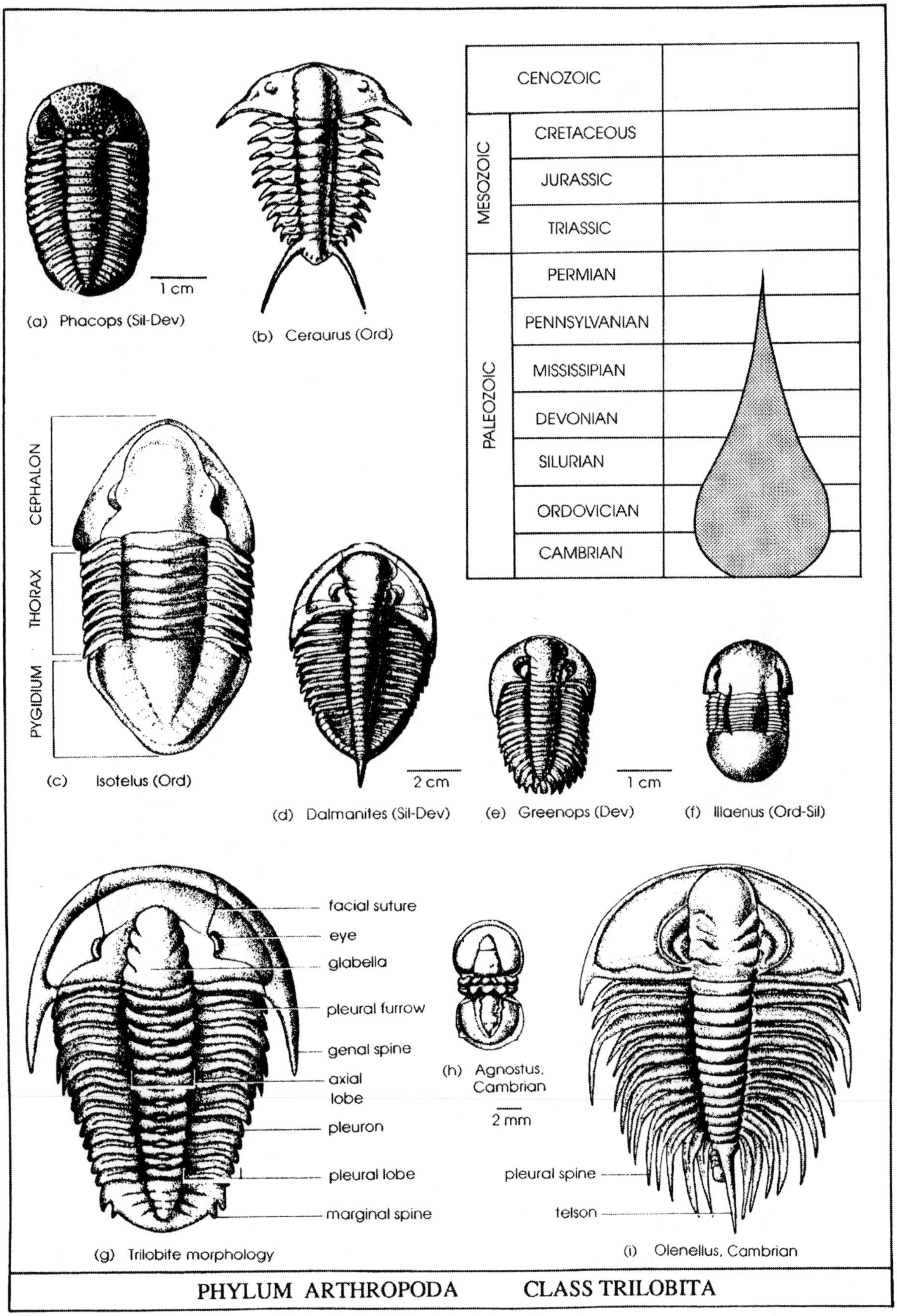

FIGURE 4.28 Phylum Arthropoda, class Trilobita: (a, b, d–f) The Illinois Geological Survey's "Guide to Beginning Fossil Hunters," by C. W. Collinson (1956) (c, g–i) Terry Chase.

walking leg and the other branch having a gill. Definite trilobite trails and burrows are found in the fossil record. Trace fossils called *Cruziana,* a trail with internal chevron-like markings, may have been made by trilobite walking and resting activities. Some trilobite trails show individual marks made by walking legs; others appear to show a sideways crablike movement.

With a ventral mouth and walking/swimming appendages, trilobites are thought to have been bottom-dwelling occasional swimmers that grazed or scavenged the seafloor, and some may have burrowed. They ranged in size from less than an inch to nearly 2 feet. Some trilobites rolled the pygidium under the cephalon for protection, as modern sowbugs do. Trilobite species were relatively short-lived and are useful as index fossils for the Lower Paleozoic.

Phylum Echinodermata (Cambrian to Recent)
Echinoderms (Figs. 4.29 and 4.30) are marine animals with calcareous skeletons composed of plates. Most have fivefold (*pentameral*) symmetry. They possess a water vascular system for locomotion, food gathering, respiration, and sensory functions and have a poorly developed circulatory system. Some have light-sensitive cells that function as simple eyes.

Subphylum Crinozoa

Class Crinoidea (Middle Cambrian to Recent) (Fig. 4.29)

Subphylum Blastozoa

Class Cystoidea (Lower Ordovician to Devonian) [Fig. 4.30(d), (e)]

Class Blastoidea (Middle Ordovician to Permian) [Fig. 4.30(f), (g)]

Crinoids, cystoids, and blastoids were most common in the Paleozoic. Only one group, the crinoids (sea lilies), continues to Recent times. The anatomy of these three classes of attached echinoderms is somewhat similar. They have a root or anchor system attached to a stem or column of circular plates, with a central hole through which run ligamentous fibers. Atop the column is a cup-shaped calyx that houses the internal organs. In the crinoids, arms covered with calcareous plates and pinnules trap food from the seawater; food grooves in the arms direct it to the mouth. However, both cystoids and blastoids feed by brachioles [Fig. 4.30(g)]. Another difference among the stalked echinoderms lies in the structure of the calyx. In crinoids, the calyx is composed of two or three groups of five plates each. Cystoids have irregular numbers of plates, and the body is not symmetrical. The plates are pierced with pores for respiration. Blastoids have three groups of five plates in the calyx, and ambulacral grooves make a five-rayed pattern.

Subphylum Echinozoa

Class Echinoidea (Ordovician to Recent)
This class includes sea urchins and sand dollars [Fig. 4.30(a), (b), (c)]. Echinoids, which fall into two categories, irregular and regular, are most common in the Mesozoic and Cenozoic.

Regular echinoids have pentameral symmetry and move in all directions across the seafloor. The mouth is found at the center of the bottom surface, and the anus is at the center of the top surface. Spines and tube feet are used for locomotion. The

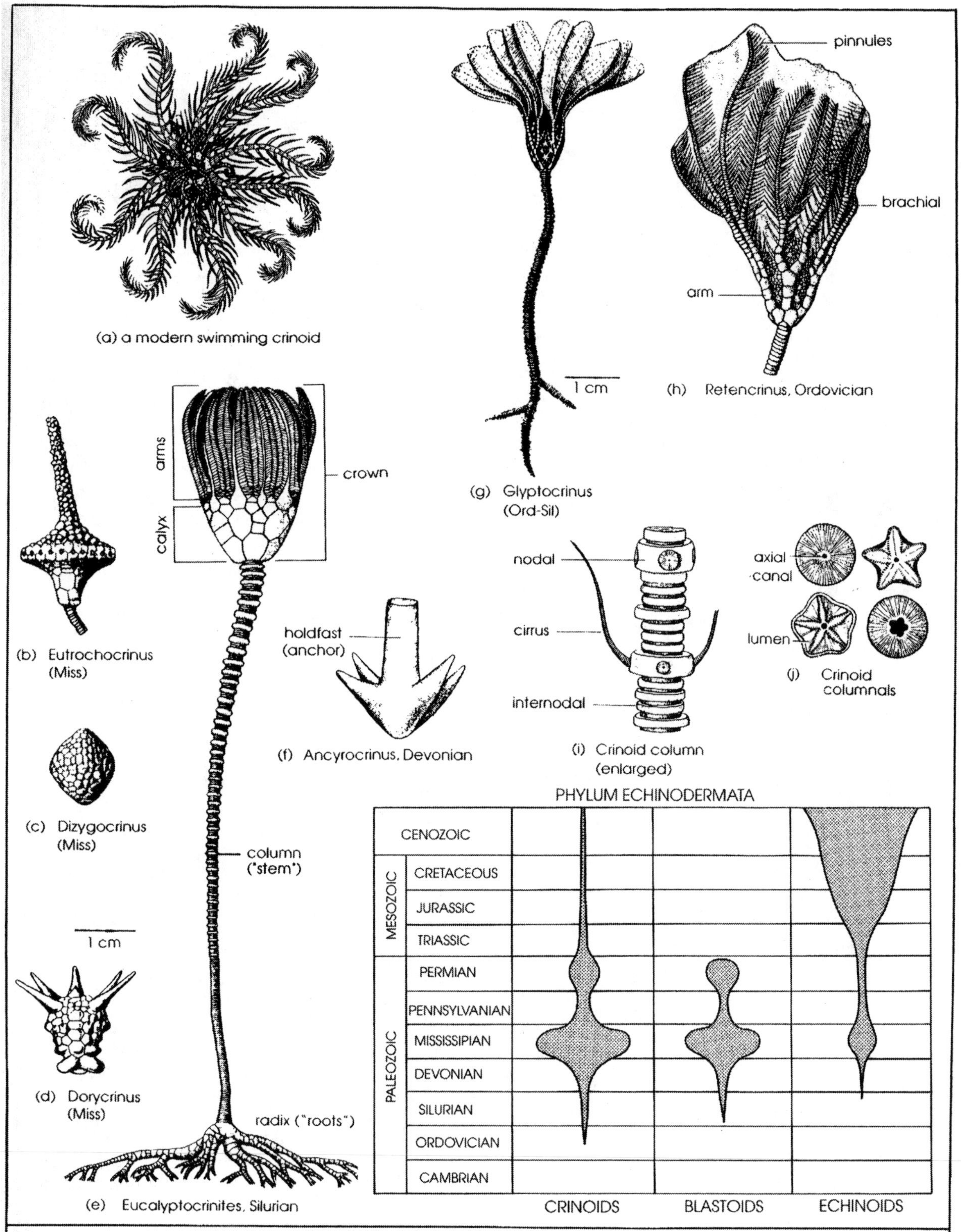

PHYLUM ECHINODERMATA

FIGURE 4.29 Phylum Echinodermata: (a) Lucy Mauger (b–d, g) The Illinois Geological Survey's "Guide to Beginning Fossil Hunters," by C. W. Collinson (1956) (e, f, h–j) Terry Chase.

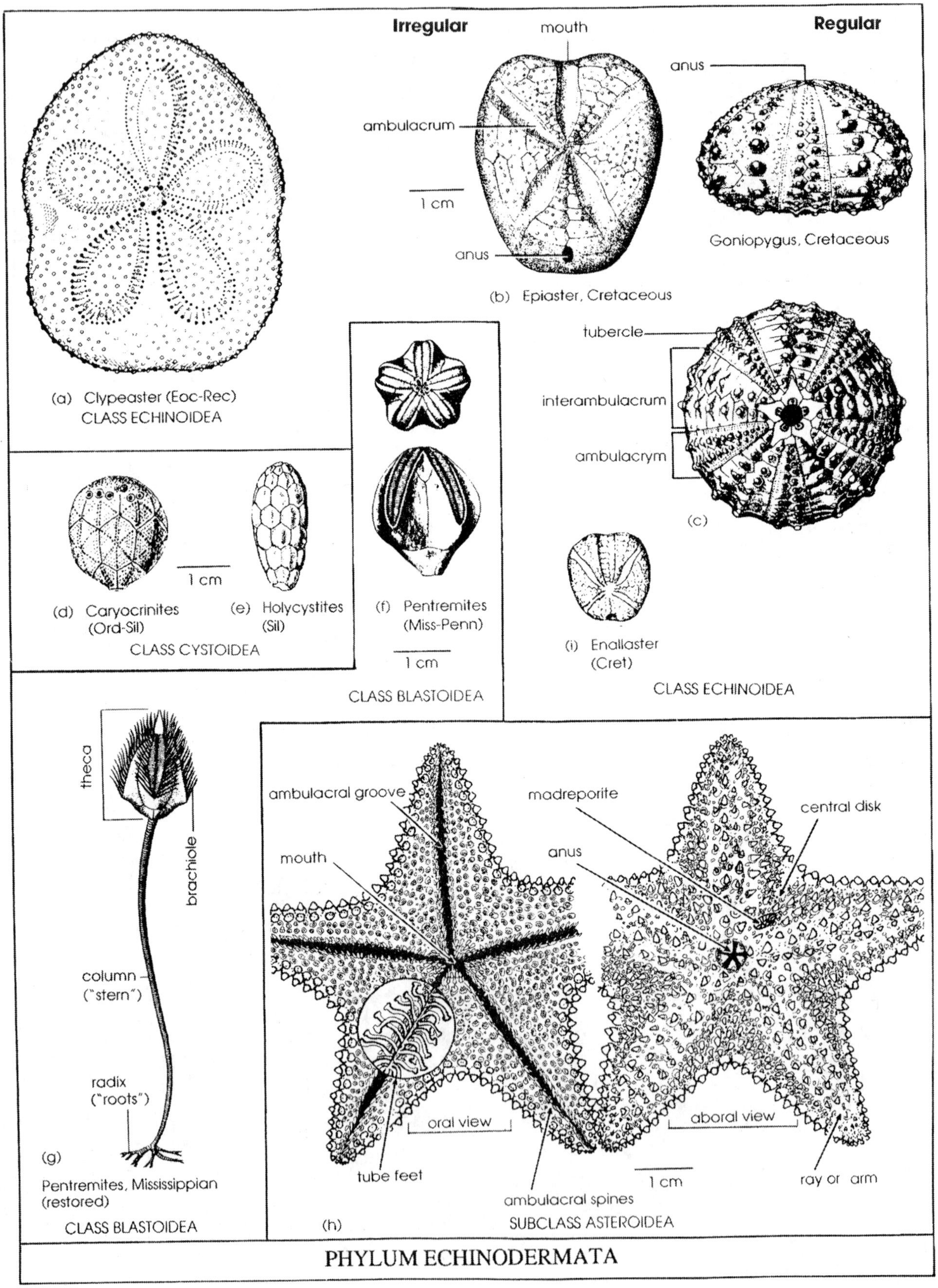

FIGURE 4.30 Phylum Echinodermata: (a, h) Lucy Mauger (b, c, g) Terry Chase (d–f) The Illinois Geological Survey's "Guide to Beginning Fossil Hunters," by C. W. Collinson (1956).

spines are attached to body plates with ball-and-socket joints, and muscles move the spines in all directions. Regular echinoids graze on algae or scavenge flesh on the seafloor. Their spines are common sediment components in some strata.

Irregular echinoids have bilateral symmetry and tend to be heart shaped or oval. They create a burrow in sediment several centimeters deep connected to the surface by a tube and obtain nourishment by suspension or deposit feeding. The mouth and anus are at opposite ends of the bottom surface. Sand dollars are flattened irregular echinoids that live under a thin layer of sediment on the seafloor.

Subphylum Asterozoa

Class Stelleroidea

Subclass Asteroidea (Ordovician to Recent)
Sea stars (starfish) [Fig. 4.30(h)] fall within this class. They possess fivefold symmetry and extended arms. The mouth is on the bottom of the body and the anus on top. The madreporite is both the entry and exit point for water to the water vascular system. Tube feet line the ambulacral grooves on the underside of the body and allow the sea stars to move freely over the seafloor. Sea stars are predators, especially favoring bivalves. They can pull the valves apart, evert their stomachs into the opening and digest the bivalve in its shell.

Phylum Hemichordata

Class Graptolithina (Middle Cambrian to Carboniferous)
An extinct group called *graptolites* are the major fossil representatives of this phylum (Fig. 4.31). They are considered hemichordates because they secreted tubes of protein, similar to the modern hemichordate, *Rhabdopleura*. Graptolites are often found preserved in black shale as compressed carbonized impressions, sometimes replaced by pyrite.

Graptolites are thought to have been colonial marine organisms. The colonies were composed of short tubes (*thecae*) connected by a common canal. There are two main orders of graptolites: (1) Dendroidea, bushlike graptolites that attached to the shallow ocean floor or to floating seaweed, and (2) Graptoloidea, colonies consisting of branching stipes from an initial chamber, which may have floated by gas bubbles or swum weakly by cilia in the open ocean.

Graptolites are good index fossils for the Lower Paleozoic. Graptoloids declined and became extinct in the Early Devonian. Dendroids became extinct in Permian times. Graptoloids are extremely valuable for stratigraphic zonation and correlation, being planktonic, widely distributed, not greatly affected by water depth or temperature, and having short geologic ranges.

Conodonts (Cambrian to Triassic)
Conodonts (Fig. 4.31) are toothlike structures composed primarily of calcium phosphate less than 1/4 inch in size. The animal with the conodonts was an elongate eel-like, soft-bodied creature with an assemblage of conodonts near the head. It was about 1-1/2 inches long and had a bilobed head, transversely segmented muscles, and long fins with ray supports. The animal shows some similarities to chaetognaths, the modern arrowworm, which is a member of the marine zooplankton.

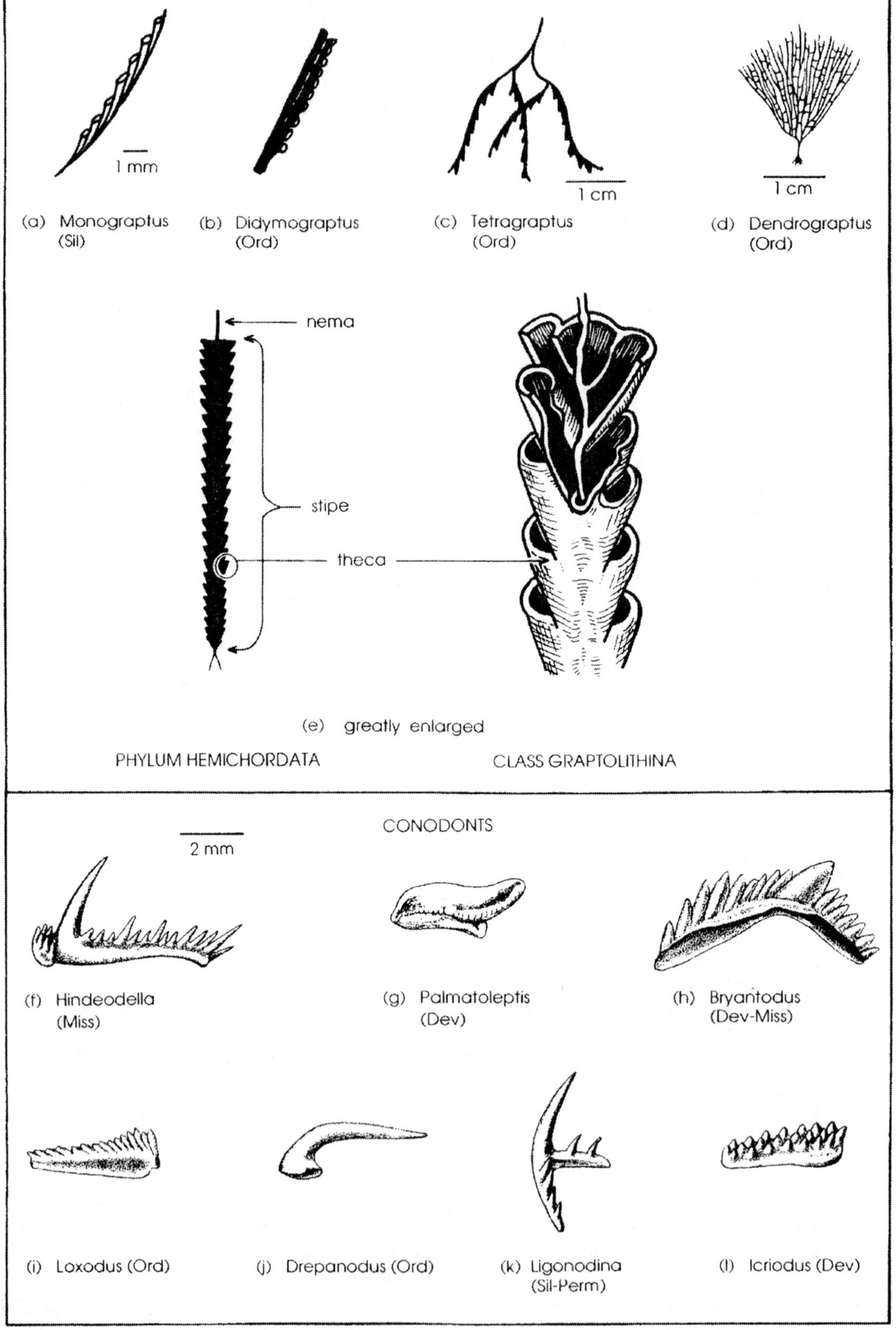

FIGURE 4.31 Phylum Hemichordata, class Graptolithina, and conodonts: (a–d, f–l) The Illinois Geological Survey's "Guide to Beginning Fossil Hunters," by C. W. Collinson (1956) (e) Lucy Mauger.

Conodonts are very useful for zonation and correlation of Paleozoic strata. In addition, the color (amber to black to clear) indicates the maximum temperature to which the conodont, and therefore the rock, was subjected after burial. The thermal history of a sedimentary rock inferred from conodonts can be very useful in the search for petroleum.

Phylum Chordata (Cambrian to Recent) (Figs. 4.32 and 4.33)
The most important chordate characteristics are a *notochord* (a dorsal hollow nerve cord) and segmentally arranged muscles. The cephalochordate called *Amphioxus,* found in modern shallow marine waters, is a good representative of nonvertebrate chordates. Because of the lack of hard parts, the fossil record of early chordates is scarce, although *Pikaia* from the Burgess Shale of Canada may belong to this group.

Subphylum Vertebrata (Cambrian to Recent)
Vertebrates possess a brain, a spinal cord enclosed by vertebrae, an internal skeleton of bone or cartilage, paired sensory organs, and well-developed nervous, circulatory, digestive, and muscular systems. The most commonly preserved portions of the internal skeleton are the bones and teeth of the animal. The teeth and bones are composed of apatite, which is a variety of calcium phosphate.

Class Agnatha (Cambrian to Recent)
This class includes modern jawless vertebrates, such as lamprey and hagfish. Fossil ostracoderms, ranging from Ordovician to Devonian, were jawless fish. Some had a flexible light covering of scales and may have had good swimming ability. Others, with heavier armored plates, were probably benthic.

Class Placodermi (Devonian to Carboniferous)
Placoderms were primitive jawed fish with platy armored skin. *Bothriolepis* is shown in Figure 4.32(a). Jaws may have evolved from gill supports in earlier, jawless vertebrates. More heavily armored placoderms were probably benthic, but those few with reduced armor thickness may have been more active swimmers. Instead of true teeth, placoderms had bony plates on the margin of their jaws. Placoderms are found in freshwater deposits as well as marine. The evolutionary relationships of placoderms to other fish groups is unclear.

Class Chondrichthyes (Silurian to Recent)
Cartilaginous fishes include modern sharks, rays, and skates. Since the skeleton is composed of nonmineralized material, usually only the teeth and spines are fossilized (Fig. 4.33). The group is primarily marine, but freshwater fossil forms have been found. Most sharks are marine predators today and have been in the past as well, but some groups have taken up a benthic life, developing shell-crushing teeth. Instead of bony armor or scales, sharks' external covering consists of small scalelike features called *denticles*.

Class Osteichthyes (Silurian to Recent)
This class includes the bony fish. Major groups are (1) extinct forms called acanthodians, (2) ray-finned fish, and (3) lobe-finned fish. Acathodians are an early group of bony fish that are difficult to classify. Their geologic range is Silurian to early Permian. The tail is prominently asymmetrical, like that of sharks. They differ from other bony fish by the structure of their teeth and the spines supporting their

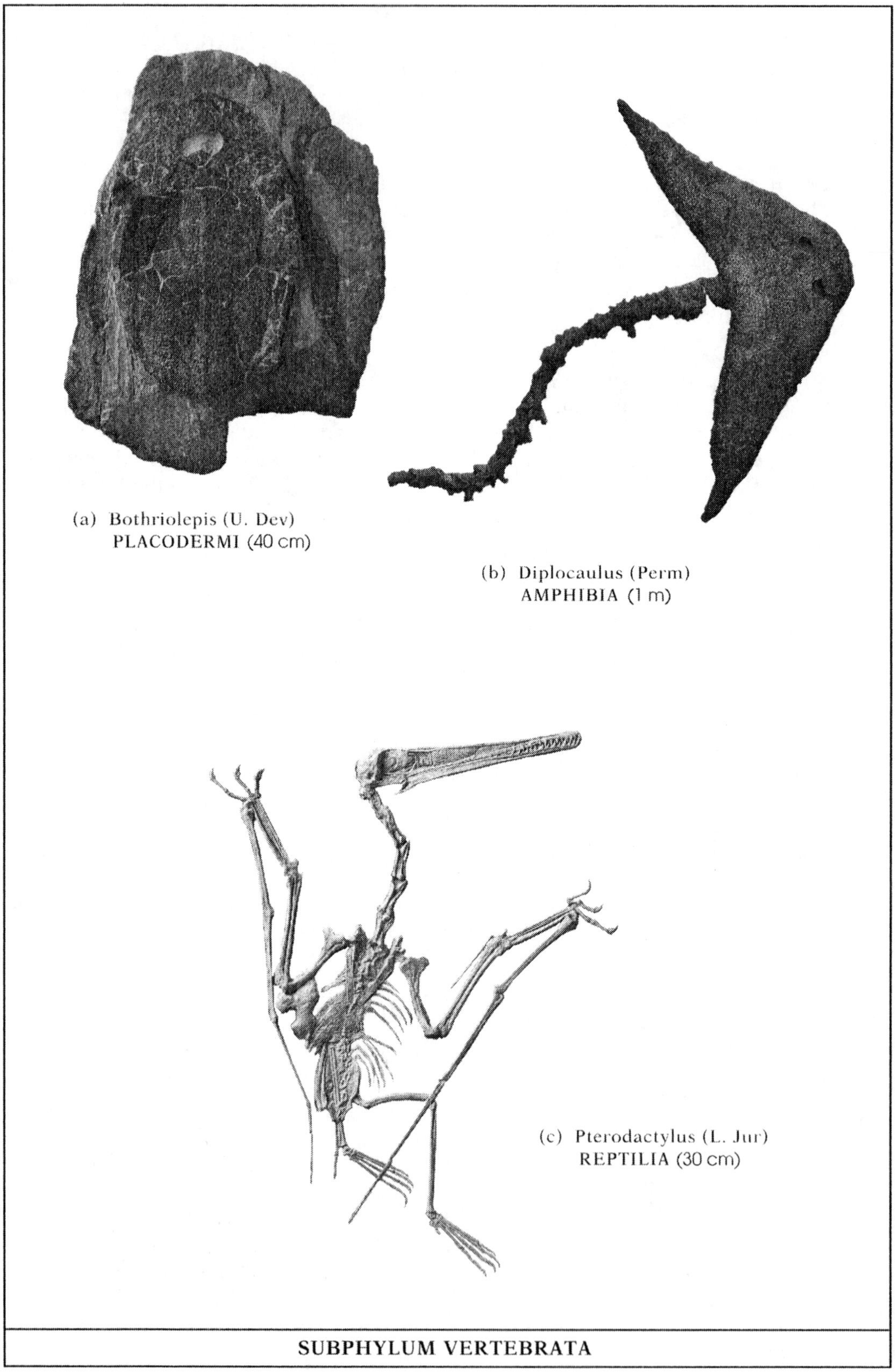

FIGURE 4.32 Phylum Chordata, subphylum Vertebrata: DK.

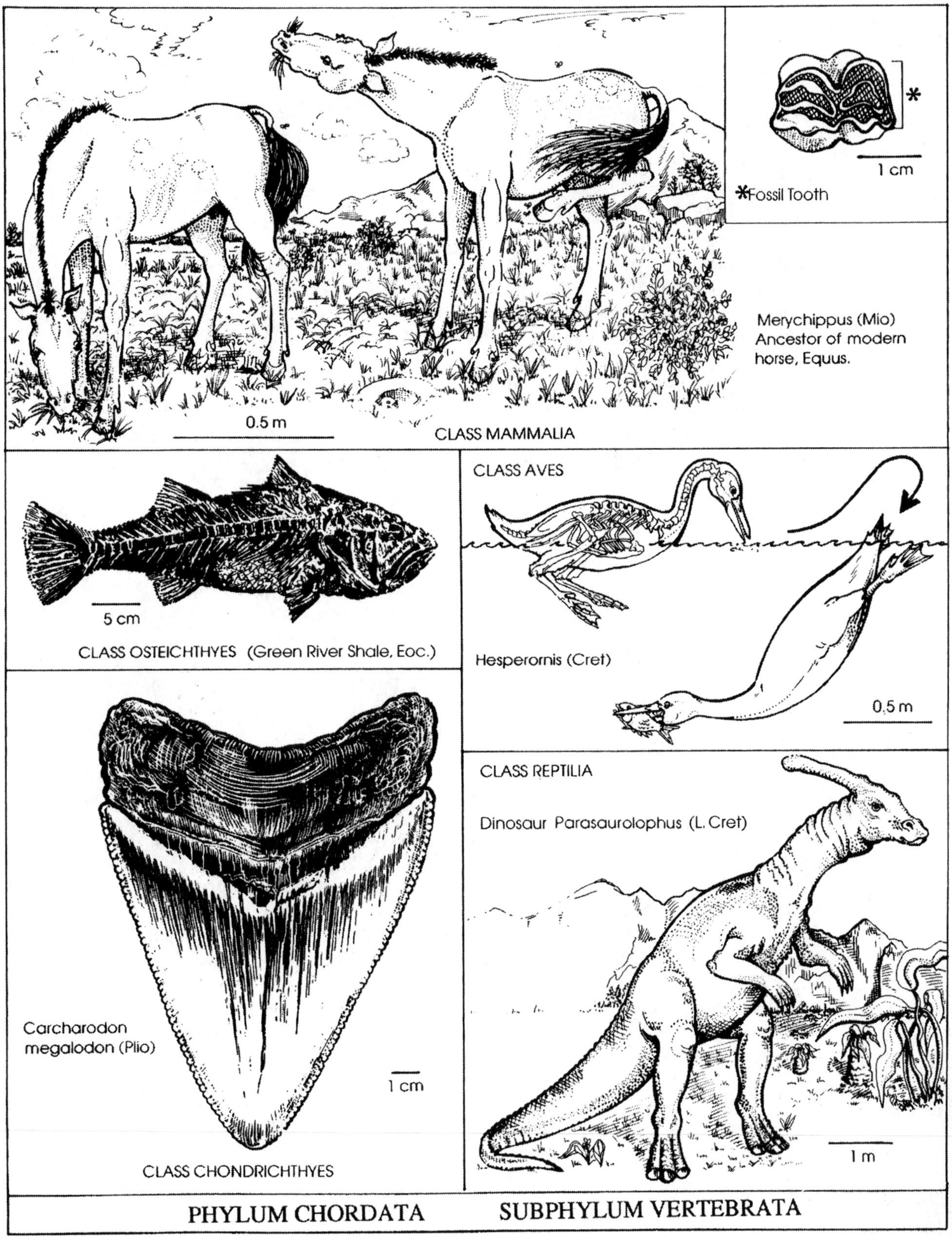

FIGURE 4.33 Phylum Chordata, subphylum Vertebrata: Lucy Mauger.

fins as well as spines intermediate between pectoral and pelvic fins. They are the first fish with jaws to appear in the fossil record. Most modern fish are ray-finned fish of the group called the *teleosts* (Fig. 4.33), which includes over 20,000 living fish species. Three major groups of lobe-finned fish are the modern lungfish and coelacanths and a group from the Paleozoic called the *rhipidistians*, which are considered ancestral to amphibians. The lobe fin has a bone structure that is similar to the limbs of other vertebrates. Modern theories suggest that the fin bone structure might have been useful for movement of the fish in shallow-water ponds. Rhipidistians and early amphibians also have great similarities in tooth structure.

Class Amphibia (Devonian to Recent)
The amphibians are the first vertebrates to be able to live on land, although they need moisture in which to lay eggs. Juveniles are aquatic with gills; adults possess lungs. Labyrinthodont amphibians were common Paleozoic vertebrate fossils (Fig. 4.32). They evolved in the Devonian from rhipidistian ancestors and achieved great diversity in the Carboniferous. Relationships between the Paleozoic amphibians and modern groups have not yet been established. Modern representatives are frogs, salamanders, and caecilians (legless amphibians). Frogs and salamanders are first found in the fossil record in the Jurassic.

Class Reptilia (Late Carboniferous to Recent)
Reptiles evolved from amphibians in the Late Paleozoic and were the first vertebrates to conduct their lives fully on land, although many returned to aquatic environments. The key to this ability was the development of an egg with membranes for fluid retention and gas exchange and a leathery or mineralized covering. Thus reproduction was not limited to watery or moist areas. Early reptiles had a lizard-like body structure. Modern representatives of Reptilia include turtles, crocodiles, snakes, and lizards. Fossil members include dinosaurs (Fig. 4.33), flying reptiles (Fig. 4.32), marine reptiles, phytosaurs, therapsids, and many others. Therapsid reptiles have many mammalian tooth and jaw features and are considered mammal ancestors.

Class Aves (Jurassic to Recent)
Birds are warm-blooded, egg-laying vertebrates with an external covering of feathers. Their metabolic rate is very high due to the energy requirements of flight. Although birds are rare as fossils, in the Jurassic Solnhofen Limestone of Germany, the earliest fossil bird, *Archaeopteryx*, provides an excellent link between birds and their reptilian ancestors. The largest members of genus *Hesperornis*, a Cretaceous diving bird (Fig. 4.33), were as much as 1 m in height. In the Tertiary of South America, Asia, North America, and Europe, giant predatory birds towered 1.5–2 m in height. Some paleontologists consider birds to be descendants of small theropod dinosaurs, and the link between dinosaurs and birds is strengthened by the discovery in China of small dinosaurs with feather-like ornamentation on their bodies.

Class Mammalia (Triassic to Recent)
Mammals have hair, are warm-blooded, and nurse their young. Their ancestors appear to have been reptiles from the Triassic therapsid group called *cynodonts*. These animals show a relatively larger brain size and may have achieved a higher metabolic rate than their reptilian cousins. Early mammals show differentiation of teeth into incisors, canines, premolars, and molars. Body size was generally small, and nocturnal habits may have prevailed among many groups of early mammals.

There are three major modern groups:

Monotremes are egg-laying mammals and include the duckbill platypus and the echidna.

Marsupials are pouched mammals and include the kangaroo, koala, and opossum.

Placental mammals are born at a more advanced stage of development than marsupials, nourished within the female by the placenta. Examples are rodents, elephants, whales, hoofed mammals such as deer and horses (Fig. 4.33), rabbits, bats, carnivores such as cats and wolves, and primates such as monkeys, apes and humans.

IV. KINGDOM PLANTAE (FIG. 4.34)

This kingdom consists of multicellular organisms capable of photosynthesis but lacking organs of sensation, digestion, respiration, or movement.

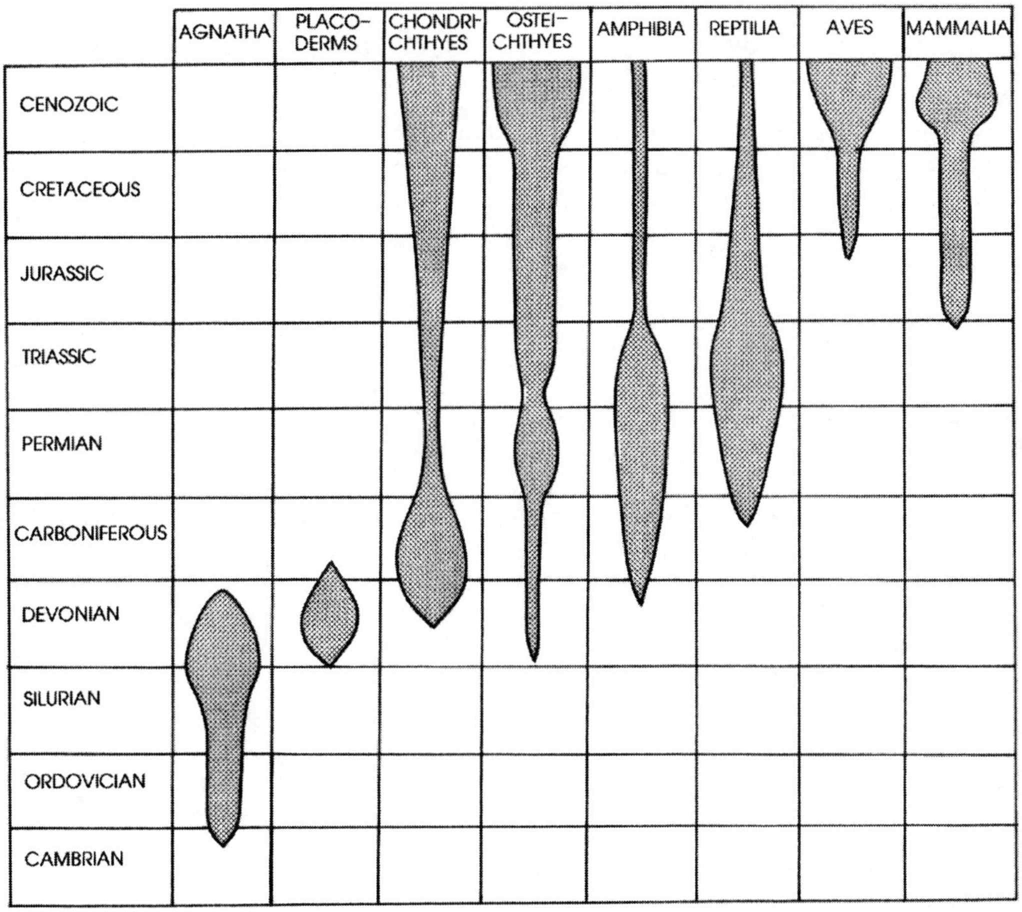

FIGURE 4.34 Diversity of selected vertebrate classes.

Divisions:

Anthocerotophyta: Hornworts can be found in tropical forests, along streams, and in disturbed areas. They are spore-bearing plants of questionable origin and relationships. Fossil hornwort spores date from the Late Cretaceous. It is possible that the group originated in the Paleozoic, but fossil plant or spore material has not yet been found.

Bryophyta: Mosses are a very diverse group of plants today but are rare in the fossil record.

Hepaticophyta: Liverworts are ancient nonvascular land plants that may date back to the Silurian. Modern liverworts are found in some of the same places as are mosses: moist shaded areas. Leafy and nonleafy types are two of several subgroups among liverworts. Modern liverworts have unique chemicals and cellular structures that distinguish them from mosses and other nonvascular plants.

Trachaeophyta: Vascular plants [Figs. 4.35 and 4.36(a) and 4.36(b)] live on land and possess several requirements for such a life: (1) a pipe system for transferring water and nutrients from the soil to the cells of the plant, (2) a waxy leaf covering for retaining moisture, (3) leaf pores for the exchange of gases, and (4) rigid stem cells to support the weight of the plant above the ground.

There are two general reproductive modes of vascular plants. Spore-bearing plants have alternation of generations: a spore-bearing sexual phase and an asexual phase. The more advanced seed-bearing plants reproduce by seeds produced after pollination (sexual reproduction). The following groups are included within the Tracheophytes.

Cladoxylopsida: This group of fossil plants, found from the Middle Devonian to the Early Carboniferous, has no living members but is thought to be related to ferns and horsetails (see later) because of features in their vascular tissue. Fossils of the group are preserved stems, and it is possible these plants had no leaves.

Lycophyta: Fossil lycophytes are first found in the Middle Silurian but are most common from the Carboniferous. As an important part of the forest flora, these trees grew to towering heights, upwards of 35 m. The most familiar of these trees are *Lepidodendron* and *Sigillaria* (Fig. 4.35). In the Late Carboniferous, extinction claimed most of the giant lycophyte groups. Modern lycophytes, such as club mosses, are small and inconspicuous. Lycophytes produce spores on special leaves called *sporophylls*. Spore-bearing plants have a two-stage life cycle. Spores germinate under moist conditions and grow into a gametophyte plant with egg and sperm. The fertilized eggs become a sporophyte plant that produces the spores for the next cycle. Lycophyte leaves are called *microphylls*; they have only one unbranched vein. This leaf structure is probably not related to the leaf structure in other tracheophytes.

Progymnosperms: These fossil plants were important in the flora of the Middle Devonian through the Late Mississippian, but left no living relatives. Many had woody trunks and grew to be tall trees (up to 20 m) but reproduced by means of spores. The relationship of these trees to the gymnosperm group is now being reconsidered, although the fact that some progymnosperms produced two different kinds of spores may indicate close relationships with seed plants.

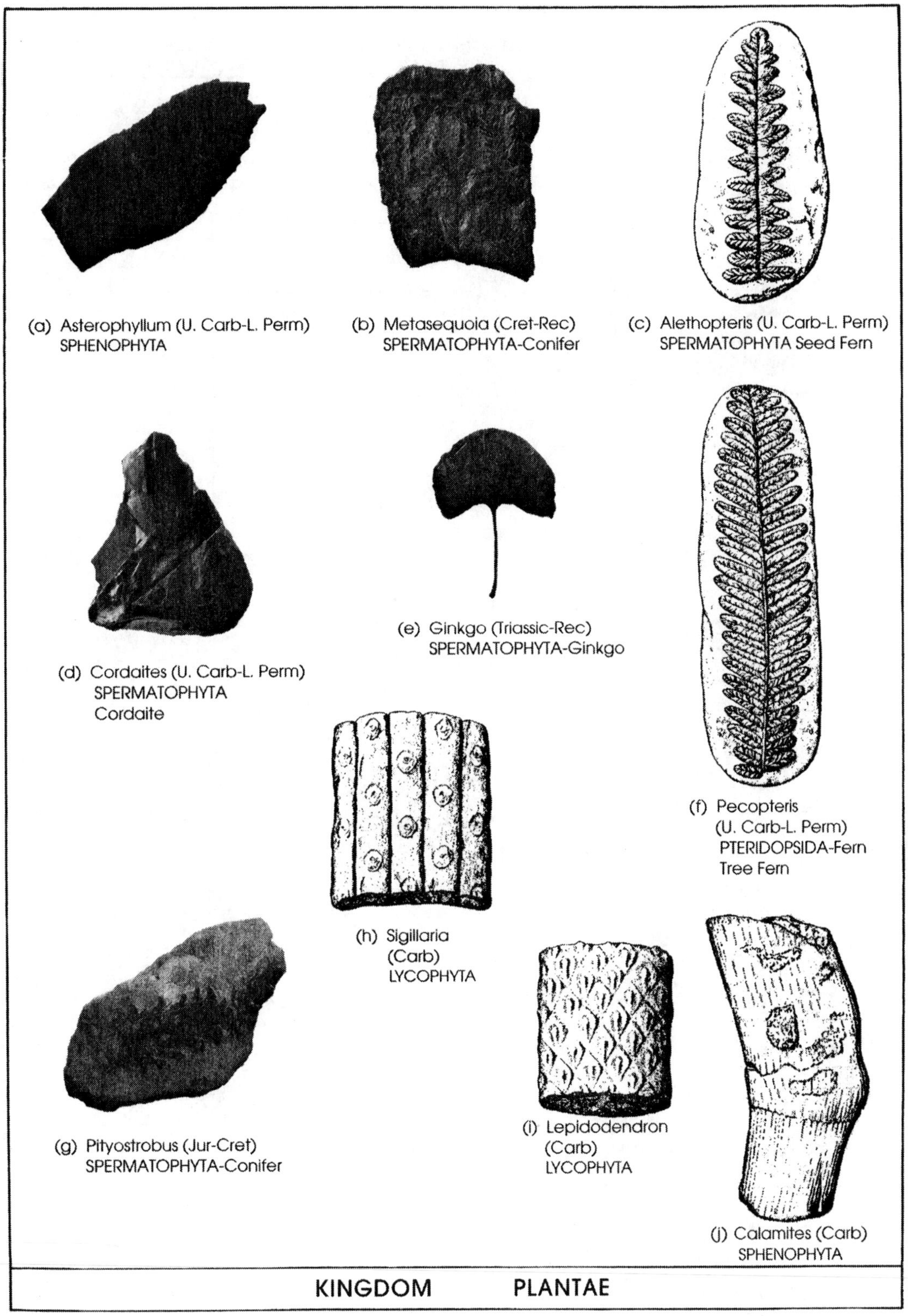

FIGURE 4.35 Kingdom Plantae: (a,b,d,e,g) DK (c,f,h,i,j) Illinois Geological Survey, "Pennsylvanian Plant Fossils of Illinois," by C. W. Collinson and R. Skartvedt (1960).

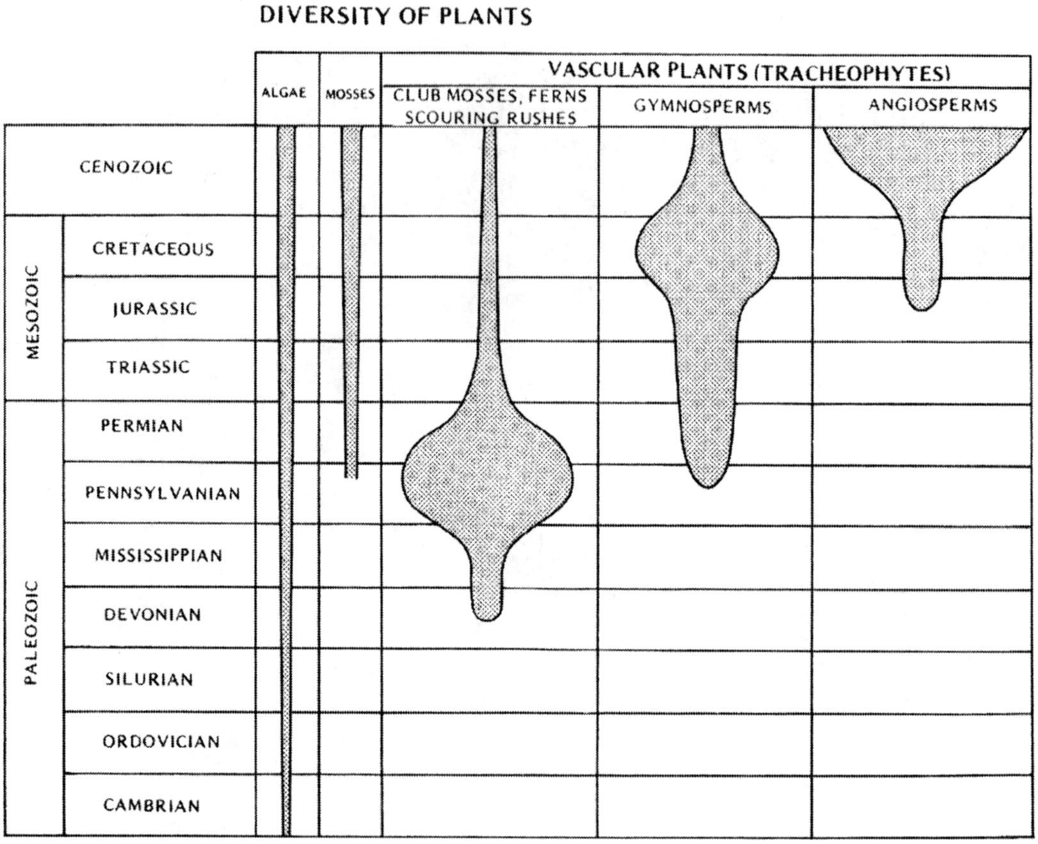

FIGURE 4.36(a) Plant diversity.

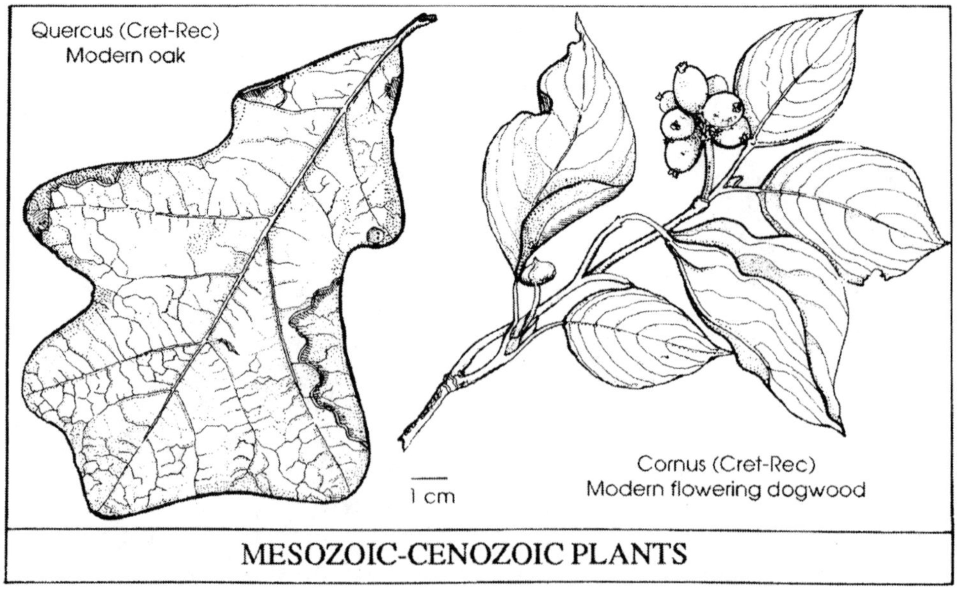

FIGURE 4.36(b) Mesozoic-Cenozoic plants. Drawn by Lucy Mauger.

Pteropsida: Ferns (Fig. 4.35) are spore-bearing plants dating back to at least the Devonian. There are four living groups of ferns and two groups now extinct. They are closely related to lycophytes and may also have evolved from the Cladoxylopsida. The most diverse modern group, the leptosporangiate ferns, are first found in the Early Carboniferous. Their two-phase life cycle is similar to that of the lycophytes described earlier. The stems of many modern ferns are *rhizomes*, stems that trail horizontally just below the ground surface, allowing them to vegetate areas rapidly. Fern leaves are referred to as *fronds*, and these fronds bear the spore-producing organs, often on the underside.

Sphenophyta: This group includes several important Carboniferous species, such as *Calamites* and *Asterophyllum* (Fig. 4.35), and the modern horsetail, *Equisetum*. Sphenopsids first appear in the Late Devonian and may have evolved from ancestors in the group Trimerophytophyta, discussed next. The Mazon Creek area of Illinois yields a great variety of sphenopsids from the Late Carboniferous, ranging from shrub size to 20-m tree-sized *Calamites*. These plants thrived in "coal swamps" as well as in drier lowland areas. The life cycle of sphenopsids is similar to that of other spore-bearing plants, alternation of generations. The modern horsetail, *Equisetum*, grows commonly along river banks and other areas with moist soils.

Trimerophytophyta: This spore-bearing group from the Devonian may be ancestral to ferns, progymnosperms, and sphenopsids. They lacked true leaves, and photosynthesis took place along the stem.

Zosterophyllophta: Another Devonian spore-bearing plant group, the zosterophylls may be ancestral to the lycophytes.

Spermatophyta: Seed plants belong in this division and are a varied group. Seed plants, unlike spore-bearing plants, have only a one-stage reproductive cycle. Production of gametes and fertilization take place within the parent plant. The most common division is between *angiosperms* (flowering) and *gymnosperms* (lacking flowers). Gymnosperms, however, are a varied group and include cycads, ginkgoes, conifers, cordaites, glossopterids, and a variety of other flowerless plants. Progymnosperms, mentioned earlier, share some characteristics with early seed plants and may be ancestral to the group. The earliest known seed plants come from the Late Devonian of West Virginia.

"Gymnosperm" groups:

Pteridospermophyta: Seed ferns are a varied and informal taxonomic group. They are seed plants with fernlike foliage. *Medullosa* is an example of one that has been closely studied. The foliage of this species is classified under the names *Alethopteris* and *Neuropteris* (Fig. 4.35). Seed ferns can have unusually large pollen grains and large, avocado-sized seeds. In general, seed ferns range geologically from the Late Devonian to the Permian.

Glossopterids. These tongue-shaped leaves were an important component of the Gondwana flora and as such joined the mass of evidence for early continental drift and later plate tectonics theories. They range from Late Carboniferous to Triassic in Australia, Africa, South America, Antarctica, and India. The relationship of glossopterids to other gymnosperms is still under investigation.

Cycadophyta. Cycads first appear in the Upper Carboniferous and experienced their greatest abundance and diversity in the Mesozoic. Over 100 species exist today. They may have descended from a group called *Medullosans*, early seed plants described earlier. Modern cycads exist in tropical and subtropical rain forests as well as drier environments. In the Mesozoic their distribution was global, from Alaska to the Antarctic, due to the warmer, more moist climate. Fossil leaves, cones, and seeds from these plants have been found worldwide.

Ginkgoes (Fig. 4.35): The earliest probable ginkgophyte dates from the Early Permian, but they attained their maximum diversity in the Jurassic. A single species, *Ginkgo biloba*, represents the group today. Ginkgo leaves are generally triangular in shape and develop a partial split, or bifurcation (hence the species name, *biloba*).

Cordaites (Fig. 4.35): This group of woody gymnosperms was most prominent in the Late Paleozoic, when they played an important role in upland forests of the Siberian platform and coal swamps of the Upper Carboniferous and Permian in Europe and North America.

Conifers (Fig. 4.35): Since the Late Carboniferous, conifers have been important components in both drier upland and moist lowland forests. As climates cooled and dried toward the end of the Paleozoic, conifers expanded their ranges. They held a dominant position in forests during the Mesozoic, yielding to the angiosperms in the Cretaceous and Cenozoic. Most conifers are woody evergreen plants with a single erect trunk and branches emerging at nearly right angles to the trunk. Reproductive structures in most conifers consist of woody cones, although in junipers the cones are more berry-like. Wind transport is the most common form of pollination.

"Angiosperm" group:
Anthophyta: Angiosperms (flowering plants) are the most prominent members of the group Anthophyta (Fig. 4.36b). Their origins and ancestors are still somewhat mysterious. The first true angiosperms appear in tropical sediments of the Early Cretaceous, although there is a possible angiosperm species found in the Upper Triassic of Colorado. Angiosperms are characterized by various aspects of their reproductive structures, pollen, and vascular tissues. The most common parts of angiosperm plants to be fossilized are the pollen and the leaves; fruits and seeds are also sometimes found. Flowers, being seasonal and delicate structures, are seldom preserved. Some angiosperms, such as the grasses, are wind-pollinated; others rely on a symbiotic relationship with insects to effect pollination. Angiosperms are by far the most common plant group in modern floras.

EXERCISES

Exercise 4-8 INVESTIGATING THE BURGESS SHALE

Go to **http://park.org/Canada/Museum/lobby.html**. Click on the following links: *Paths to Knowledge; Virtual Paleontological Museum; Museum Tour; Burgess Shale*. The Burgess Shale is a famous fossil locality in western Canada. It is unusual because of the exquisite preservation of fossils 530 million years old. At the Web site, explore the information about the Burgess Shale, and then answer the following questions.

a. Describe the sedimentary environment of the Burgess Shale and the associated formations. How does that environment differ from the environment in this part of Canada today?

b. How were the Burgess organisms preserved, and why is that preservation unique?

c. Go to the fossil menu. Using it and the information on fossil groups in this manual, create a Burgess Shale ecosystem. Find one or more representatives of each of the following categories:

- Producers
- Benthic
  - mobile grazers
  - sessile filter feeders
  - burrowers
- Nektonic
  - predators
  - scavengers
  - filter feeders
- Planktonic

d. Which Burgess Shale species seem particularly peculiar or even bizarre to you? Why?

e. Are there some Burgess Shale species that seem to have no living relatives? What might account for that?

f. What does the discovery of well-preserved faunas such as the Burgess Shale suggest about the "completeness" of faunas from fossil localities elsewhere?

Exercise 4–9 VENDIAN "ANIMALS"

Go to **http://www.ucmp.berkeley.edu/**. From the home page, follow the links *History of Life; Special Exhibits; Vendian Life* to investigate these interesting and enigmatic creatures. Answer the following questions.

a. List and describe three Vendian fossils whose affinities (classification) are fairly well established; in other words, three fossils on which there is some agreement about what they are.

b. List and describe three Vendian fossils whose affinities are in doubt or simply unknown.

c. After reading the descriptions of these "critters," would you be inclined to consider them as a fauna, that is, a group of animal species?

Exercise 4-10 SILURIAN REEF

Go to **www.mpm.edu/**. Follow these links: go to *Collections and Research; geology; K-12 Education Virtual Silurian Reef.* This Web site does a good job exploring reefs in the present and in the geologic past. As you explore the Web site, answer the following questions.

a. What is a reef? What environmental factors are common to reefs throughout geologic time? Why would central North America be a good place to find Silurian reefs?

b. List the five major types of organisms and their roles in the life of a reef.

c. Look at organisms of the Silurian reef, and click on each listed. Determine into which of the five categories (from question b) each organism fits.

d. What caused the demise of Silurian reefs in Wisconsin? What kinds of environmental factors threaten reefs today?

Exercise 4-11 THREE SOLUTIONS TO FLIGHT

Go to the Web site **www.ucmp.berkeley.edu/**. Follow these steps: go to *History of Life; Special Exhibits; Vertebrate Flight*. Here you'll find a very interesting discussion on the evolution of flight and how three different groups of vertebrates "solved" the problem of flight: the pterosaurs, birds, and bats. Look at the various sections in the Web site and then answer the following questions.

a. Describe the differences in the flight structures, particularly the wing structures and the thoracic structures, for each group of vertebrate fliers. How are the wings supported? What structures in the chest power the flight?

b. Look at the *Origins of Flight* section. What are the two main scenarios for how flight evolved?

c. Decide which scenario each of the three flying vertebrate groups might have used in initiating flight.

Exercise 4-12 EXAMINING FOSSILS

The following 10 questions require a simple set of fossils to be available in the laboratory.

a. Using the axis of coiling and the animals' physical features, compare specimen 1 and specimen 2 to determine which is a gastropod and which is a cephalopod.

1. ____________________ 2. ____________________

b. Demonstrate, by drawing simple diagrams, where the location of the symmetry plane would be in specimen 3 (rugose coral), specimen 4 (bivalve), and specimen 5 (trilobite).

c. What type of suture pattern is present in specimen 6 (cephalopod)?

d. Using specimen 7, draw an illustration showing the location of the bivalve's muscle scars inside the valve. Are these muscles used for opening or closing the shells, or both?

e. Using a sand dollar or sea urchin as specimen 8, sketch the specimen and show evidence of its symmetry.

f. Draw specimen 9 (spiriferid brachiopod) and specimen 10 (bivalve) and show where the plane of symmetry for each would be located. Why is only one valve of a bivalve often found, and why are both valves often found in common brachiopods?

g. A piece of fossiliferous limestone bearing five different "things" with circular outlines was collected by a student and brought into the laboratory. How might a paleontologist determine which particular outline belonged to the following organisms: a coral, a crinoid, a bryozoan, a belemnite, and a nautiloid.

h. What is the main difference between the modern-day shells and fossil shells in tray B?

i. To what kind of vertebrate do the teeth in tray C belong? Why are the teeth most often the only part of this animal preserved as fossils?

j. The fossil wood in tray D has been replaced by silica (SiO_2). Suggest three ways in which this fossil wood could be distinguished from recent wood.

Exercise 4–13 USING FOSSILS FOR PALEOENVIRONMENTAL INTERPRETATION

Refer to Figure 4.37 showing a paleoenvironmental interpretation.

Zone A. Soft weathered limestone. The upper surface is a product of karst topography with abundant sinkholes. The basic fauna present are sea urchins and abundant spines, abundant bivalves, and bryozoans.

Zone B. Infill of sinkholes, grey micaceous shaly sand. Typical fauna present: abundant shark teeth (Tertiary and mid-Cretaceous species), large crocodile teeth and vertebrae, fish vertebrae and spines, ray and skate teeth, turtles, and large pieces of carbonized wood. In the upper part of the zone are oyster banks (*Ostrea thirsae*) and *Ophiomorpha* burrows. *Ophiomorpha* is the term given to a specific type of preserved burrow. The modern analog is the burrow made by the marine decapod shrimp called *Callianassa*. For a geologist, the important feature is that these modern animals live

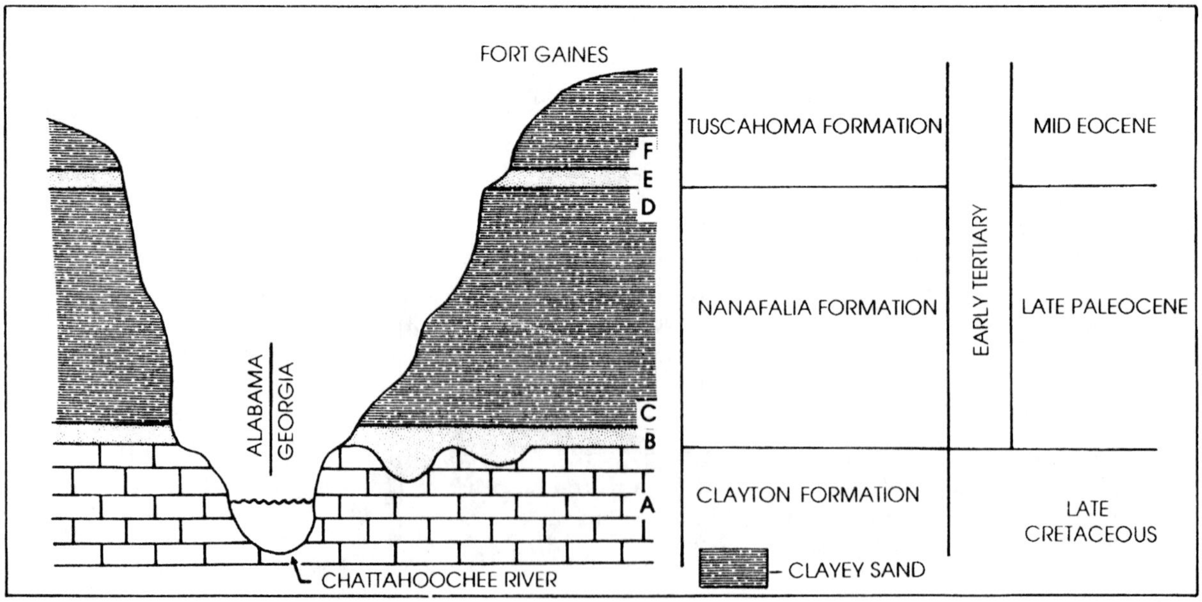

FIGURE 4.37 Paleoenvironmental interpretation.

only in the foreshore or beach face and thus their fossils can be used as excellent indicators of the ancient shoreline.

Zone C. Grey shaly sand with some large concretions grading to clean sand. Basic fauna present are open marine bivalves and gastropods.

Zone D. Grey laminated shaly clay. The uppermost part of the Nanafalia has very abundant burrowing by *Ophiomorpha* and scattered large clams, shark teeth, and crab claws. An interval just below the intensely burrowed zone has abundant small gastropods and bivalves mixed with abundant carbonized leaves and plant remains.

Zone E. Coarse conglomeratic sand containing some heavy minerals, such as kyanite, and large pieces of petrified wood (trunk sections and limbs).

Zone F. Fine sandy clay to clay. Typical fauna present are continental-shelf bivalves and gastropods.

Questions

a. There is an unconformity between the Clayton Formation and the Nanafalia Formation. What evidence can you cite to support this fact?

b. What environment and water depth could be envisioned for zone B?

c. Give a brief summary of the depositional history of the outcrop from zone A to zone F using the faunal and sedimentary evidence available.

Exercise 4-14 FOSSILS AND RADIOMETRIC DATING

Refer to Figure 4.38 for this exercise on radiometric dating.

a. Using the decay curve in Figure 2.26 on page 81, determine the ages of the two igneous intrusions.

b. During what geologic period did mystery fossil Z probably live? ________________

c. Is intrusion Q older or younger than unit B? ________________________________

d. Stratigraphic unit C was deposited during the ______________________ period.

e. Note the contact metamorphism associated with igneous intrusion R. Using this evidence, is unit D younger than, older than, or the same age as intrusion R?

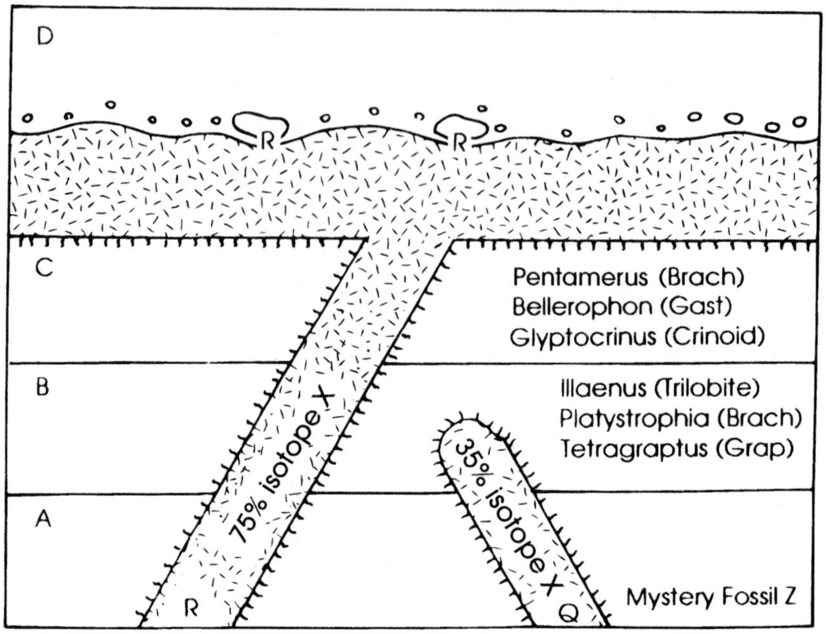

FIGURE 4.38 Fossils and radiometric dating.

f. What is the specific nature of the contact between unit R and unit C?

g. Give a brief narrative summary of the geologic history represented by this geologic cross section.

Exercise 4-15 QUESTIONS

1. List the three requirements that must be met before a fossil can be classified as an index fossil.

2. What are two major prerequisites that must be met for most organisms to be fossilized?

3. Are both the genus and species name of a fossil capitalized? Give an example.

4. Why are there so few Precambrian fossils?

5. Would the open prairie country of the midwestern United States be a better place to preserve a vertebrate animal than a lake? Explain.

6. Would you expect better preservation of very thin-shelled bivalves in an oceanfront environment or in a sheltered lagoon? Explain.

7. Describe a natural environment near your home or college where fossil preservation may be taking place. What does this environment provide to enhance preservation?

8. What characteristics might distinguish a radiolarian from a foraminiferan?

9. Spicules are a means of structural support for the skeleton of a ________________.

10. Sponges are sessile benthic organisms; therefore, they live in trees close to swamps. True or false; if false, explain why.

11. How would you distinguish among rugose, tabulate, and scleractinian corals?

12. What can corals tell us about past climates and environments?

13. Corals are not found growing at the mouth of the Mississippi River, even though the water is warm and ample nutrients are available. Why are the corals absent from this environment?

14. How would you distinguish a bryozoan from a colonial coral?

15. If a fossil brachiopod were found to be composed entirely of pyrite, would you suspect that replacement had taken place? Why?

16. What distinguishes the coiling in many cephalopods from that of the gastropods?

17. Ostracods, brachiopods, and clams all have two valves. How would you differentiate each of these groups?

18. Trilobites are known to have molted their exoskeletons. How would this affect their preservation potential?

19. Why are so many subdivisions of the arthropods poorly represented in the fossil record?

20. Describe the symmetry of the phylum Echinodermata.

21. What is an endoskeleton, and what is an exoskeleton? Which of the following groups have endoskeletons and which have exoskeletons? brachiopods, bryozoans, belemnites, ammonoids, corals, crinoids, trilobites, gastropods, sponges, vertebrates

Exercise 4-16 FOSSILS AS PALEOENVIRONMENTAL INDICATORS

In a modern marine environment, the distribution of certain species of animals or plants is dependent upon environmental variables mentioned in this chapter (light, salinity, temperature, bottom sediments, and water motion). Different depositional environments are characterized by distinctive faunal assemblages. Using the principle of uniformitarianism, one should be able to interpret ancient depositional environments by comparing the fossil assemblages with their modern analogs.

Data 1

The following data are given:

1. A cross section of a modern environment.

2. A partial list of the common species occurring within the environment (Table 4.2).

3. From the partial list the following species were chosen to represent each subdivision because they were environmentally controlled and maintained a high abundance:

Subdivision I	(A) *Littorina irrorata*	(B) *Crassostrea virginica*
Subdivision II	(C) *Callianassa* sp.	(D) *Cyrtopleura costata*
Subdivision III	(E) *Moira atropos*	

4. The percentage distribution of these key species (A, B, C, D, E) within the modern environment is illustrated in Figure 4.39. These data are also plotted as frequency distribution diagrams in Figure 4.39.

PERCENT OF KEY SPECIES IN EACH SUBDIVISION OF THE MODERN ENVIRONMENT.

	ENVIRONMENT SUBDIVISIONS		
SPECIES	I	II	III
A	50	5	0
B	38	5	0
C	7	40	0
D	5	50	10
E	0	0	90

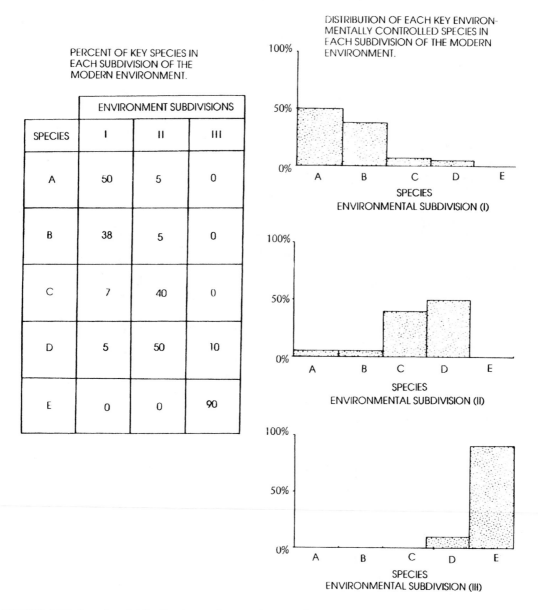

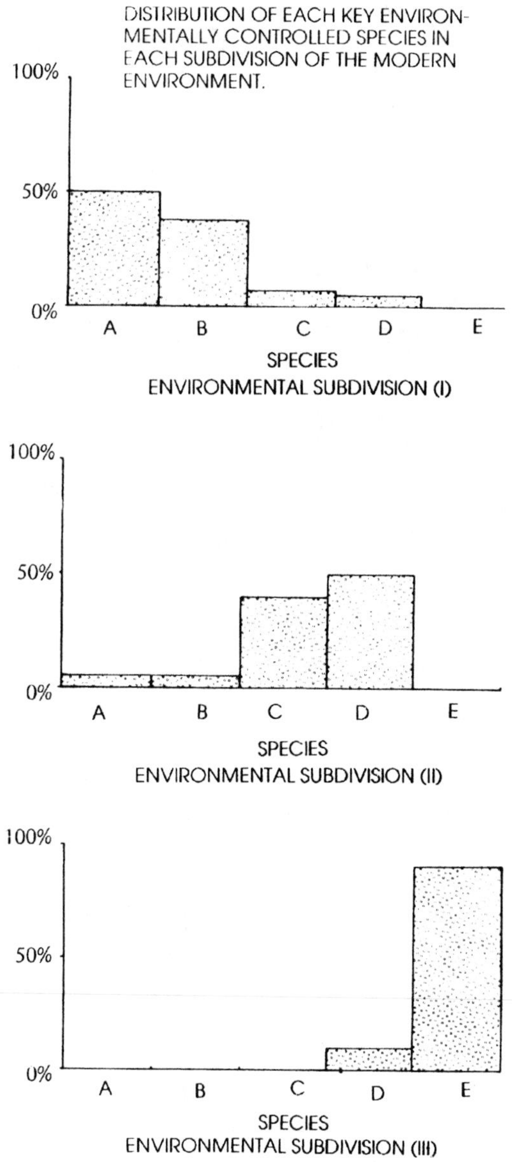

FIGURE 4.39 Species distribution in modern environments.

TABLE 4.2 Partial List of the Biologic Components in the Modern Environment

Component	*Subdivision occurrence*
WORMS	
Diopatra sp.	I
Pectinaria sp.	I
Spiochaetopterus sp.	I
CRUSTACEANS	
Ocypode sp. (ghost crab)	II
Uca pugilator (fiddler crab)	I, II
Sesarma reticulatum (marsh crab)	I
Eurytium limosum (mud crab)	I
Callianassa atlantica (shrimp)	I, II
Callianassa major (shrimp)	I, II
Squilla sp. (mantis shrimp)	I
ECHINODERMS	
Amphipolis sp. (brittle star)	II, III
Mellita quinquisperforata (keyhole urchin)	II, III
Moira atropos (heart urchin)	III
Lytechinus variegatus (urchin)	II, III
Clypeaster subdepressus (sand dollar)	II
MOLLUSKS (BIVALVES)	
Anadara ovalis (blood ark)	II, III
Donax variabilis (Florida coquina)	II
Ensis directus (Atlantic jackknife clam)	I, II, III
Crassostrea virginica (common oyster)	I
Cyrtopleura costata (angel wing)	II
Tagelus plebius (stout tagelus)	II, III
Tellina alternata (alternate tellin)	II, III
Mercenaria mercenaria (northern quahog)	I, II
Dinocardium robustum (great Atlantic cockle)	II, III
Solen viridis (green jackknife clam)	II, III
Atrina serrata (sawtooth pen shell)	II, III
Petricola phaladiformis (false angel wing)	II
MOLLUSKS (GASTROPODS)	
Polinices sp. (sand-collar snail)	II
Busycon caricum sliceans (southern knobbed whelk)	I, II
Busycon canaliculatum (channeled whelk)	I, II
Littorina irrorata (marsh periwinkle)	I
Oliva sayana (lettered olive)	II
Terebra discolorata (Atlantic auger)	II, III
GRASSES	
Spartina alterniflora	I
Spartina patens	I
Salicornia sp.	I
Juncus sp.	I
Uniola paniculata (sea oats)	

Data 2

A stratigraphic section of Cretaceous age is composed of six distinctive fossiliferous formations. After a detailed study, six fossils (P, Q, W, X, Y, Z) were found to be uniquely distributed within this stratigraphic section. The results of a laboratory analysis are presented in Figure 4.40.

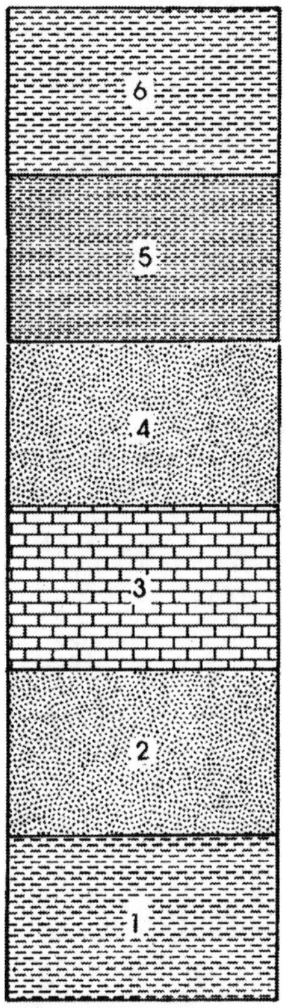

STRATIGRAPHIC SECTION OF FOSSILIFEROUS SEDIMENTARY ROCKS OF CRETACEOUS AGE.

DISTRUIBUTION OF KEY SPECIES IN EACH FORMATION (IN PERCENT).

	ENVIRONMENTAL INDICATOR FOSSILS					
Fomations	P	Q	W	X	Y	Z
6	50	40	3	6	1	0
5	30	25	10	20	15	0
4	5	10	40	30	5	10
3	0	0	20	20	40	20
2	25	25	10	20	20	0
1	60	30	10	0	0	0

FIGURE 4.40 Distribution of Cretaceous species.

Instructions and Questions:

1. Plot the frequency distribution of all species in each of the six formations using the following groupings: fossils P + Q, fossils W + X, and fossils Y + Z. (Use the blank frequency diagrams provided in Figure 4.41.)

2. Assume that the environmental subdivisions in the stratigraphic section can be directly compared to the subdivisions in the modern example; then answer the following questions.

a. Which rock layers (1–6) belong to the offshore, barrier island, and lagoon/marsh facies? Cite evidence.

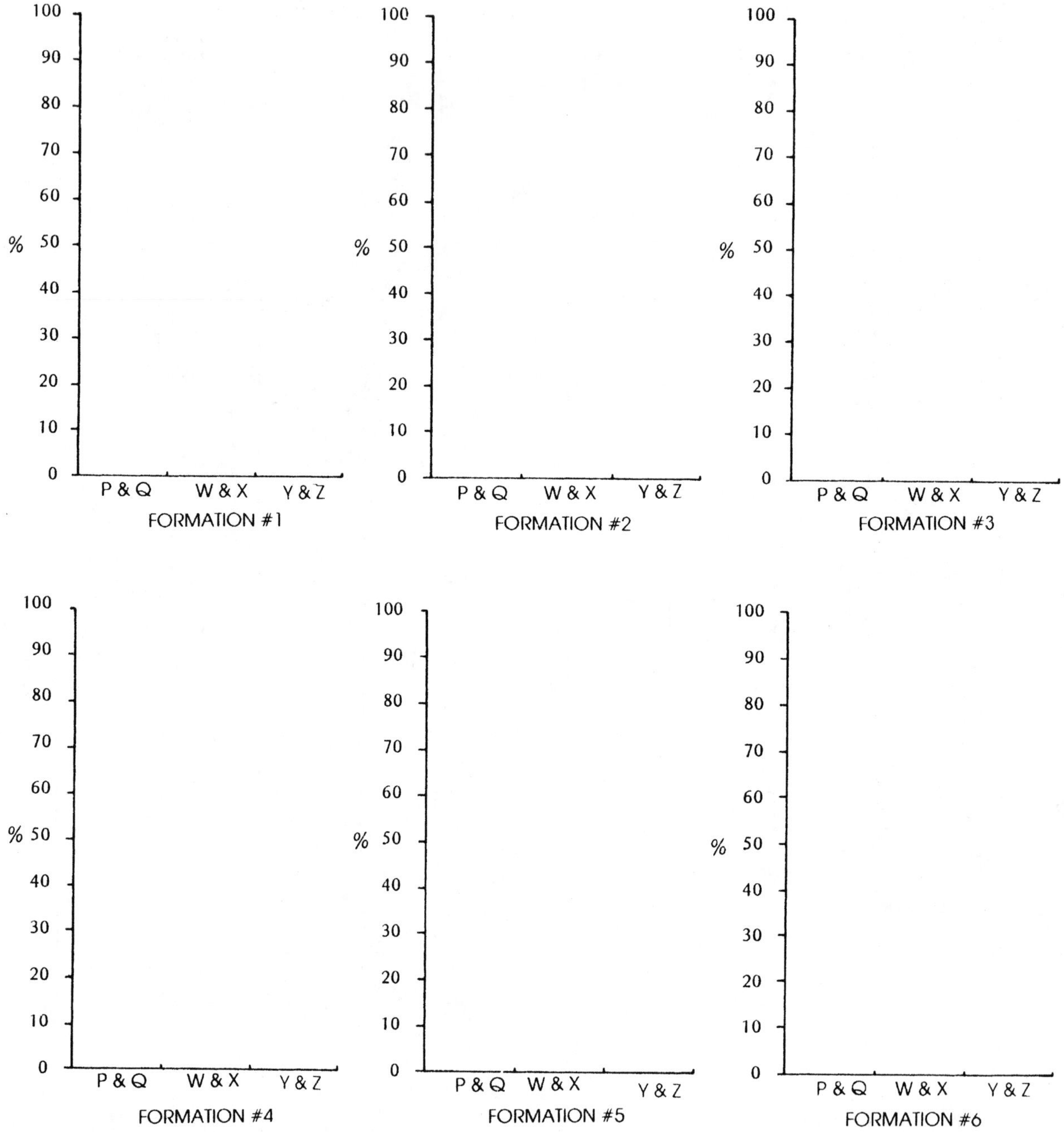

FIGURE 4.41 Frequency diagrams.

b. Which set of fossils—(P + Q), (W + X), or (Y + Z)—goes with each sedimentary environment?

c. Considering your answers to the first two questions, which of the following events would you say had taken place in the region represented by the stratigraphic section: a transgression, a regression, both, or neither? Discuss your answer.

d. Marsh deposits will often have a distinctive dark grey or black color. What causes this color, and how is it related to the presence of marsh grasses such as *Spartina* and *Salicornia*? What does this suggest about the completeness of the fossil record?

e. The small shrimp *Callianassa* lives primarily in the relatively clean sand portions of the beach face within the intertidal zone. Could *Callianassa* burrows be of value in interpreting sea level change?

f. What might the recognizable effects in the geologic and biologic record be if this modern environment were hit by a hurricane?

CHAPTER

5

Geologic Map Interpretation

In your study of geology you have learned a great deal about the rock cycle, surface processes and topography, structural geology, and interpretation of earth history. Now you are equipped to study and interpret geologic maps.

Geologic maps are very useful tools for interpretation of earth history. Geologic maps differ from paleogeographic and lithofacies maps in one fundamental aspect: whereas paleogeographic maps are reconstructions of the depositional environments or the geography present within a region during a specific interval of the geologic past, geologic maps show the distribution of geologic strata of all ages presently exposed on the earth's surface. Because of modern erosion, the earth's surface is being truncated, which exposes many existing older structures produced by past tectonic activity. Geologic maps are very useful in delineating the existence of exposed structural features such as folds and faults, the location of horizontal strata, and the geographic distribution of igneous and metamorphic terrains.

The basic geologic map is most often compiled by a geologist doing detailed field mapping of a local area and using a topographic map as a base. Modern photogrammetric and satellite imagery is now used to map much of the earth's surface, but even these maps are still subject to ground verification by geologists. Most detailed geologic maps, often quadrangle maps, are ultimately compiled into regional continental maps. Large-area geologic maps are useful for interpreting the extent and configuration of very large structural provinces, extensive fault systems, mountain ranges, and distinctive areas of sedimentary accumulations.

The present surface of the North American continent, and all of its various landforms, is a direct function of the distribution of different rock types, the structural configuration of these rocks, and the ongoing modification of these rocks and structures by the major weathering processes. The current geological surface of the United States has been divided into a number of physiographic provinces (see Fig. 5.1). Each physiographic province has within it a distinctive set of landforms, drainage systems, rock types, geologic structures, and numerous other geologic variables that set it apart from surrounding provinces. The uniformity and similarities of the overall geology within a physiographic province suggest that processes of geologic change have affected the area as a unit and that all parts of the region have a similar geologic history.

The present landforms of the United States are relatively young; most were developed during the Late Cenozoic. Pleistocene alpine glaciation modified the shape of the mountain ranges of the western United States, just as Pleistocene continental glaciation was a major factor in shaping landforms in the north-central Great Lakes portion of the United States and Canada. The major river systems and extensive Great Plains also attained their present configurations during the Late Cenozoic.

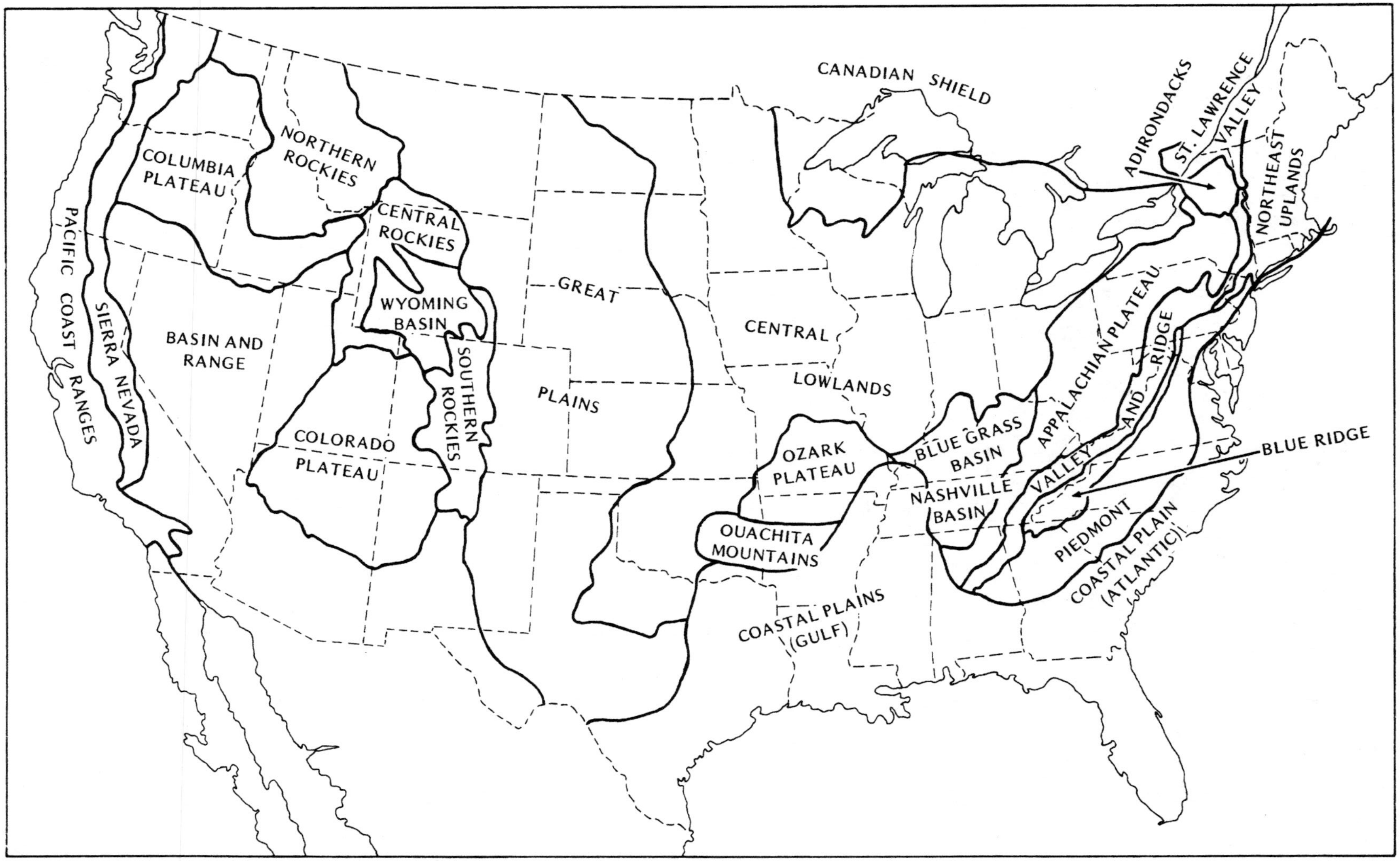

FIGURE 5.1 Physiographic provinces of the United States. From Nevin M. Fenneman, *Physical Divisions of the United States* (USGS, 1964).

GEOLOGIC MAP INTERPRETATION

The following paragraphs will present an overview of the basic methods and symbols required to understand how to "read" a geologic map.

In the geological profession, geologic maps usually have a standard set of reference symbols, patterns, or colors and a scale to guide one's interpretation of the map. The geologic symbols primarily denote surface and subsurface structural associations and configurations of the geologic units. The structural symbols depicted include strike and dip, faults, axes of folds, and others (see Fig. 5.2).

The U.S. Geological Survey has also adopted a set of standard colors and symbols to signify different periods of geologic time. The specific colors and symbols that are used on a particular map will be presented in the map's explanation, or legend. (For examples of these standard colors and patterns, see Table 5.1.) These standardized colors and patterns are useful in compiling and interpreting regional geologic maps. Any geologist with a trained eye can "read" the geologic history of an area by interpreting the distribution of colors and patterns on a map of the area.

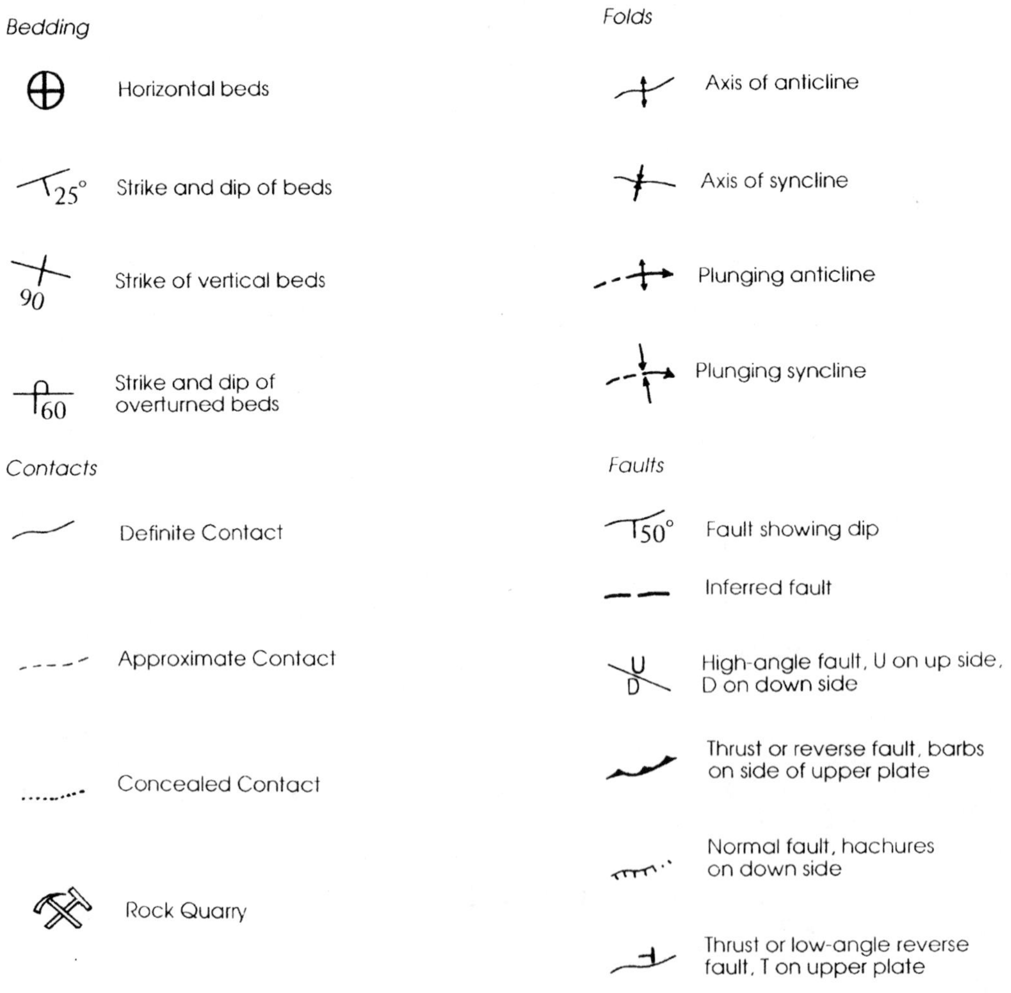

FIGURE 5.2 Structural symbols for geologic maps.

TABLE 5.1 USGS Standard Colors and Patterns

Geologic system	*Map symbol*	*Color*
Quaternary	Q	Brownish yellow
Tertiary	T	Yellow
Cretaceous	K	Green
Jurassic	J	Blue green
Triassic	₮	Peacock blue
Permian	P	Blue
Pennsylvanian	IP	Blue
Mississippian	M	Blue
Devonian	D	Blue gray
Silurian	S	Blue purple
Ordovician	O	Blue purple
Cambrian	Є	Brick red
Precambrian	p-Є	Brownish red

The smallest rock-stratigraphic unit normally included on a detailed geologic map is the formation and its members (if so subdivided). A formation is usually represented on a geologic map by a specific pattern or color. The width of the color pattern will vary on a geologic map due to several controlling factors, for example:

- the dip of the rock unit as it intersects the surface [see Fig. 5.3(a)].
- the true thickness of the stratigraphic unit [see Fig. 5.3(b)].
- the slope of the land surface as it cuts across the surface outcrop.

Figures 5.3(a) and (b) illustrate a few of the many possible combinations of dip, formation thickness, and outcrop patterns.

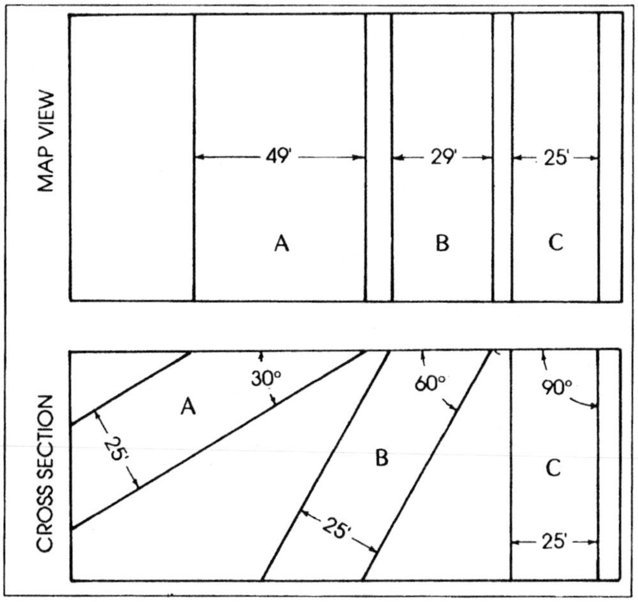

FIGURE 5.3(a) Outcrop thickness variations. Varying outcrop width produced by a stratigraphic unit of the same thickness but of varying degrees of dip.

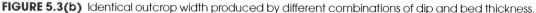

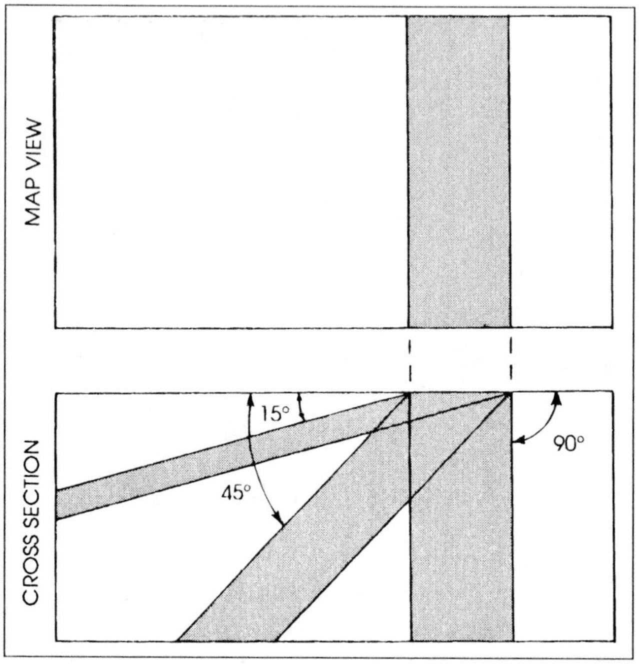

FIGURE 5.3(b) Identical outcrop width produced by different combinations of dip and bed thickness.

Geologic Map Information

The first capitalized letter of a geologic map symbol stands for the geologic system (Table 5.1). The second portion of the symbol usually designates the name of a specific mapped unit (such as a group or formation) and is shown in lowercase letters. For example, the Stark Formation of the Mississippian system would be symbolized as Ms. Occasionally a formation will span parts of two systems. For example, the Ellenberger limestone is of Upper Cambrian–Lower Ordovician age. On a geologic map, it would be designated O-Ꞓe. Since the Tertiary is subdivided into epochs, the symbols commonly begin with a "T" for the Tertiary period; the epoch is put into lowercase letters along with the formation name.

Complete the map symbols for the following geologic formations:

Formation name	System/Epoch	Symbol
Fort Wayne	Mississippian	________
Winchell	Pennsylvanian	________
Conasauga	Cambrian	________
Big Horn	Ordovician	________
Austin	Cretaceous	________
Vishnu Schist	Precambrian	________
Red Mountain	Silurian	________
Chinle	Triassic	________
Town Mountain	Precambrian	________
Claiborne	Tertiary/Eocene	________
Yorktown	Tertiary/Pliocene	________

EXERCISES

Exercise 5-1 ASYMMETRICAL FOLD

Use the given dip and strike to determine the subsurface configuration of an asymmetrical fold (see Fig. 5.4).

a. Draw in the cross-sectional view of this fold.

b. Draw in the axial plane of the fold on the cross section of the fold.

c. Draw the position of the axis on the map view.

d. What type of fold is drawn?

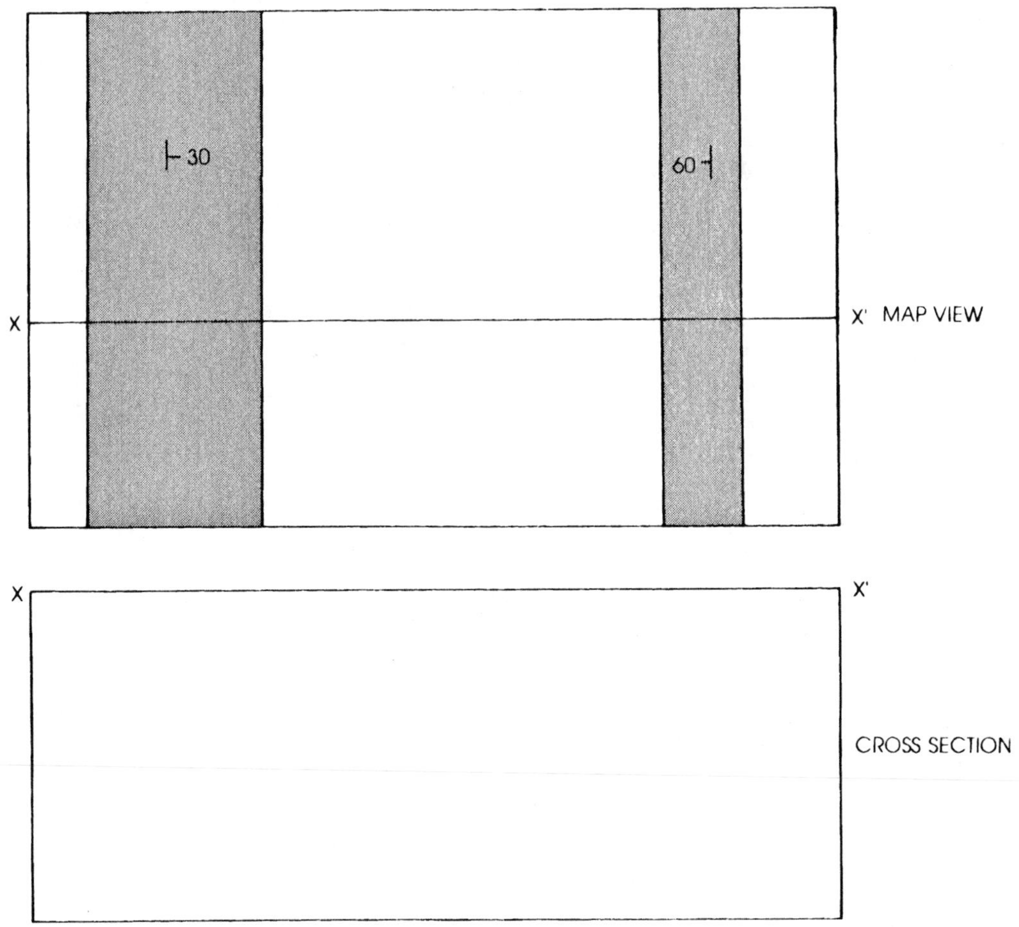

FIGURE 5.4 Diagram for Exercise 5-1.

Exercise 5-2 DETERMINING STRIKE AND DIP

Determine strike and dip (see Fig. 5.5).

a. The lower fold involves at least four stratigraphic units. Draw the dip angles on the structure section.

1. What is the angle of dip for the top of the limestone on the east side?

2. What is the structure?

b. The upper figure is a map view of the beds in the structural section.

1. What is the direction of strike?

2. Place the strike and dip symbols within the sedimentary layers on both the east and west sides of the Map View figure.

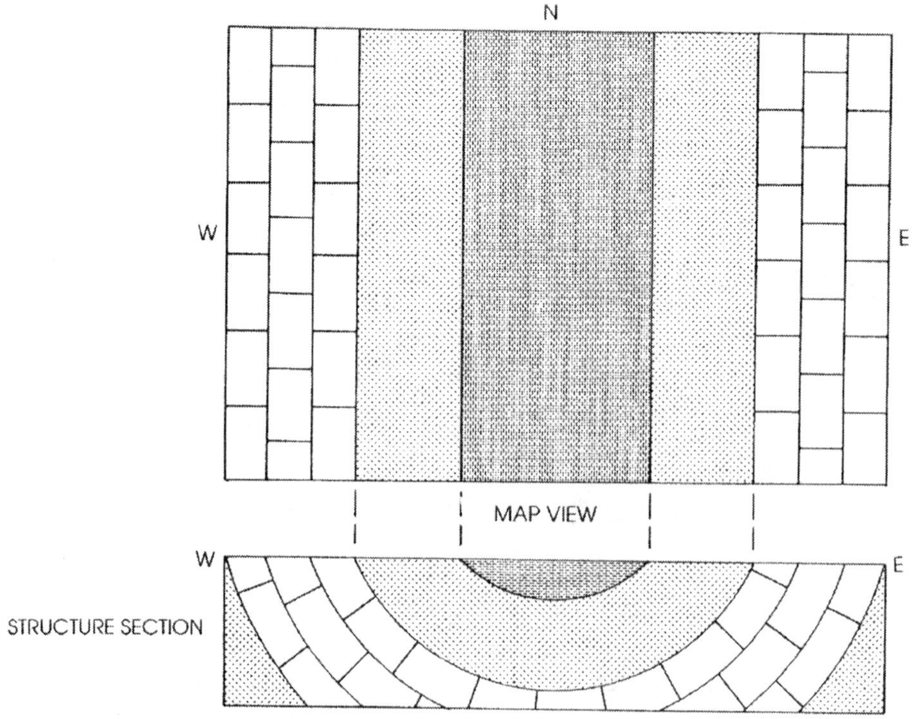

FIGURE 5.5 Diagram for Exercise 5-2.

Exercise 5-3 STRUCTURAL CROSS SECTION

Complete the structural cross section from point A to point D in Figure 5.6 and answer the following questions.

a. The three major structural features on the geologic map are ______________________,

______________________, and ______________________.

b. Where the fault is marked, what kind of fault is present? Cite your evidence.

c. Both of the central structural configurations are plunging folds. The strike of the axis is approximately ______________________. (Assume both are the same.)

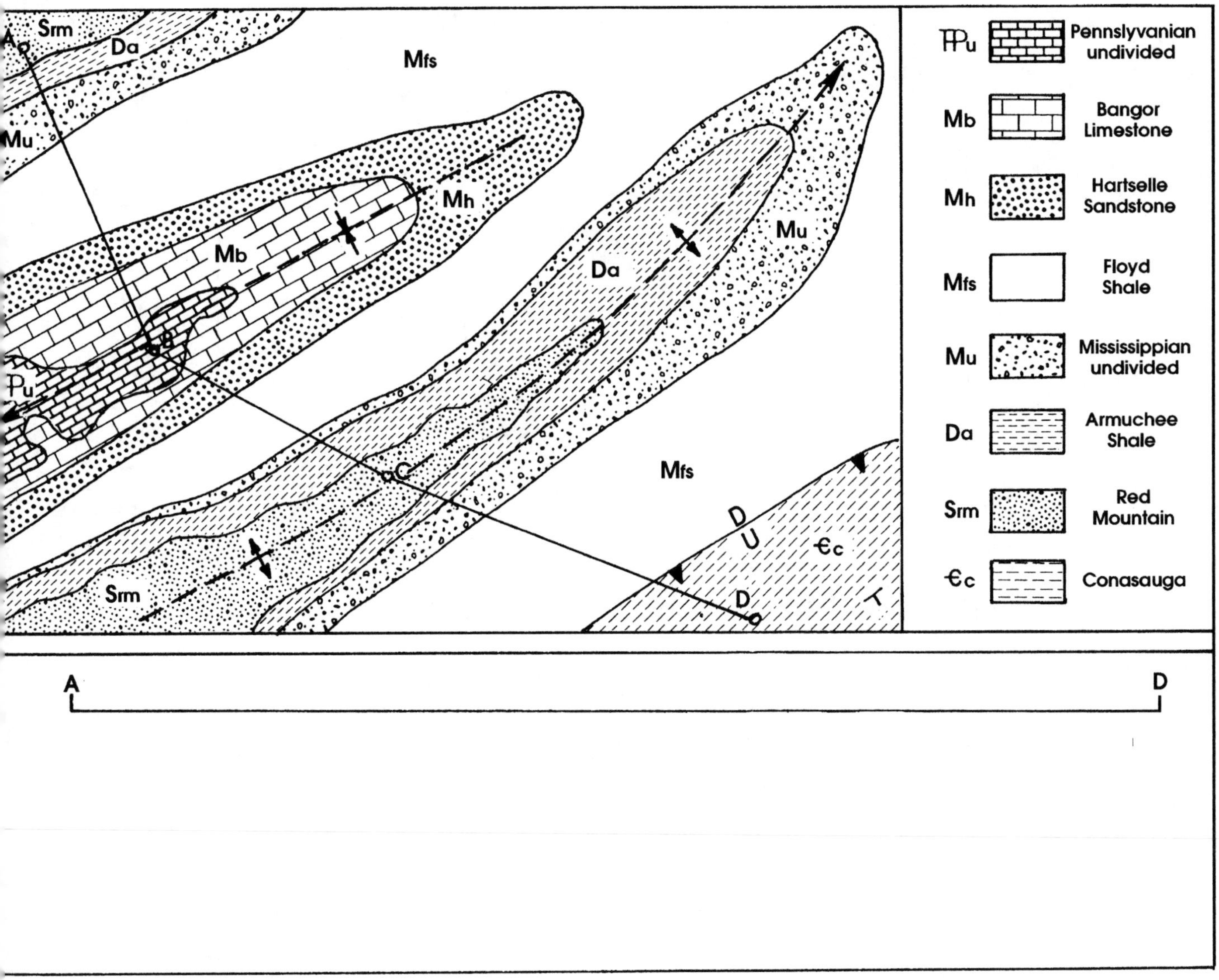

URE 5.6 Structural cross section, for Exercise 5–3.

Exercise 5-4 ARBUCKLE MOUNTAINS, SOUTH CENTRAL OKLAHOMA

Complete the following items:

a. Complete the explanation (legend) of the geologic map that covers a portion of the Arbuckle Mountains in south central Oklahoma (see Fig. 5.7).

b. Construct a structural cross section along line X-X′.

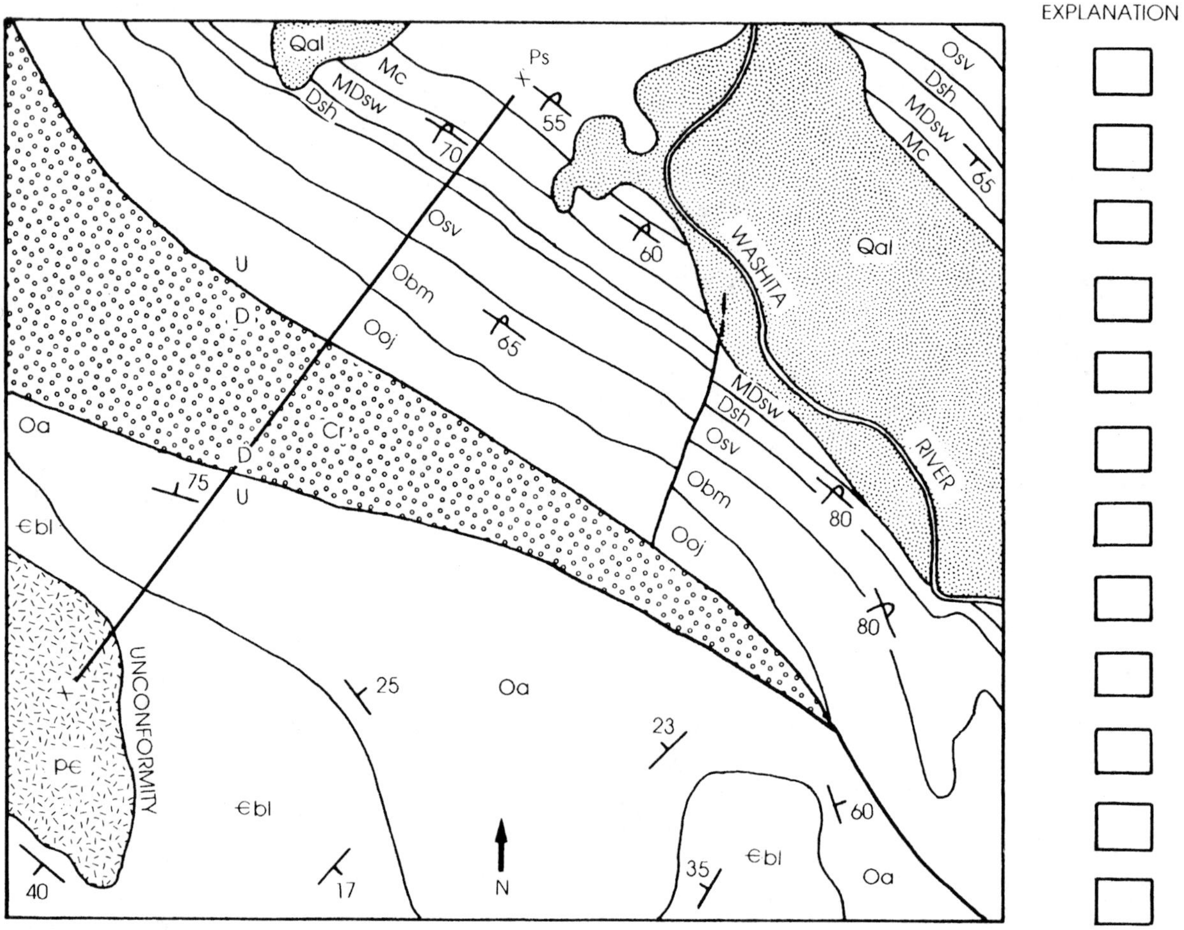

FIGURE 5.7 Geologic map of Arbuckle Mountains, for Exercise 5-4.

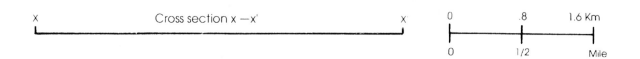

Exercise 5-5 PERRY, KANSAS QUADRANGLE

Construct a geologic map by using the Perry topographic map as a base. See box in Figure 5.8 for more information.

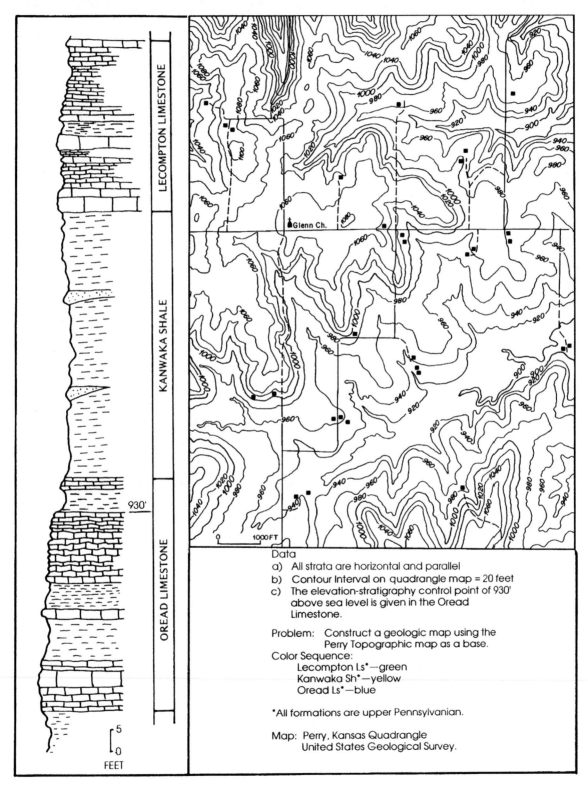

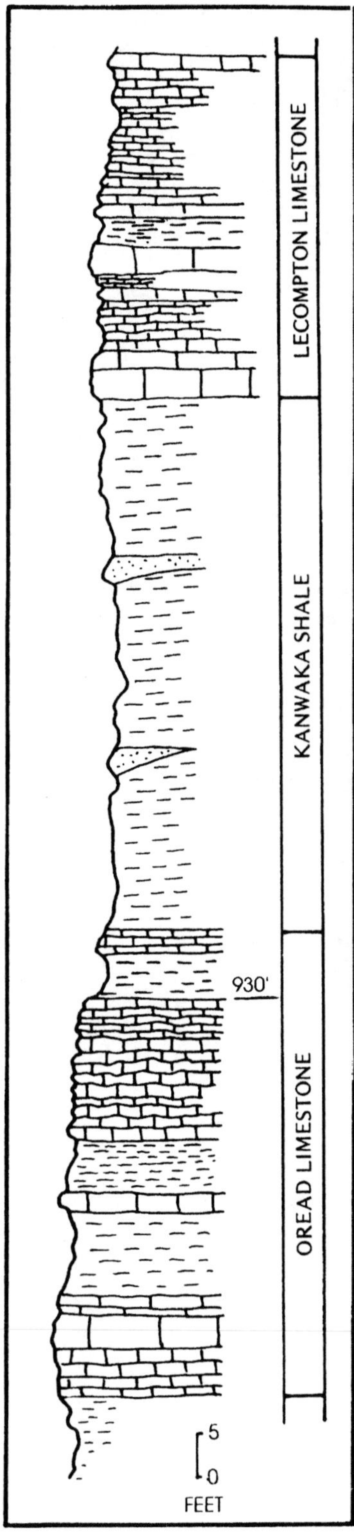

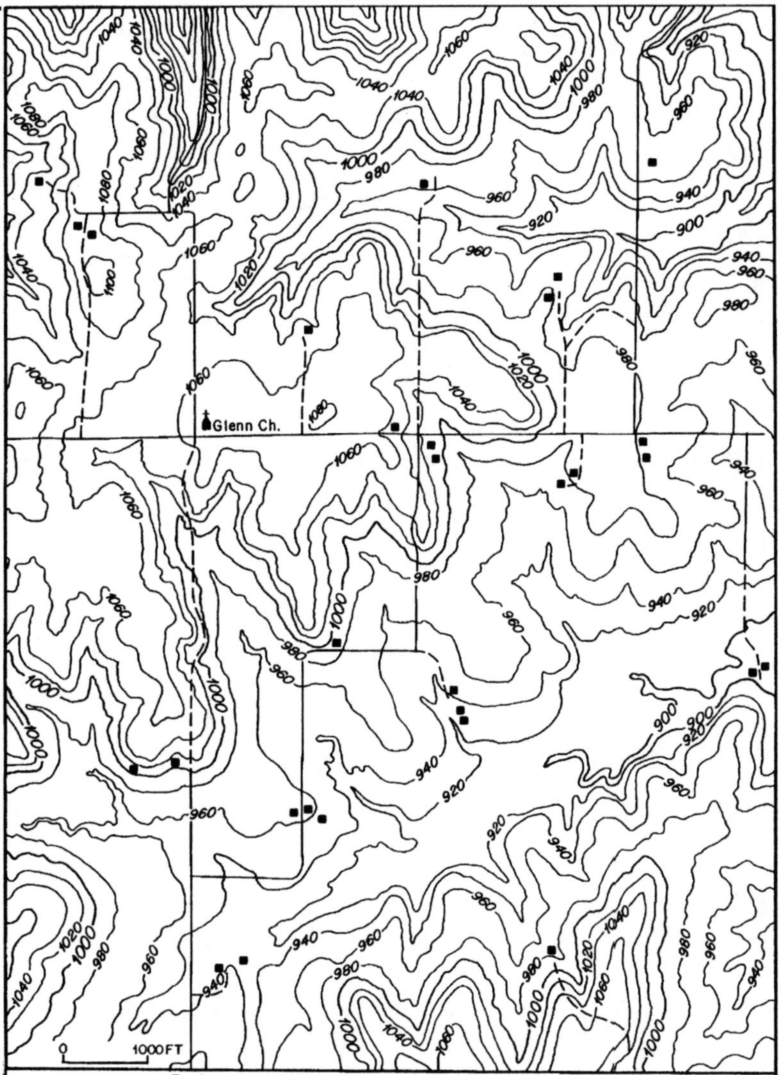

Data

a) All strata are horizontal and parallel
b) Contour Interval on quadrangle map = 20 feet
c) The elevation-stratigraphy control point of 930' above sea level is given in the Oread Limestone.

Problem: Construct a geologic map using the Perry Topographic map as a base.

Color Sequence:

Lecompton Ls*—green
Kanwaka Sh*—yellow
Oread Ls*—blue

*All formations are upper Pennsylvanian.

Map: Perry, Kansas Quadrangle
United States Geological Survey.

FIGURE 5.8 Perry, Kansas Quadrangle and stratigraphic column, for Exercise 5-5.

Exercise 5-6 HURON QUADRANGLE, SOUTH DAKOTA

Answer the following questions using Figure 5.9.

a. Which deposits are younger, the stream deposits or the glacial deposits? How did you determine this relationship?

b. Are these glacial deposits from alpine or continental glaciation?

c. What kind of moraine is the Antelope Moraine?

d. What kind of stream pattern is exhibited by the James River?

e. The floodplain alluvium (Qal) is more recent than the older stream deposits (Qod). The Qod forms flat, raised areas above the floodplain. What kind of stream features are these?

f. What do you think was the source of sediment for the Qal and Qod stream deposits before the river eroded, transported, and deposited them?

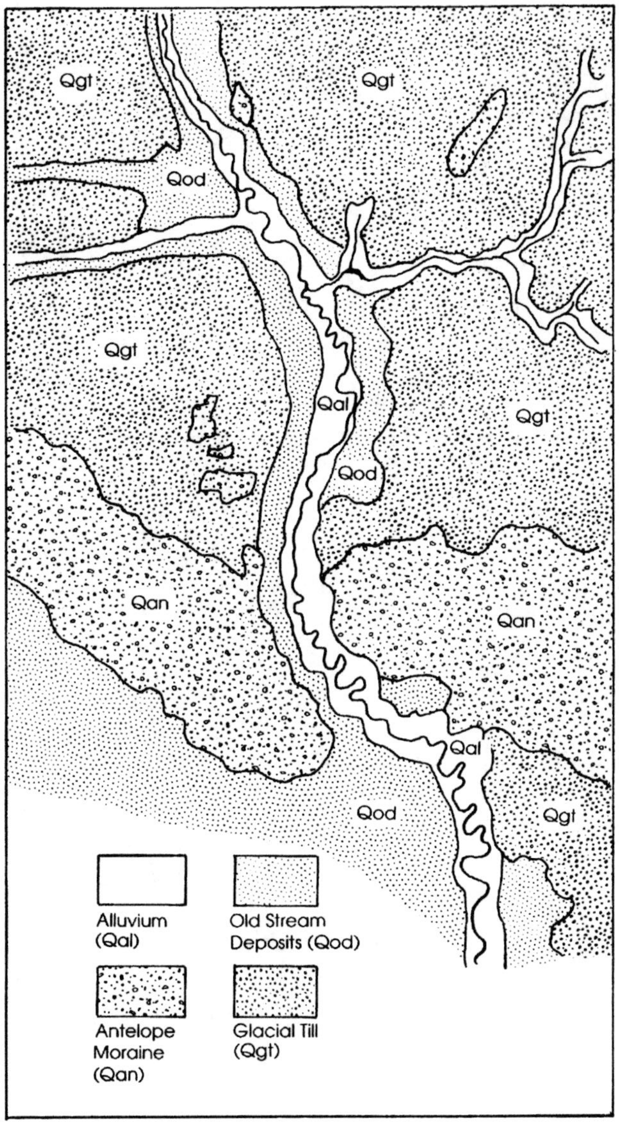

FIGURE 5.9 James River area, Huron Quadrangle, South Dakota, for Exercise 5-6.

Exercise 5-7 GLENN CREEK QUADRANGLE, MONTANA

Answer the following questions using Figure 5.10.

a. Which type of fault is the most common in the Glenn Creek area?

b. Are the faults that border Slategoat Mountain the same as or different from the majority of faults on the map?

c. Did the Glenn Fault occur before or after the fault between the Kk and Mm?

d. Look at the Renshaw Thrust. Give the name and geologic period of the formation on the upper block. Do the same for the formation on the lower block.

e. Based on the occurrence of folding and thrust faulting, what major force was acting on this area?

f. After the folding and thrust faulting events, what other type of force acted on this area?

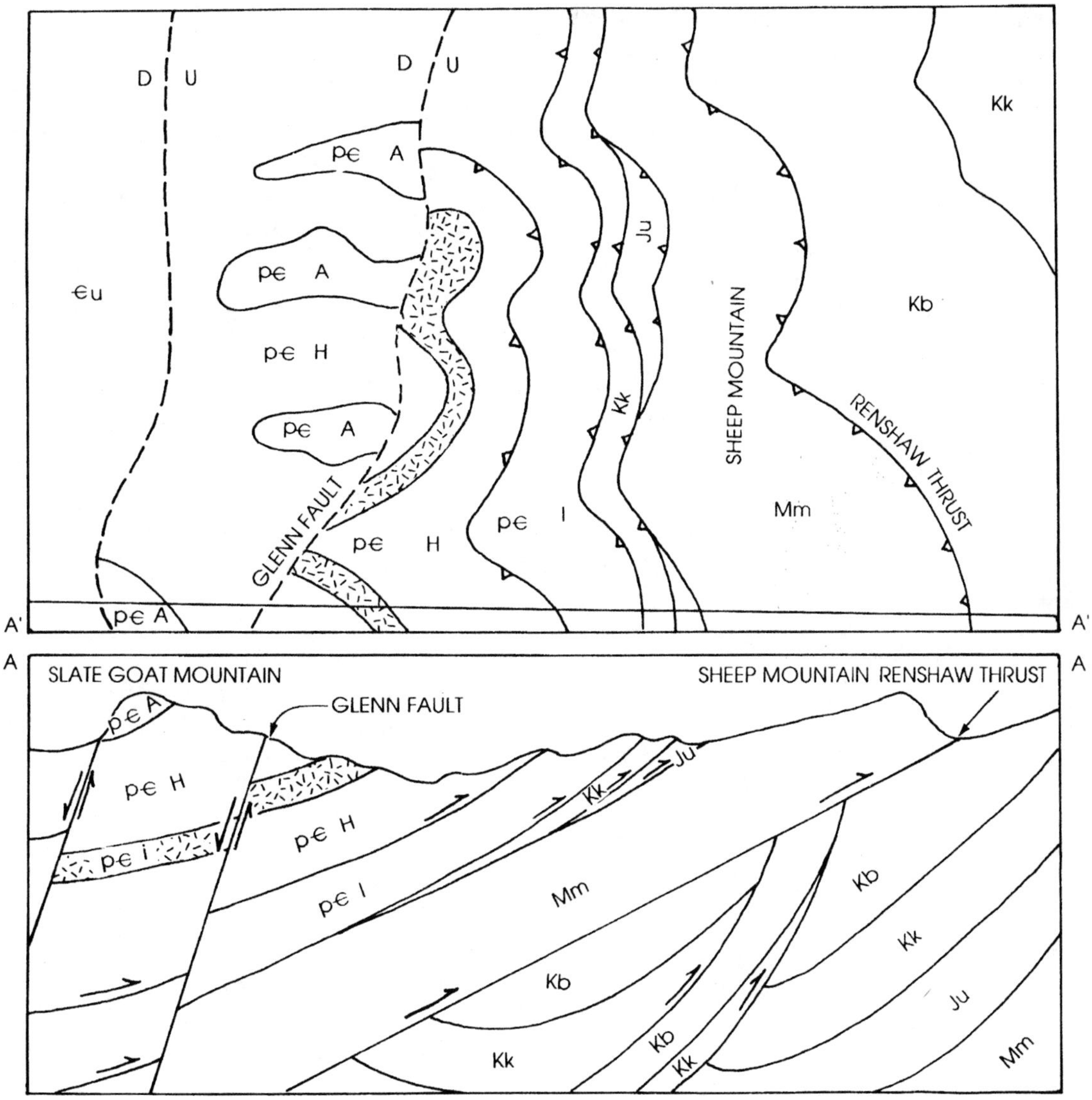

Kb Blackleaf Fm (Cret)
Kk Kootenai Fm (Cret)
Ju Jurassic undifferentiated
Mm Madison Group (Miss)
Єu Cambrian undifferentiated
pЄ i Precambrian intrusive
pЄ A Precambrian Ahorn Fm
pЄ H Precambrian Hadley Fm
pЄ I Lower Precambrian undifferentiated

FIGURE 5.10 Geological map and cross section of Glenn Creek Quadrangle, Montana, for Exercise 5–7.

GEOLOGIC MAPS AND CROSS SECTIONS

Exercise 5-8 GEOLOGIC MAPS AND CROSS SECTION EXERCISES

The color geologic maps have been chosen from various physiographic provinces in the United States and illustrate a variety of geologic structures, tectonic histories, rock types, and rock ages. Most are taken from either 15-minute or 7.5-minute quadrangles published by the U.S. Geologic Survey or by the various state geologic surveys.

Each geologic map set contains the following: (1) a geologic map showing rock age and type exposed at the surface, with colors and symbols keyed to (2) an explanation that contains the superpositional order, geologic period or epoch, and name of each rock formation shown on the map; and (3) cross sections (except for Map D) showing "slices" through several parts of each map. The cross sections show subsurface relationships and make the geology more visual and easier to interpret. In order to answer the questions with each map, you may need to examine any combination of the three parts of the map set.

PHYSIOGRAPHIC PROVINCES

For each of the maps on pages 230 through 243, determine the physiographic province by referring to Figure 5.1.

Map	Physiographic Province
A—Cross Mountain, California	Sierra Nevada
B—Devils Fence, Montana	____________
C—Williamsville, Virginia	____________
D—Stable Interior of U.S.	____________

GEOLOGIC MAPS AND CROSS SECTIONS, WITH QUESTIONS

A Map Set: Cross Mountain, California, pp. 228 to 230

B Map Set: Devils Fence, Montana, pp. 232 to 234

C Map Set: Williamsville, Virginia, pp. 236 to 237

D Map Set: Stable Interior of the United States, pp. 239 to 241

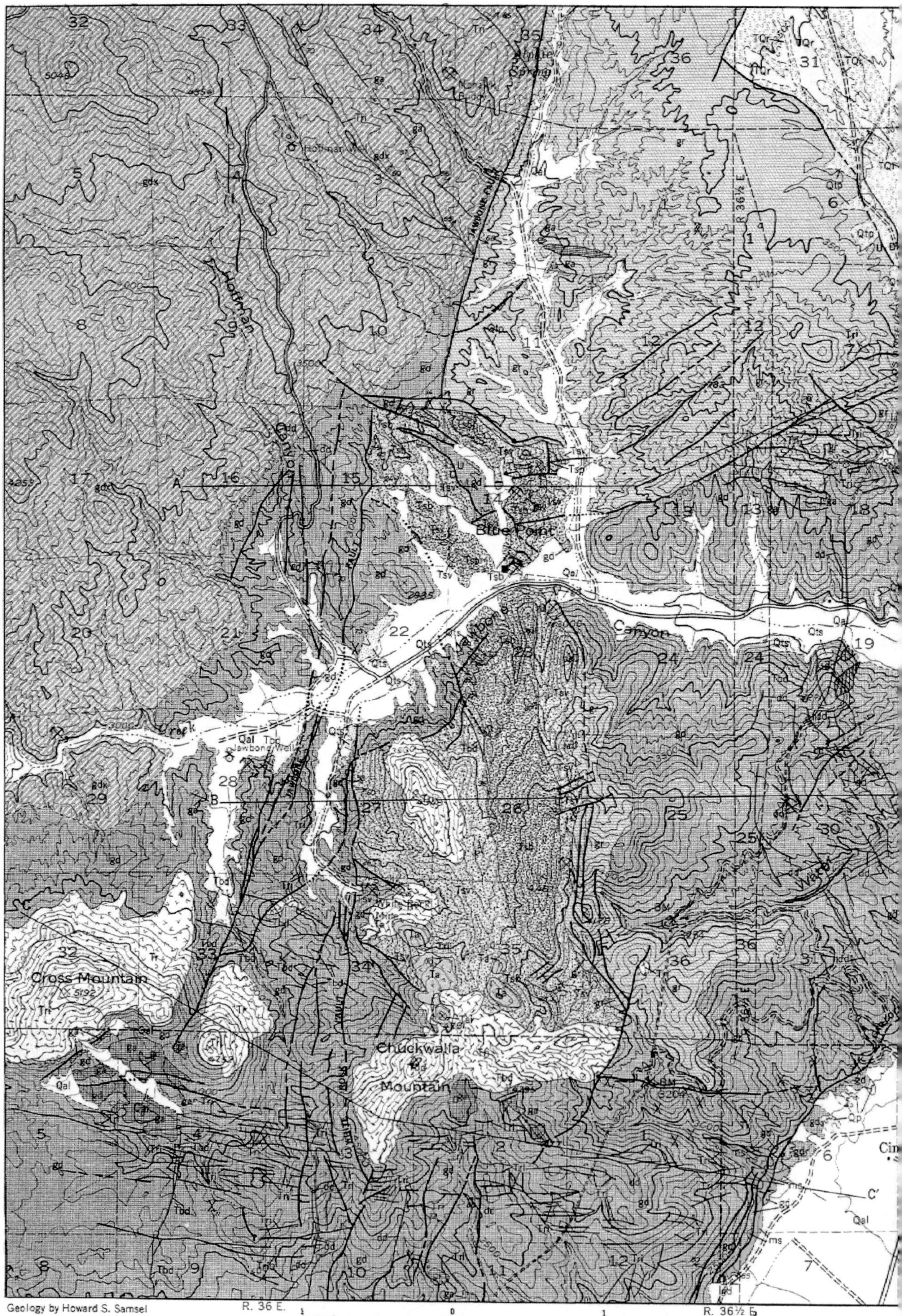

Geology by Howard S. Samsel

R. 36 E. R. 36½ E.

1 0 1

CONTOUR INTERVAL 100 FEET DATUM IS MEAN SEA LEVEL

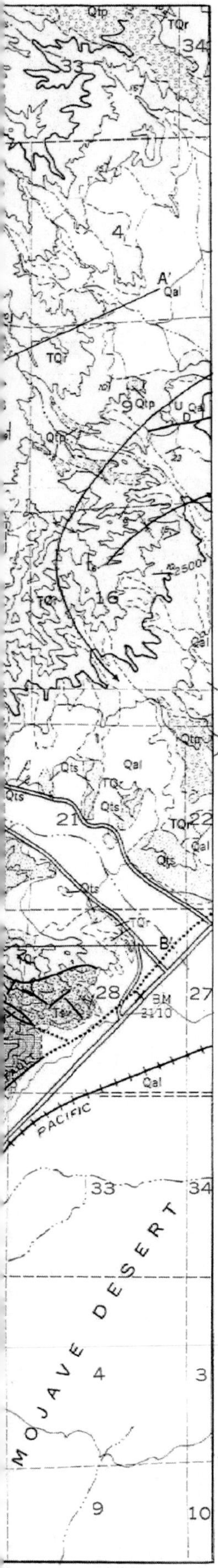

Base from U.S.G.S. 7½' quadrangles

EXPLANATION

QUATERNARY

Qal — Alluvium

Arkosic sand and gravel, volcanic debris, and fine-grained clastic and chemical playa sediments.

Qts — Stream terraces

Arkosic stream sands and gravels. Mostly remnants of an older alluvial cover.

Qtp — Pediment terraces

Arkosic sands and gravels; locally cemented with caliche.

Qp — Playa deposits

Buff clay, silt, and arkosic stream sands and gravel. Locally contain saline layers.

UNCONFORMITY

TERTIARY

TQr — Ricardo Formation

Reddish to buff, lenticular, calcareous, arkosic conglomerate and sandstone, clay and siltstone. (Members 6, 7, and 8 of Dibblee, 1952)

Ta / Tai — Hornblende andesite

Reddish to purple, prophyritic to aphanitic hornblende andesite. Ta = flows; Tai = intrusive

UNCONFORMITY

Tr / Tri — Rhyolite

Tr — *Buff, reddish to pale purple, massive, vesicular, porphyritic rhyolite in flows and welded tuff breccia.*

Tri — *White to buff, yellow and reddish-purple porphyritic to aphanitic dacite to rhyolite in plugs and dikes. Locally have glassy borders.*

Tbd — Basalt

Dark gray, porphyritic, zenolithic hornblende augite basalt dikes.

Tsv / Tsb — Sandstone, tuff and basalt flows

Tsv — *Red to buff arkosic sandstone and conglomerate; green to white tuff, silicified tuff and tuff breccia; green to pale purple rhyolite tuff breccia.*

Tsb — *Dark gray, hypocrystalline, trachytoidal, porphyritic, zenolithic, hornblende augite basalt flows.*

UNCONFORMITY

MESOZOIC

ap — Aplite

Fine- to medium-grained, equigranular granite aplite stocks and dikes.

gr — Granite

Coarse-grained, rusty equigranular biotite granite.

dd — Dacite

Pale gray, porphyritic dacite dikes.

gd / gdx — Granodiorite

gd — *Coarse- to medium-grained, equigranular to porphyritic biotite granodiorite.*

gdx — *Undivided biotite granodiorite and zenoliths of schistose and gneiseose rocks.*

PRE-MESOZOIC (?)

ms — Metasedimentary rocks

Undivided metasedimentary rocks. Consists of fine-grained, white to dark gray, thin-bedded limestone, dolomitic limestone, and fissile argillaceous layers; fine-grained, alternately layered, pale and dark gray quartz-sericite schist. Gray to white, banded, hornblende-andesine gneiss.

U — D

Fault, showing dip
Dashed where approximately located
U, *upthrown side;* D, *downthrown side*

Contact
Dashed where approximately located

Axis of anticline showing plunge

Axis of syncline showing plunge

Strike and dip of beds

Strike and dip of overturned beds

Prospect

Mine

Quarry

Mill

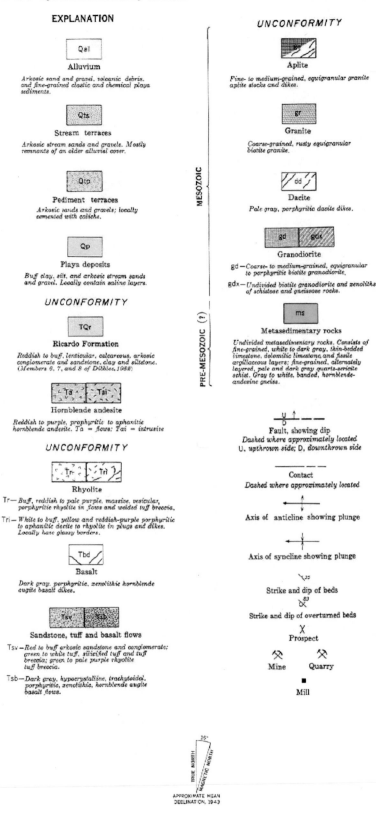

APPROXIMATE MEAN DECLINATION, 1943

STRUCTURE SECTIONS SOUTHEAST QUARTER OF THE CROSS MOUNTAIN QUADRANGLE, CALIF. MAP NO. 6

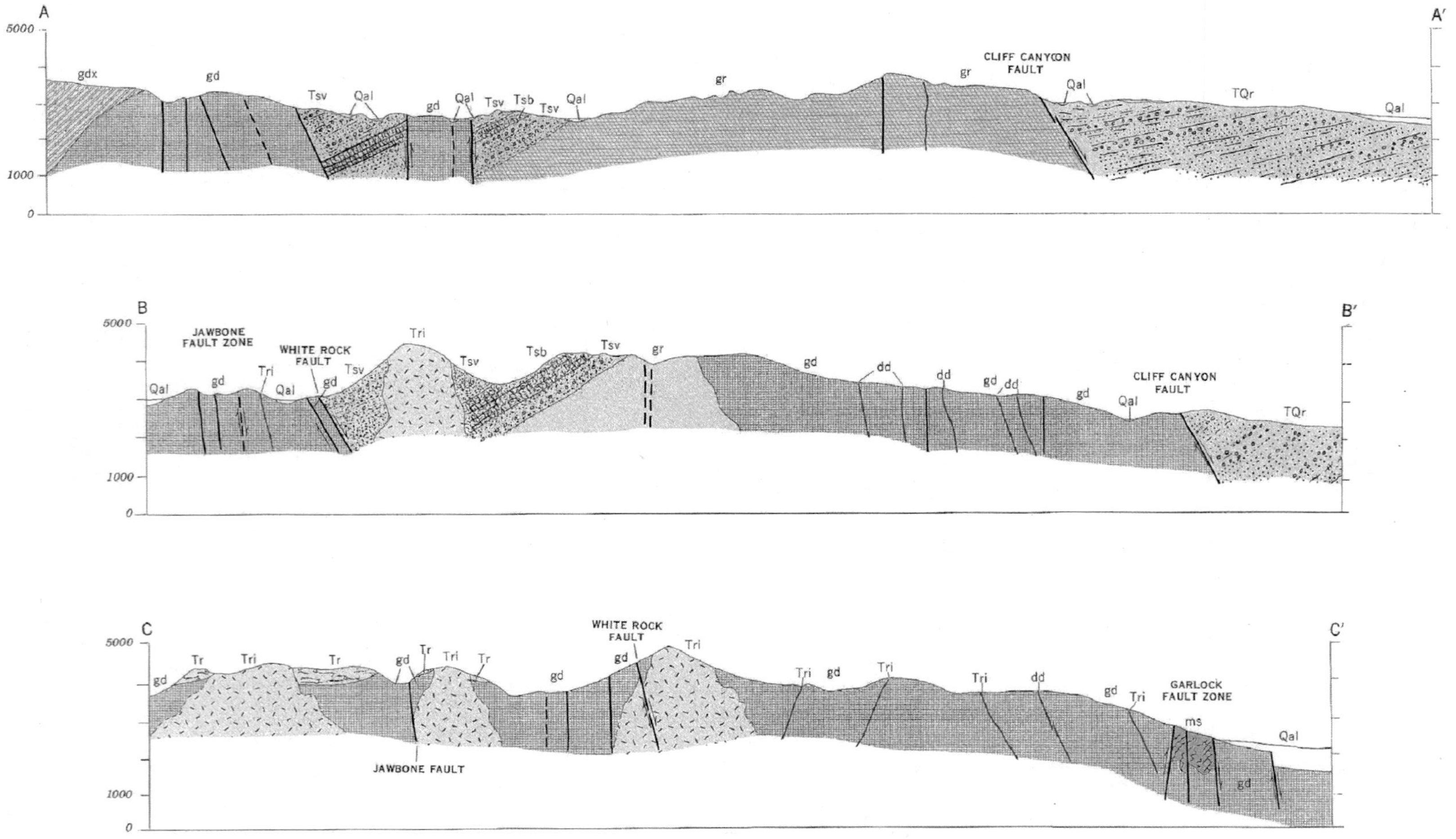

MAP A: SOUTHEAST QUARTER OF CROSS MOUNTAIN QUADRANGLE, CALIFORNIA

1. What is the most common general rock type in this area?

2. How does the dominant Mesozoic rock type differ from the dominant Tertiary rock type?

3. Examine the Cliff Canyon and White Rock faults. What type are they and what stress caused them?

4. Does faulting appear to have been active geologically recently? Cite your evidence.

5. On cross section B-B′ east of the White Rock Fault, explain the relationship of the following units: gr, Tsv, Tri.

6. What type of sediment is present in the TQr deposits? What is its origin?

7. When did dikes form in this area and what types of igneous rock intruded?

8. Which rock units appear to be alluvial fan and playa deposits from different ages?

9. Write a generalized geologic history of this area using the explanation and the cross sections as a guide.

From: U.S.G.S. PROFESSIONAL PAPER 292 PLATE 2

GEOLOGIC MA

Geology by M. R. Klepper, R. A. Weeks and E. T. Ruppel

Scale 1:31,250 1 0 2 Miles

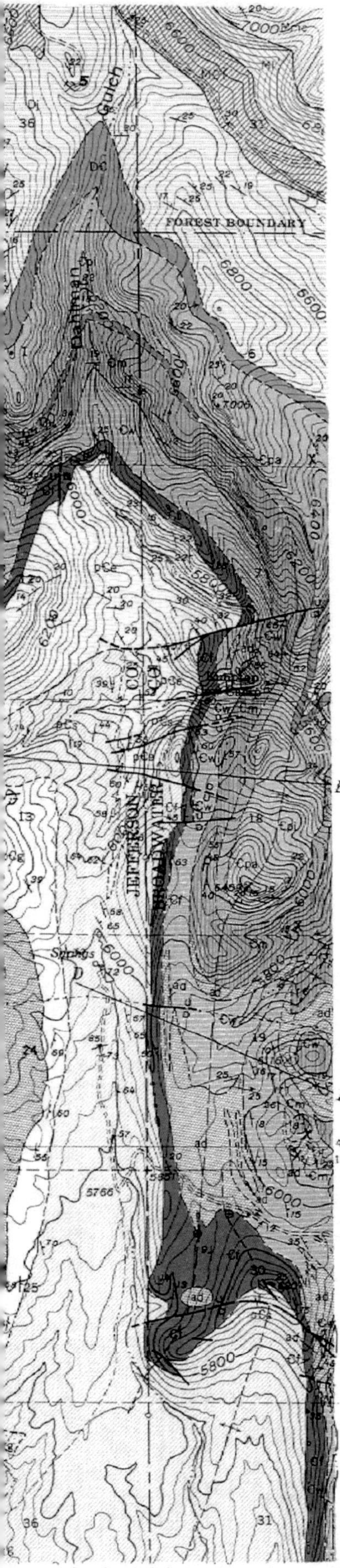

Base from U. S. Geological Survey from map of Devils Fence quadrangle, Montana.

E X P L A N A T I O N

QUATERNARY — Recent

Cal

Alluvium

TERTIARY — Pliocene (?)

Tfg

Fan gravel and pediment gravel

Unconsolidated or weakly consolidated fan deposits and gravel veneer on pediments

TERTIARY — Miocene (?)

Tt

Tuff and gravel

Light-colored rhyolitic tuff with a few interbeds of stream gravel

UNCONFORMITY

CRETACEOUS — Lower and Upper Cretaceous

Kcm / Kcl

Colorado formation

Upper black shale unit not mapped. Middle siliceous mudstone and sandstone unit, Kcm, is intertongued marine and nonmarine sandstone, mudstone, and siliceous mudstone, in part of volcanic origin. Lower black shale unit, Kcl, is a basal tan-weathering quartz sandstone, drab and olive-gray blocky siltstone, very dark gray to black shale, and an upper dark-gray carbonaceous, limonitic speckled to mottled sandstone

EROSIONAL UNCONFORMITY

CRETACEOUS — Lower Cretaceous

Kk

Kootenai formation

Comprises three units. Upper unit is 10 to 25 ft of gastropod-bearing limestone overlain by a few feet to 30 ft of drab mudstone; middle unit is red and green mudstone and shale with concretions and lentils of limestone; lower unit is crossbedded "pepper-and-salt" sandstone and interbedded shale and mudstone

JURASSIC — Upper Jurassic

Jm

Morrison and Swift formations

Morrison formation: variecolored nonmarine shale, mudstone, and siltstone, with thin beds of limestone and sandstone, and near the top a unit of black shale. Upper hundred feet locally contains thick lenticular "pepper-and-salt" sandstone and grades into overlying Kootenai formation

Swift formation: grayish-brown punky calcareous marine sandstone, 25 to 35 ft thick, with a basal chert-pebble conglomerate

EROSIONAL UNCONFORMITY

PERMIAN

Pp / PPq

Phosphoria formation

Brown and gray chert and sandstone, in part phosphatic; may locally contain one or two thin beds of phosphate rock, Pp. In places mapped with the Quadrant formation, PPq

CARBONIFEROUS — PENNSYLVANIAN

Pq

Quadrant formation

Light-colored quartzitic sandstone and interbedded light-gray sugary-textured sandy dolomite

CARBONIFEROUS — PENNSYLVANIAN and MISSISSIPPIAN

PMa

Amsden formation

Red to grayish-red mudstone, shale, and subordinate amounts of carbonate rock with interbeds of gray, brown, or yellow argillaceous sandstone in upper and lower parts; middle part of medium- to dark-gray thick-bedded dolomite

EROSIONAL UNCONFORMITY(?)

CARBONIFEROUS — MISSISSIPPIAN — Madison group

Mmc

Mission Canyon limestone

Medium-gray to light-gray medium-grained thickly and indistinctly bedded limestone, with a few thin siliceous layers in lower 200 ft and sparse gray chert nodules and lentils in upper half. A breccia unit, br, about 200 ft below top of formation has been mapped locally

Ml

Lodgepole limestone

Upper part of medium-gray fine- to medium-grained limestone in distinct beds as much as 5 ft thick alternating with zones of much thinner beds containing more mudstone partings; lower part of medium-gray limestone in beds 1 in. to 1 ft thick with partings and interbeds of yellow to red calcareous mudstone; grades into Mission Canyon limestone through a 150- to 250-ft zone

DEVONIAN AND MISSISSIPPIAN — Upper Devonian and Mississippian

MDt

Three Forks shale

Predominantly greenish-gray and brown shale with subordinate amounts of interbedded sandstone and limestone. Dolomitic siltstone at top. Locally a 20- to 25-ft fossiliferous limestone unit, l, has been mapped

DEVONIAN — Upper Devonian

Dj

Jefferson dolomite

Dark-gray granular-weathering fetid well-bedded dolomite with subordinate amounts of dark-gray limestone and light-gray dolomite

Upper Cambrian and Upper Devonian

EROSIONAL - - - - DC - - - - UNCONFORMITY

Maywood and Red Lion formations undifferentiated

Varicolored, generally in shades of red and yellowish-brown, argillaceous, dolomitic, and calcareous rocks; poorly exposed

CAMBRIAN — Upper Cambrian

Cpi

Pilgrim dolomite

Comprises three units. Upper unit is light-gray thick-bedded dolomite commonly mottled medium-gray near base. Middle unit is light- to medium-gray crystalline limestone irregularly ribboned with yellowish-gray silty dolomite. Lower unit is mottled light- and dark-gray dolomite with sparse intraformational conglomerate, locally, basal 6 to 16 ft is bluish-gray limestone

CAMBRIAN — Middle Cambrian

Cpa

Park shale

Olive-gray, gray, and light-brown shale with minor amounts of argillaceous limestone, siltstone, and sandstone

Cm

Meagher limestone

Comprises three units. Upper and lower units are medium-gray limestone irregularly ribboned or mottled with yellowish-orange, yellowish-brown, and yellowish-gray dolomite. Middle unit is thickly and indistinctly bedded medium-gray limestone, commonly with oolitic beds

Cw

Wolsey shale

Upper half is interbedded gray argillaceous limestone and greenish- and yellowish-gray calcareous mudstone and shale. Lower half is greenish-gray and drab shale with some interbeds of sandstone and limestone; many beds are micaceous, some are glauconitic

Cf

Flathead quartzite

White to pale shades of gray, pink, brown, and purple medium- to thick-bedded homogeneous even-grained quartz sandstone; most beds are cemented to vitreous quartzite; thin, discontinuous sparse pebble zones in lower part; crossbedding common

UNCONFORMITY

PRECAMBRIAN — Upper Precambrian — Belt series

pCe

Empire shale

Gray, greenish-gray, and brown, siliceous mudstone or argillite with interbeds of quartzite sandstone and shale. Intertongues with Spokane shale

pCs

Spokane shale

Grayish-red mudstone, shale, and sandstone, with a few thin beds of limestone near base

pCg

Greyson shale

Gray and brown mudstone or shale alternating with sandstone or quartzite. Base not exposed. Grades into Spokane shale

INTRUSIVE ROCKS

YOUNGER INTRUSIVE ROCKS

TERTIARY(?) — Lower Tertiary(?)

gd

Granodiorite and quartz diorite

In Sagebrush Park stock

ic

Composite or hybrid intrusives

Small plutons containing diverse and unusual rocks including olivine-rich and quartz-rich types

OLDER INTRUSIVE ROCKS

CRETACEOUS — Upper Cretaceous

ba

Basalt and related rocks

Dark-gray to greenish-black fine- to medium-grained rocks, mainly as sills

ad

Andesite porphyry, diorite porphyry, and related rocks

Greenish-gray to dark-gray porphyritic rocks with phenocrysts of plagioclase and hornblende or augite; mainly as sills

hl

Hornblende lamprophyre

Very fine grained gray rock with conspicuous hornblende phenocrysts

67

Contact, showing dip

Dashed where approximately located; length of dashes indicates relative degree of accuracy; long dash most accurate

90

Vertical contact

Overturned contact

U 39 ?
D

Fault, showing dip

Dashed where approximately located; length of dashes indicates relative degree of accuracy; long dash most accurate. Dotted where concealed, queried where doubtful. U, upthrown side; D, downthrown side

F

Preintrusion fault

Forming boundary along margin of intrusive body; dashed where approximately located

?

Anticline

Showing crestline and plunge of axis. Dashed where approximately located; length of dashes indicates relative degree of accuracy; long dash most accurate. Dotted where concealed; queried where doubtful

?

Syncline

Showing position of trough and plunge of axis. Dashed where approximately located; length of dashes indicates degree of accuracy; long dash most accurate. Dotted where concealed; queried where doubtful

Overturned anticline

12

Strike and dip of beds

90

Strike of vertical beds

Horizontal beds

50

Strike and dip of cleavage

17

Strike and dip of joints

90

Vertical vein, approximately located

85

Vein, approximately located, showing dip

Vertical shaft

Adit

Prospect pit

Mine

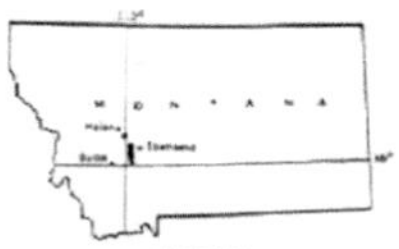

INDEX MAP

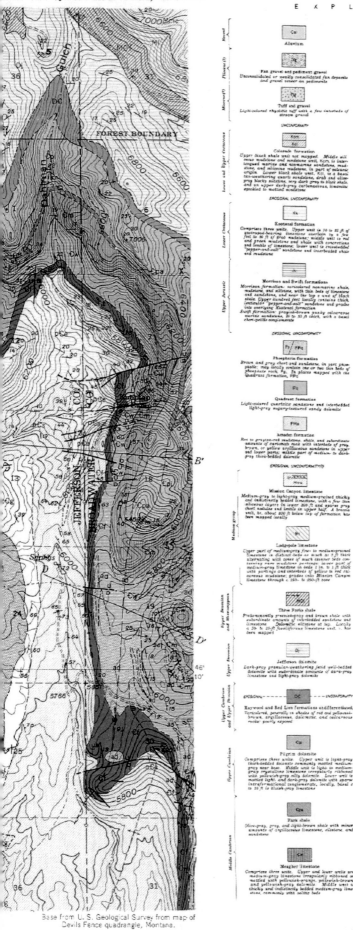

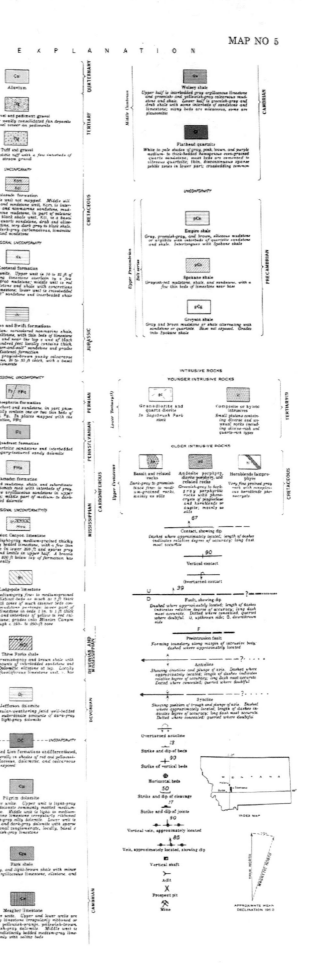

APPROXIMATE MEAN DECLINATION 1950

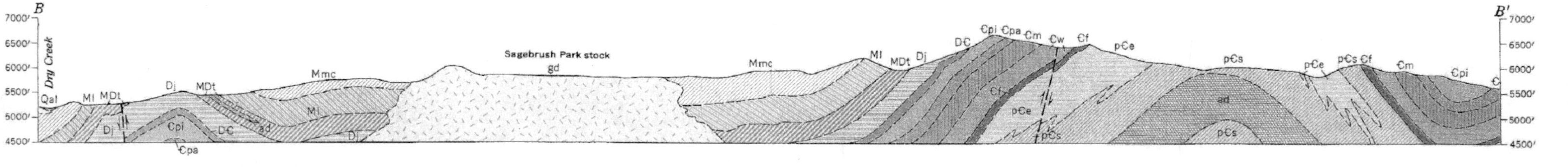
B
B′
Dry Creek
Sagebrush Park stock
gd
Qal
Ml
MDt
Dj
Mmc
€pi
€pa
D€
ad
€m
€w
€f
p€e
p€s
7000′
6500′
6000′
5500′
5000′
4500′

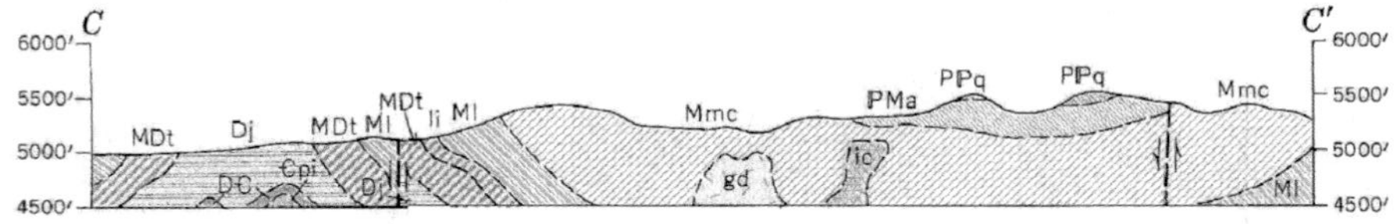
C
C′
MDt
Dj
Ml
li
Mmc
PMa
PPq
gd
ic
D€
€pi
6000′
5500′
5000′
4500′

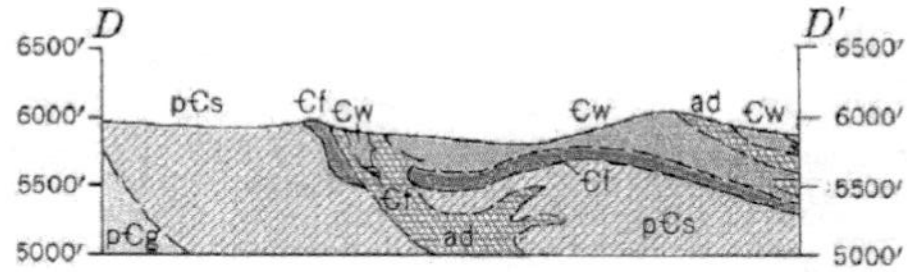
D
D′
p€s
€f
€w
ad
p€g
6500′
6000′
5500′
5000′

MAP B: DEVILS FENCE, MONTANA

1. Describe the large structure on the eastern side of this map. Notice the relative age of the rock in the center and the many strike and dip symbols on the structure. Also examine the cross section B-B′. What is the term for such a structure?

2. Contrast the structure in Question 1 with the similar-looking structure in the north central part of the map. What type of stress produced these two structures and how do the two structures differ?

3. Examine the orange patterned "ad" unit that intrudes the Grayson shale in the eastern structure. What kind of intrusion conforms to the sedimentary layering as this does? What other sedimentary units does this unit intrude?

4. How does the shape of the "gd" intrusive body differ from the "ad" unit in Question 3?

5. During which geologic periods did intrusion occur in the area of this map?

6. What geologic periods are not represented by rock in this area?

7. Approximately how many million years are missing in the unconformity between the Phosphoria Formation and the Morrison/Swift Formations? Refer to the map explanation and the geologic time chart on the inside back cover.

8. What appears to be happening to sea level in the first three formations in the Cambrian period?

9. In which physiographic province is this area located?

10. Write a generalized geologic history of this area, including geologic events such as sedimentary deposition, folding, faulting, intrusion, and the geologic periods when these events occurred.

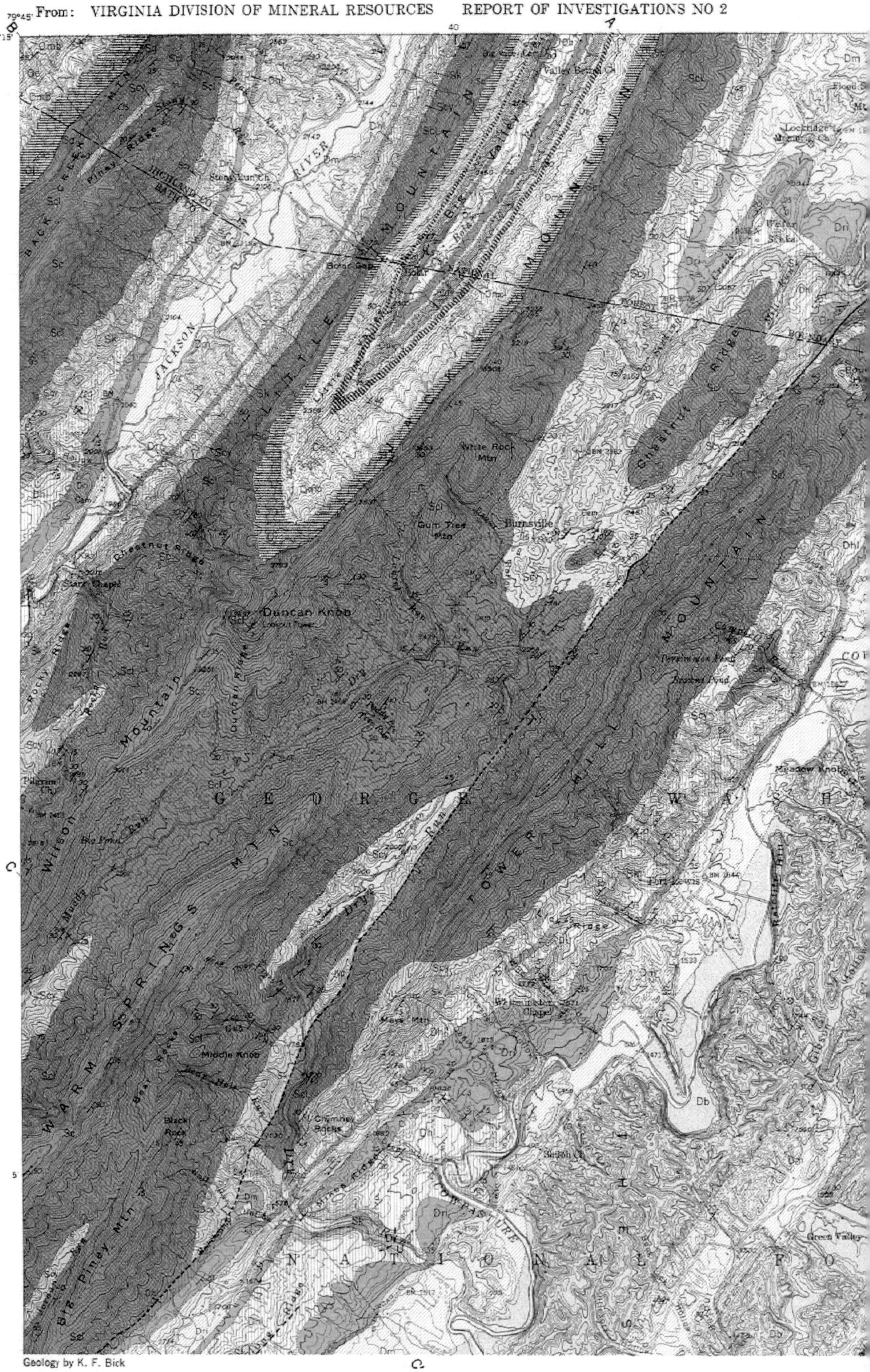

Geology by K. F. Bick

GIC MAP OF THE WILLIAMSVILLE QUADRANGLE, VA.

MAP NO 3

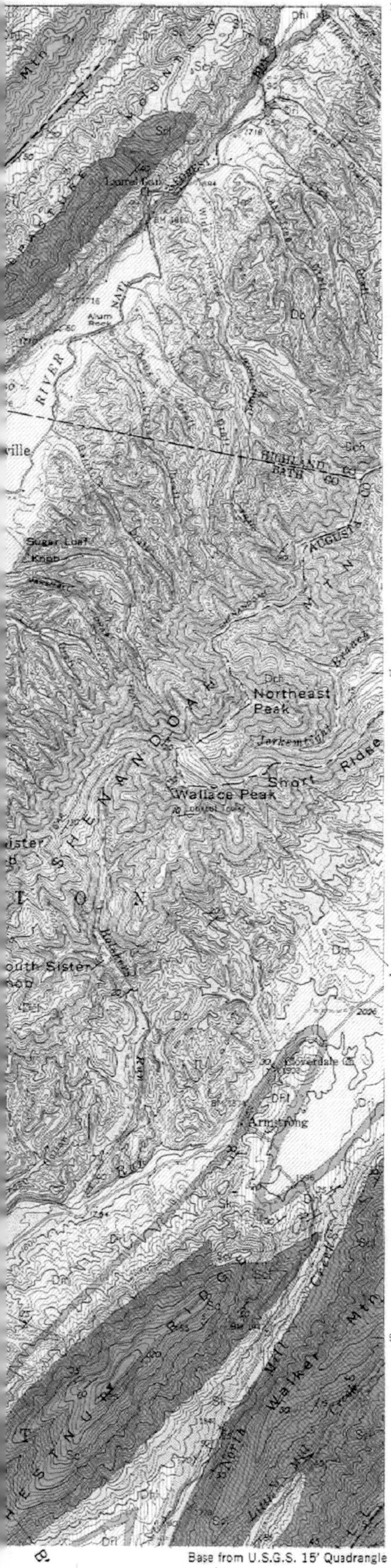

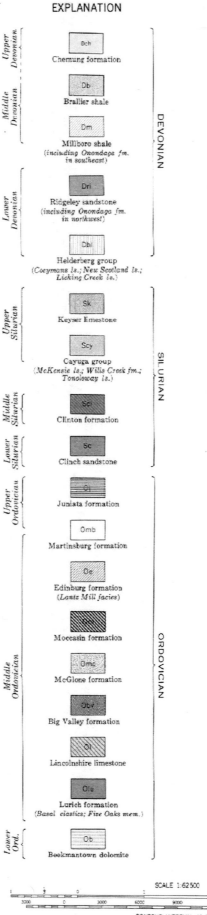

EXPLANATION

DEVONIAN

Upper Devonian

- Dch — Chemung formation

Middle Devonian

- Db — Brallier shale
- Dm — Millboro shale (*including Onondaga fm. in southeast*)

Lower Devonian

- Dri — Ridgeley sandstone (*including Onondaga fm. in northwest*)
- Dhl — Helderberg group (*Coeymans ls.; New Scotland ls.; Licking Creek ls.*)

SILURIAN

Upper Silurian

- Sk — Keyser limestone
- Scy — Cayuga group (*McKensie ls.; Wills Creek fm.; Tonoloway ls.*)

Middle Silurian

- Scl — Clinton formation

Lower Silurian

- Sc — Clinch sandstone

ORDOVICIAN

Upper Ordovician

- Oj — Juniata formation
- Omb — Martinsburg formation

Middle Ordovician

- Oe — Edinburg formation (*Lantz Mill facies*)
- Om — Moccasin formation
- Omc — McGlone formation
- Obv — Big Valley formation
- Ol — Lincolnshire limestone
- Olu — Lurich formation (*Basal clastics; Five Oaks mem.*)

Lower Ord.

- Ob — Beekmantown dolomite

CONTACTS

- exposed
- approximate
- covered

FAULTS

NORMAL AND REVERSE

- exposed
- approximate

U - *upthrown side*

D - *downthrown side*

THRUST

- exposed
- approximate

T overthrust side

ATTITUDE OF ROCKS

- 50 Strike and dip of beds
- 50 Strike and dip of overturned beds
- Strike of vertical beds
- Horizontal beds

QUADRANGLE LOCATION

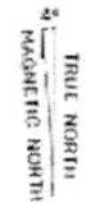

APPROXIMATE MEAN DECLINATION, 1946

SCALE 1:62500

1 ½ 0 1 2 3 4 MILES

3000 0 3000 6000 9000 12000 15000 18000 21000 FEET

CONTOUR INTERVAL 40 FEET
DATUM IS MEAN SEA LEVEL

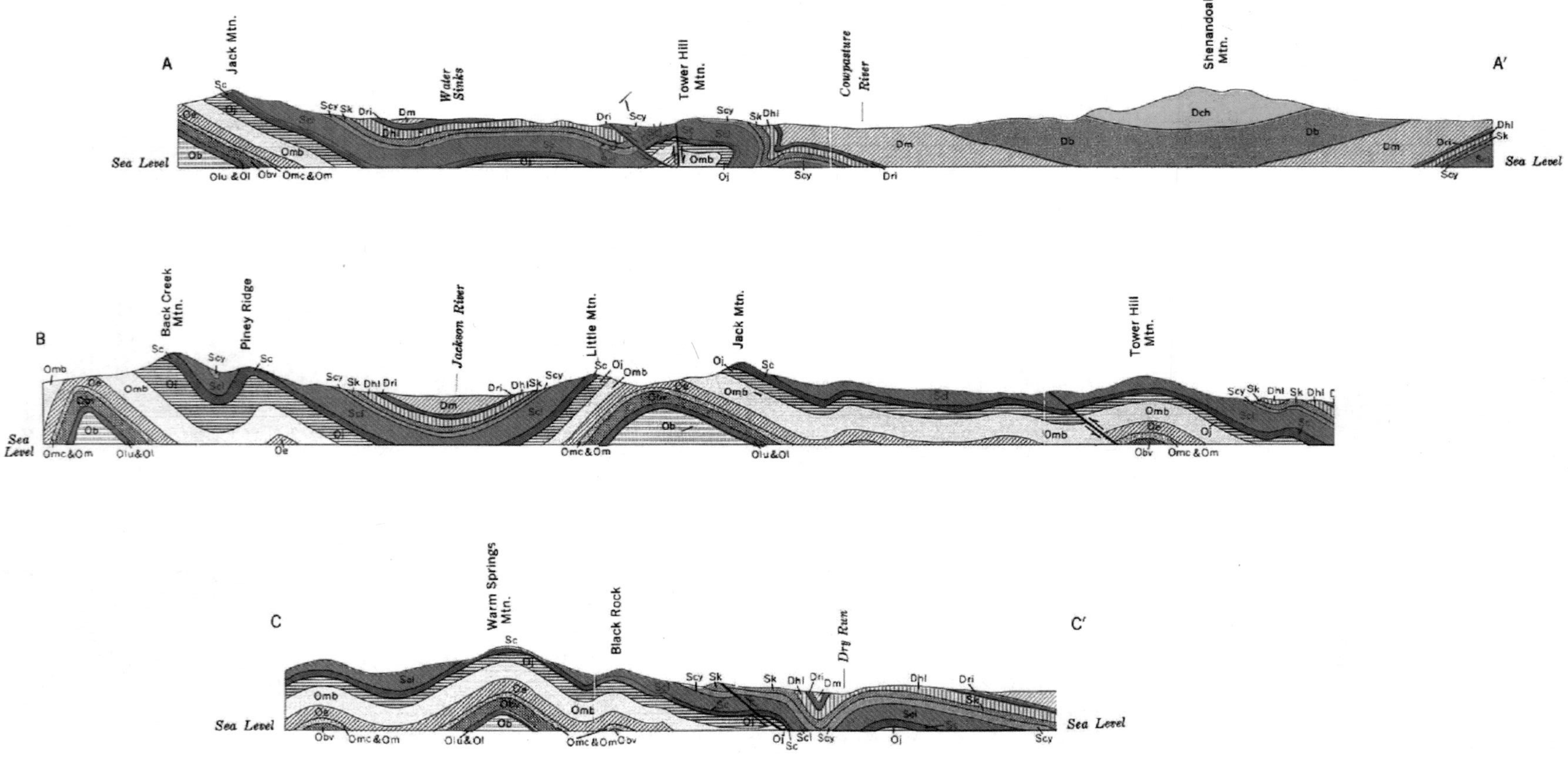
A
Jack Mtn.
Water Sinks
Tower Hill Mtn.
Cowpasture River
Shenandoah Mtn.
A′
Sea Level
B
Back Creek Mtn.
Piney Ridge
Jackson River
Little Mtn.
Jack Mtn.
Tower Hill Mtn.
Sea Level
C
Warm Springs Mtn.
Black Rock
Dry Run
C′
Sea Level

MAP C: WILLIAMSVILLE, VIRGINIA

1. "Anticlines create mountains and synclines create valleys." Examine the cross sections of this area and evaluate this statement. If this statement is not true for all cases, then what other principle governs mountain and valley formation in this area?

2. What specific kind of anticlines and synclines are present in this area?

3. In the northwest corner of the map, Jack Mountain, Big Valley, and Little Mountain form a breached anticline. Examine the rock types present and explain how this breached anticline formed. (See the Glossary for more information.)

4. The Cowpasture River flows between Tower Hill Mountain and Shenandoah Mountain. Predict what rock type would be found in this valley and why. Then check your answer with the key in the map explanation.

5. Which one of the other three geologic maps does this map most closely resemble? What plate tectonic situation must have existed in both of these areas? (See Chapter 6, pages 244 and 245 for information on plate boundaries.)

6. Compare the times of geologic deformation on the two maps from Question 5. In which area does the deformation appear to have occurred earlier in geologic time? When did mountain building occur in Virginia? When in the other area?

7. What other structure on the Virginia map confirms the action of compressive stress in this area?

8. What geologic event occurred in the other map area (see Question 5) but not in Virginia?

9. Write a generalized geologic history of this area using the explanation and the cross sections.

GEOLOGIC MAP OF STABLE INTERI

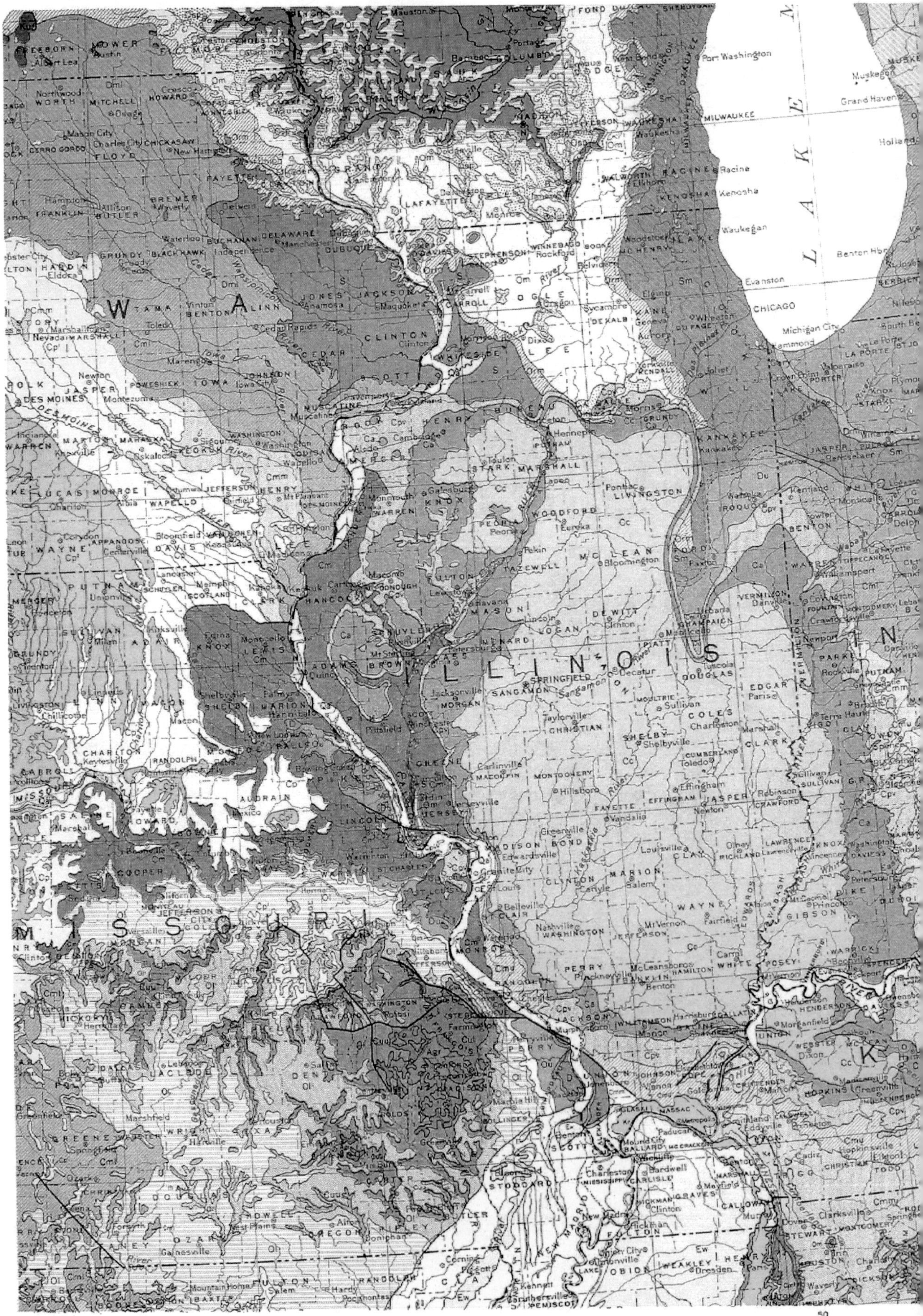

Compiled by George W. Stose and O. A. Ljungstedt

Base from U.S. Geological Survey topographic quadrangles

SCALE 1:2,500,000

50

0 100 150 MILES

EXPLANATION

QUATERNARY
- Qa — Alluvium

TERTIARY
- Ew — Wilcox group
- Em — Midway group

CRETACEOUS
- Ku — Upper Cretaceous rocks
- Kr, Ks — Ripley formation and Selma chalk
- Ke — Eutaw formation
- Kt — Tuscaloosa formation

CARBONIFEROUS

Permian
- Cd — Dunkard group

Pennsylvanian
- Cmh — Monongahela formation
- Cc — Conemaugh formation
- Ca — Allegheny formation
- Cpv — Pottsville group
- Cp³ — Upper-middle Pennsylvanian rocks
- Cp² — Lower-middle Pennsylvanian rocks
- Cp¹ — Lower Pennsylvanian rocks

Mississippian
- Cmu — Rocks of Chester age
- Cmm — Rocks of Meramec age
- Cml¹, Cml — Rocks of Osage and Kinderhook age
- Cm — Mississippian rocks

DEVONIAN
- Du — Rocks of Portage to Catskill age
- Dm¹, Dm — Rocks of Hamilton to Onondaga age
- Dl — Rocks of Helderberg age
- D — Devonian rocks

SILURIAN
- Su — Rocks of Cayuga age
- Sm — Rocks of upper Niagara (Lockport) age
- Sml — Rocks of lower Niagara (Clinton) and Medina age
- S — Silurian rocks

ORDOVICIAN
- Orm — Shale of Richmond age
- Ou — Maquoketa shale to Joachim limestone
- Om — St. Peter sandstone and Everton limestone
- Osp — St. Peter sandstone
- Ol, Ol¹ — Rocks of Beekmantown age

CAMBRIAN
- CO — Upper Cambrian and Lower Ordovician age
- Cuu — Upper part of Upper Cambrian rocks
- Cu — Upper Cambrian or St. Croixan rocks
- Cul — Lower part of Upper Cambrian rocks
- Cl — Limestone and shale of Lower and Middle Cambrian age
- Cq — Lower Cambrian quartzites and slates

PRE-CAMBRIAN
- Awb — Oligoclase-biotite schist
- Aw — Albite-chlorite schist and garnetiferous phyllonite
- Agg — Mylonitized granite gneiss and hornblende gneiss
- Abm — Middle Huronian rocks
- Agr — Granite, porphyry, and gabbro

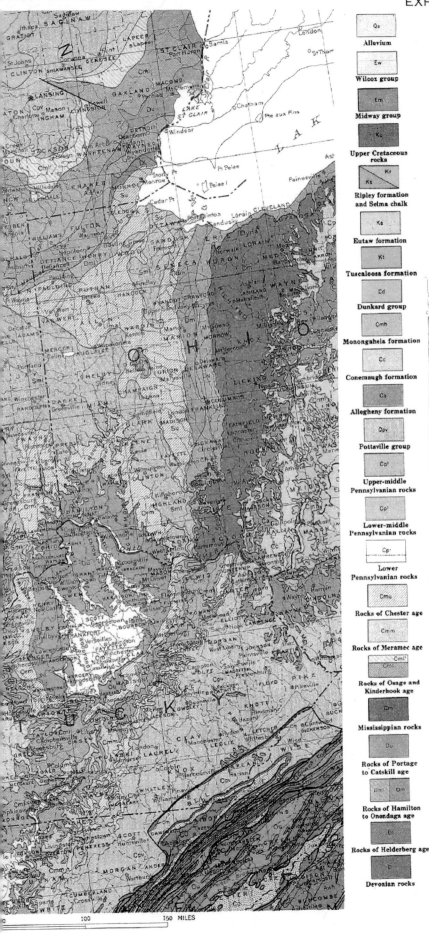

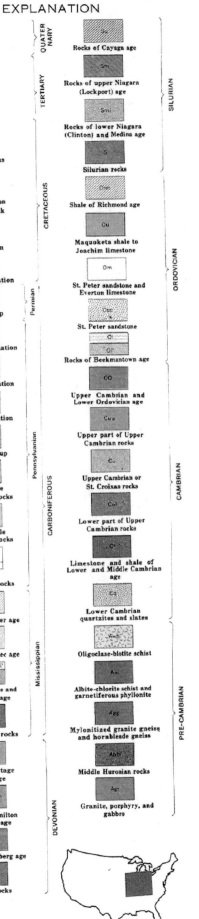

LOCATION OF AREA

MAP D: STABLE INTERIOR

1. The stable interior of the United States is an area of gently "warped" sedimentary rock layers overlying Proterozoic basement rock. Broad up- and down-warped areas form domes and basins. Domes are like anticlines in that they contain older rock at their cores. Basins are just the opposite. Determine whether each of these areas is a dome or basin:

a. the state of Illinois

b. the area of Tennessee centered on Nashville

c. southeastern Missouri

d. the state of Michigan

e. northern Kentucky centered on Frankfort

2. The area in the far southeastern corner of this map is structurally very different from the rest of the map. Notice the many narrow bands of different colors. Describe this area's structure, and compare it to the structure on the rest of the map.

3. What causes the branching pattern of the Cuu and Ol′ units in southern Missouri?

4. What is the agent of deposition for the Qa alluvium on this map?

5. Estimate the approximate percentage of rocks of the following ages on this map:

a. Precambrian

b. Lower Paleozoic (Cambrian to Silurian)

c. Upper Paleozoic (Devonian to Permian)

d. Cretaceous and Tertiary

6. Determine the system (geologic period) of rocks on which each of the following important cities was built:

a. Des Moines, IA

b. Jefferson City, MO

c. Milwaukee, WI

d. Springfield, IL

e. Lansing, MI

f. Cleveland, OH

g. Indianapolis, IN

7. Notice the Cretaceous- and Tertiary-age rocks in the vicinity of Paducah, Kentucky. These rocks are in contact with and overlie rocks of Mississippian (Cmu) age. Describe the nature of this contact. Approximately how much geologic time is missing at this contact?

8. Why do you think the map of this area has been titled "Stable Interior"? What physiographic provinces are present here? (See physiographic map, Fig. 5.1.)

CHAPTER 6

Plate Tectonics

Seafloor spreading and plate tectonics are major concepts in geology. In 1915 Alfred Wegener suggested that today's continents had originally been part of a "supercontinent" called Pangaea that had, at some point in geologic time, fragmented into pieces (the continents) and gradually drifted to their present positions. Very few additional technical studies were undertaken to explain or refine this concept of continental drift until the early 1950s. Scientists like Harry Hess, J. Tuzo Wilson, Robert Dietz, Patrick Hurley, and Frederick Vine, among others, began to synthesize the extensive oceanic geological data on a global scale to evaluate further the viewpoint current at that time that the oceanic crust was still closing and opening. Abundant new hydrographic and geophysical data derived from studies of ocean ridges, ocean trenches, earth geomagnetism, earthquakes, and the like were amassed and analyzed utilizing computer synthesis. Thus many geologists began to reevaluate Wegener's ideas in terms of a dynamic earth on which surface plates composed of continental and ocean-floor rock move as large units.

Figure 6.1 is a composite illustration of the modern concept of plate tectonics. The rigid upper layer of the earth, known as the *lithosphere*, includes the continental and oceanic crust and the upper part of the mantle. Because of stresses set up, most likely by movement of mantle material, the lithosphere is often broken by a system of fractures that occur in specific zones called *plate boundaries*. These plate boundaries are often marked by the locations of volcanoes and earthquake epicenters (see Fig. 6.1). Modern plate tectonics theory postulates that plate boundaries can be included in one of three categories.

The first type of plate boundary is the *divergent boundary*, also known as a *spreading center*, exemplified by the East Pacific Rise and the Mid-Atlantic Ridge. Impelled by the movement of heated mantle material upwelling from below the lithosphere, the older lithospheric rock is spread apart, and newly generated volcanic material fills the fracture at the spreading center.

As new ocean-floor rock is created at the midocean ridges, there must be a compensating destruction of old ocean-floor crust, because no evidence suggests that the earth is constantly expanding. This destruction of earth's old crust appears to happen in the areas where it is subducted into ocean trenches. A subduction zone is an area in which a cold slab of seafloor is forced back into the mantle beneath another plate. Plates are moving toward each other at this, the second type of plate boundary, known as a *convergent boundary*. Geothermal heat and friction increase the temperature of the down-going plate, and water decreases the melting point to create magma, which erupts at the surface as a chain of andesitic volcanic islands (for example, the Mariana Islands) or andesitic volcanoes along a continental margin (for example, the Andes of South America). Continental collisions produce mountain chains, such as the Himalayas, where the continents are sutured together. Ocean plates can be subducted beneath other ocean plates and continents, but continents cannot be subducted. When one ocean plate is subducted beneath another, it is the older, colder ocean crust that submits to subduction. Earthquakes occur at various depths along the subducted plate.

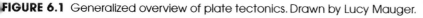

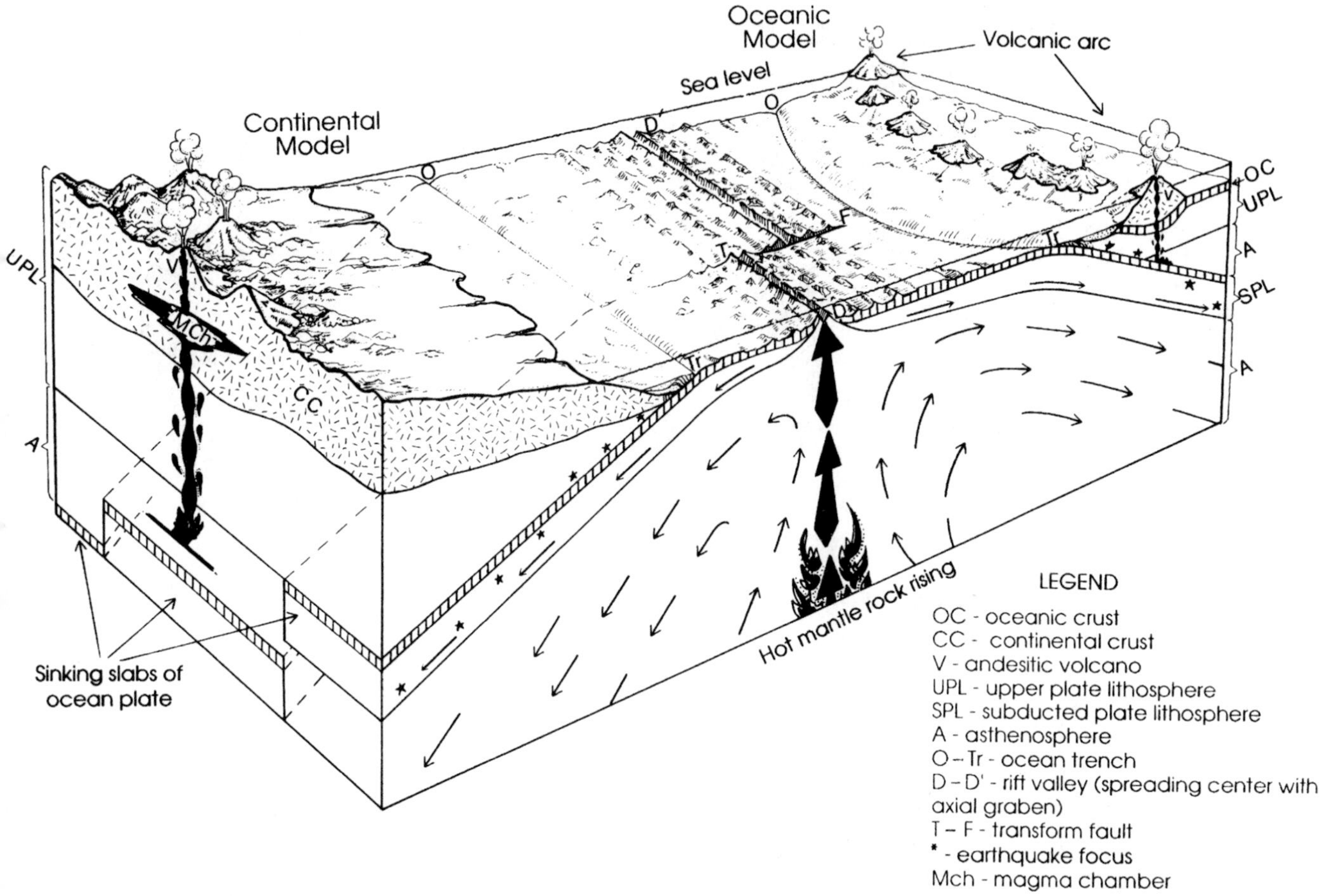

FIGURE 6.1 Generalized overview of plate tectonics. Drawn by Lucy Mauger.

The third common type of plate boundary is formed in areas where lithospheric plates slip horizontally past each other; it is called a *transform boundary*, which follows a transform fault. Transform faults are formed either in the ocean or on a continent by a system of large lithosphere fractures that laterally connect two spreading ridges, two subduction zones, or a subduction zone and a spreading ridge. The transform faults are developed as a consequence of differential spreading rates caused by the curvature of the earth's spherical surface. The San Andreas fault system in California is an example of a transform fault connecting the East Pacific Rise in the Gulf of California to the Juan de Fuca Rise off the coast of Oregon and Washington.

EXERCISES

Exercise 6-1 SEAFLOOR SPREADING IN THE SOUTH ATLANTIC

Rates of plate movement are slow, to be sure. They are calculated in terms of centimeters per year. Two of the exercises in this chapter will allow you to calculate and compare the rates of movement of two different plates. This exercise covers seafloor spreading in the South Atlantic. It is designed to demonstrate how the original concept of continental drift was envisioned by Alfred Wegener and also to calculate the rate of seafloor spreading taking place in the South Atlantic. (Cut out continents, Fig. 6.5.)

a. Wegener used the geographic shape of the coastal outlines of Africa and South America to suggest that the two continents were once united as one large landmass. In 1960 Bullard suggested that a better continental fit could be obtained by including the continental shelves as part of the continents.

1. Why should the continental shelves be considered part of the continents? (See example cross section, Fig. 6.2.)

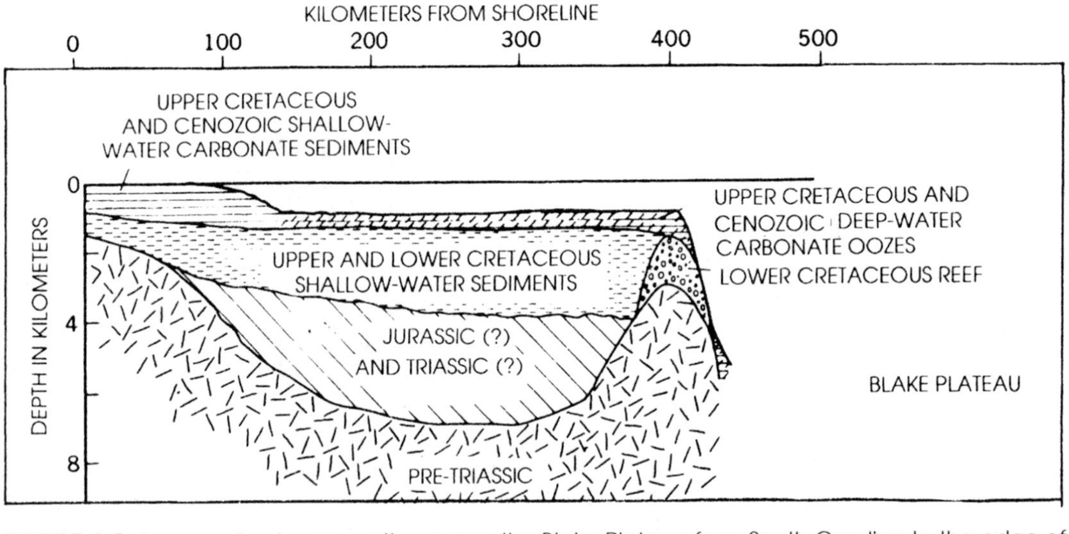

FIGURE 6.2 A generalized cross section across the Blake Plateau from South Carolina to the edge of the continental shelf.

2. After fitting the continents together by using the overlays and the map (see Fig. 6.3), decide whether Bullard's or Wegener's approach to continental fit works better. Cite your evidence.

b. The approximate distance between the continents in their present position is shown on Figure 6.3. Allard and Hurst[1] indicate that rocks of Aptian (late early Cretaceous) and older ages in the coastal portion of Brazil north of Bahia are very similar to the rocks in Gabon, Africa. This would suggest that the spreading of Africa from South America could have occurred during the last 120 million years.

1. From these data, the calculated spreading rate in the South Atlantic is approximately ______________________ centimeters per year.

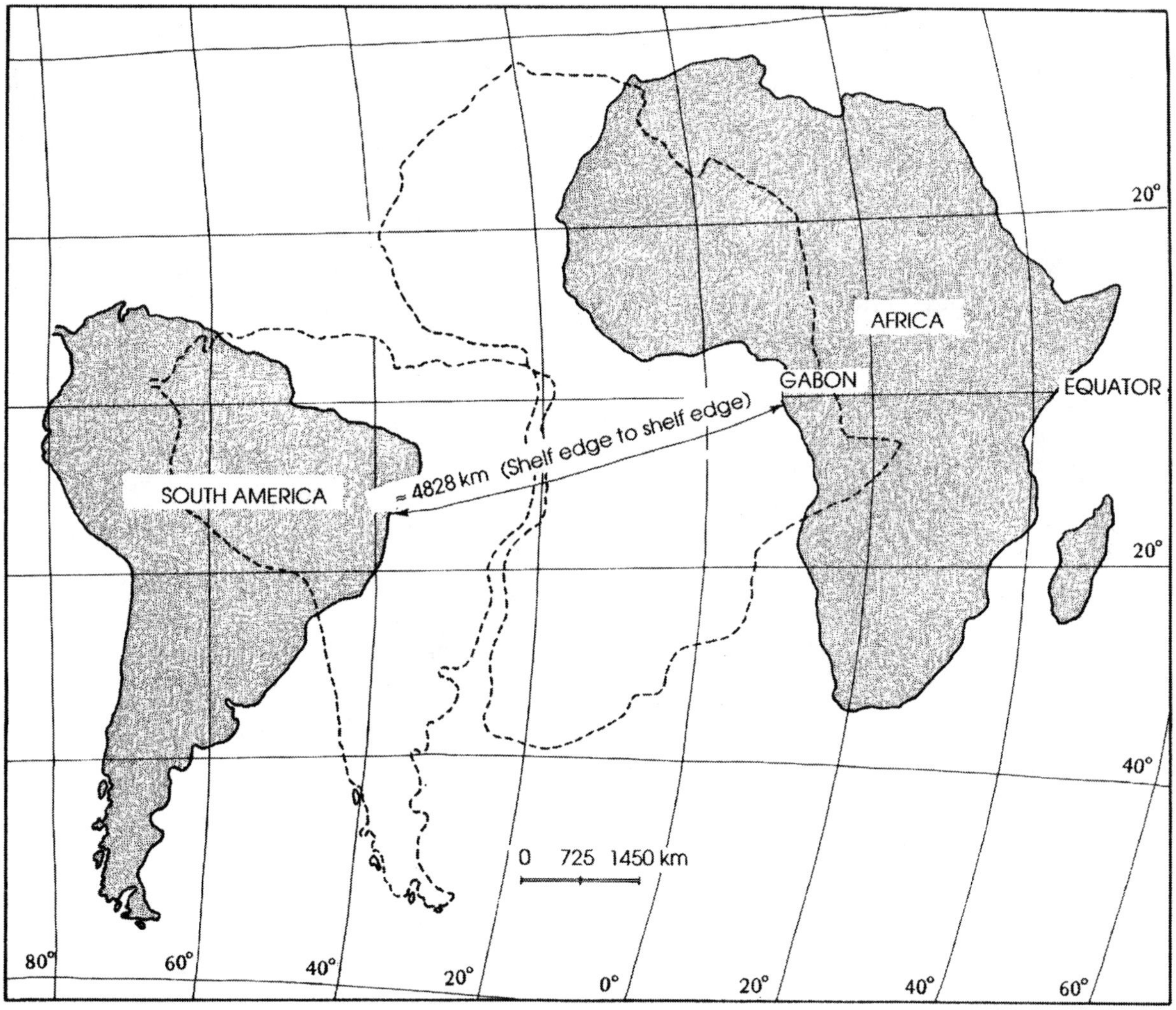

FIGURE 6.3 Data worksheet: Present position of Africa and South America (stippled) and predrift location of continents (dashed outline).

[1]Gilles O. Allard and Vernon J. Hurst, "Brazil–Gabon Geologic Link Supports Continental Drift," *Science*, Vol. 163, No. 3867, pp. 528–532, illus. (incl. sketch maps), 1969.

2. Figure 6.4 shows the age of ocean crust plotted relative to the distance from the Mid-Atlantic Ridge. How does the rate calculated in the previous question compare to the spreading rates obtained during the Deep Sea Drilling Project (Joides) Leg 3 (Fig. 6.4) in the South Atlantic? Discuss your answer. (Remember that the plate movement rate of a single continent is one-half the spreading rate of two continents moving away from each other.)

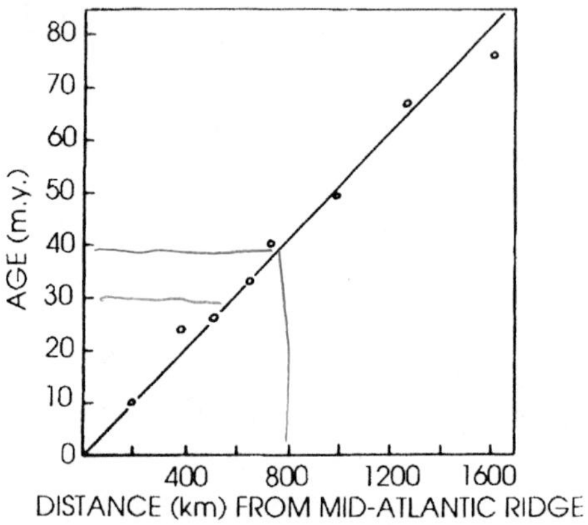

FIGURE 6.4 Plot of age of sediment immediately above basaltic crustal rocks (after Fig. 8, p. 463, "Initial Reports of Deep Sea Drilling" Project, Volume 3, National Science Foundation, April 1970) (JOIDES-South Atlantic).

c. Place your continental overlays of Africa and South America (Fig. 6.5) in the position they are in today (solid outline on the worksheet). Next, slowly move them back through time to their respective predrift position (dotted outline on the worksheet).

1. As the continents were moved back to their predrift positions, was the movement parallel to lines of latitude? Discuss the apparent motion required of each continent during its drift (for example, movement east–west, movement north–south, rotational movement). Has the drift rate been uniform throughout the South Atlantic? Explain.

2. After placing the continental overlays in their predrift position, tabulate all of the evidence you can, from the data given, that indicates that the continents were once part of a single large landmass.

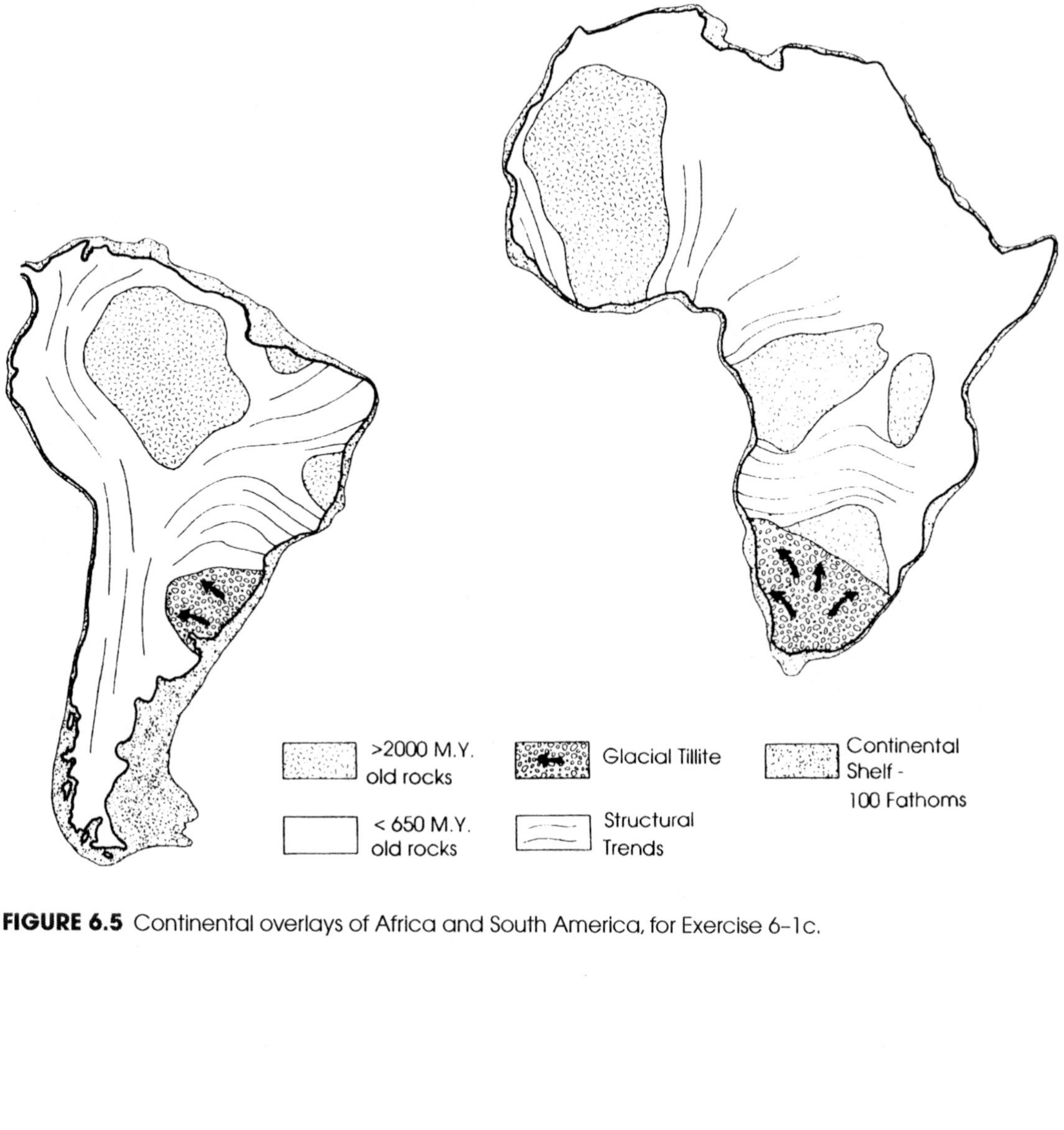

FIGURE 6.5 Continental overlays of Africa and South America, for Exercise 6-1c.

Exercise 6-2 RATE OF MOVEMENT OF THE PACIFIC PLATE

The Pacific Plate is moving away from the East Pacific Rise in a generally northwest direction. This plate consists mostly of oceanic floor or crust and its associated features, including a great number of islands and seamounts. Many of these islands and seamounts (for example, the Hawaiian Islands and the Emperor Seamount Chain) form linear chains called *aseismic ridges*. The prevailing theory for the origin of these linear ridges suggests that spreading has caused the oceanic lithosphere to move over the top of a chimney-like plume of magma generated deep within the mantle. After a long episode of eruptions of lava from the hot spot, a volcanic island is formed. Eventually, movement of the plate slides the island off the hot spot and another volcano grows up from the seafloor over the hot spot (Fig. 6.6). Continuing volcanic eruptions create other islands or seamounts as neighbors to the original. Gradually, over millions of years, as the plate moves, a chain of islands is formed and aligned in the direction of the overall plate motion. Older islands are subject to erosion and cooling, and isostatic sinking causes them to submerge relative to sea level. Many islands become submerged flat-topped features called *guyots*.

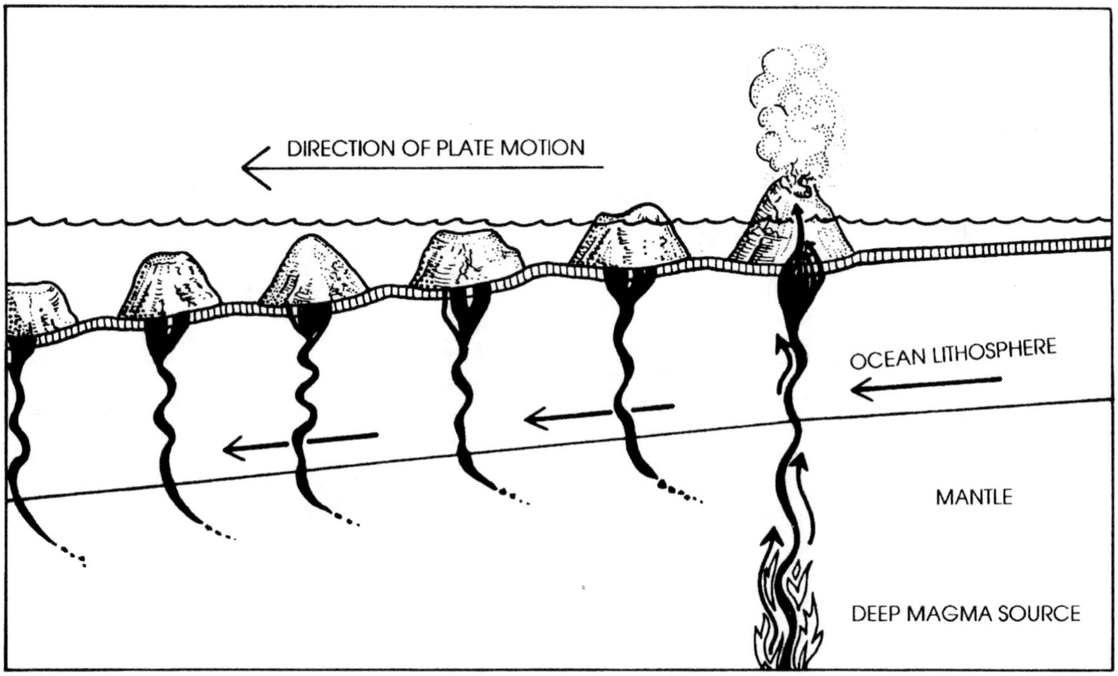

FIGURE 6.6 Hawaiian Islands hot spot. Deep magma source builds volcanoes on plate above, and plate motion moves volcanoes away from magma source. Drawn by Lucy Mauger.

The ages of rock on the Hawaiian Islands and on nearby islands to the northwest have been determined by radiometric dating. Figure 6.7 shows the present position of the various Hawaiian islands. The age of the rock is shown in Table 6.1. Since the big island of Hawaii is volcanically active today, the hot spot seems to be located beneath this island. As you would expect, Hawaii has the youngest volcanic rock as well. By finding the distance of other islands from Hawaii and by knowing the age of rock on these other islands, you can calculate a rate of plate movement. In the following exercise you will be calculating the rate of plate movement today and then calculating an average movement rate over the past 20 million years.

Fill in Table 6.1.

a. Note the rock ages, shown in the second column. The difference between Maui and Hawaii rock ages is 1.3 million years. Calculate plate movement segments at each island by subtracting the age of the rock and the distance from the figures for the previous island and divide the difference in distance by the difference in time. Remember that your distances are in kilometers and your time is in millions of years, but your answer for the rate of movement should be in units of centimeters per year. Enter your answers in Table 6.1 in the "Rate of movement" column.

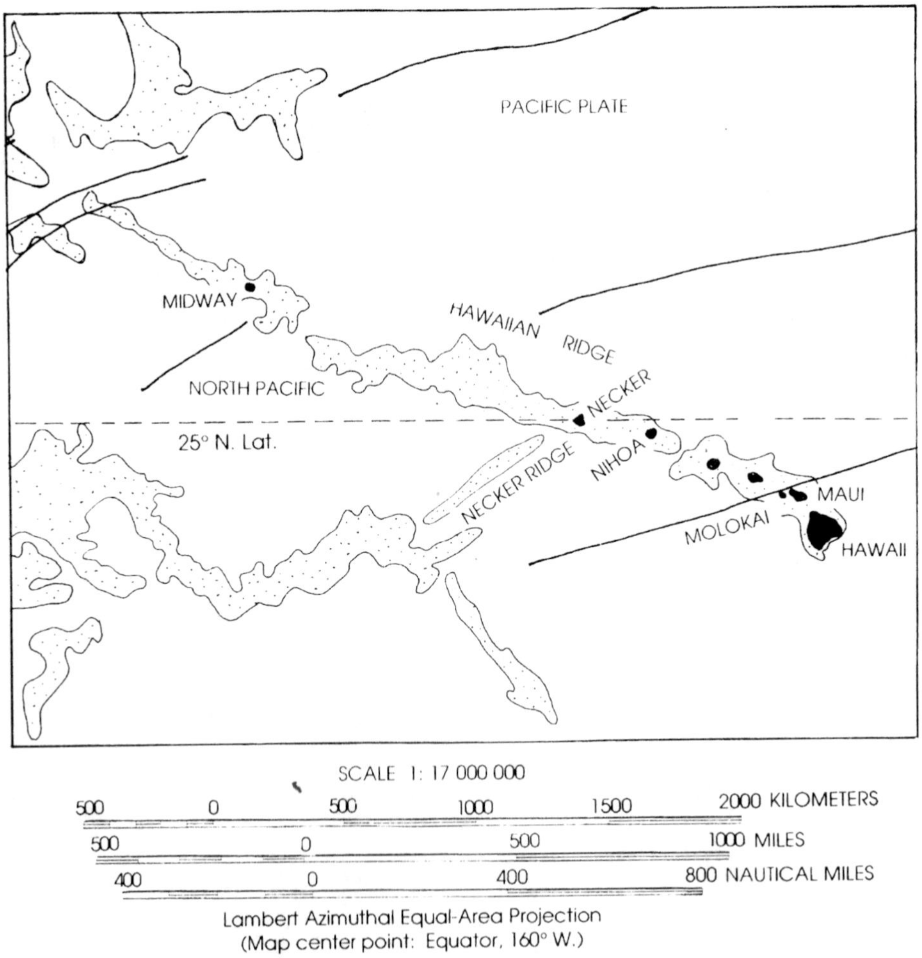

FIGURE 6.7 Map of Hawaiian Ridge. From *Preliminary Tectonostratigraphic Terrane Map of the Circum-Pacific Region.* Howell, D. G., Schermer, E. R., Jones, D. L., Ben-Avraham, Z., Scheibner, E., © 1985. Reprinted with permission from AAPG.

TABLE 6.1 Rates of Plate Movement of Pacific Plate over the Past 27.7 Million Years

Island	*Age of rock (m.y.)*	*Distance from Kilauea (km)*	*Rate of movement (cm/yr)*
Hawaii	0.0	0.0	
Maui	1.3	221.0	17.0
Molokai	1.6	250.0	
Oahu	2.5	350.0	
Kauai	5.0	519.0	
Nihoa	7.2	780.0	
Necker	10.3	1,058.0	
Midway	27.7	2,432.0	
Average plate movement rate over past 27.7 million years			

b. After your calculations are complete, average your answers and enter the number in Table 6.1.

c. Answer the following questions.

1. When was the rate of plate movement most rapid?

2. When was the rate of plate movement slowest?

3. How does the overall rate of movement of the Pacific Plate compare with the rate of movement of South American and African plates in Exercise 6–1?

Exercise 6-3 CONVERGENT BOUNDARIES AND SUBDUCTION ZONES

As stated previously, if ocean floors are spreading in areas of midocean ridges, they must be converging in other areas if the surface area of the earth is to stay constant. Apparently, deep-ocean trenches represent zones of subduction where ocean-floor rock sinks into the mantle. One line of evidence for this theory lies in the distribution of earthquakes of intermediate and deep focal depths. Earthquakes occur only in rigid crustal rock material, not in plastic or molten mantle or core material.

Most earthquakes are the shallow-focus type that occurs within the upper 70 kilometers of the earth (within the lithosphere). Only near trenches do earthquakes of intermediate (70–350 kilometers) focal depth occur. This fact would indicate that something solid and rigid is being forced deep into the mantle in these areas. Since the earthquakes are occurring within this subducted plate, it can be profiled by locating earthquakes and their various focal depths relative to the position of the trench. Figure 6.8 shows an area

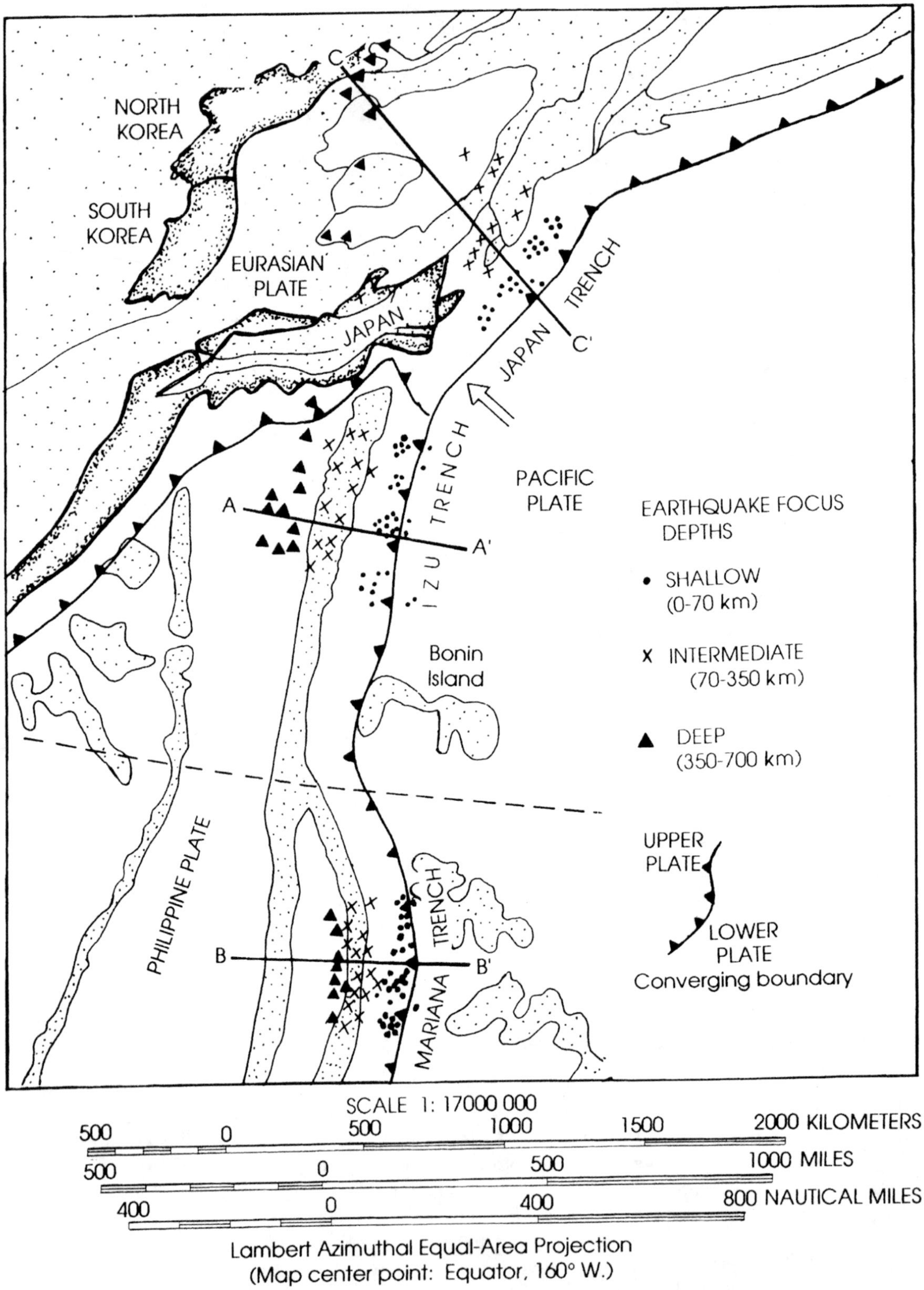

FIGURE 6.8 Map of the Japan, Izu, and Mariana trenches. From *Preliminary Tectonostratigraphic Terrane Map of the Circum-Pacific Region*. Howell, D. G., Schermer, E. R., Jones, D. L., Ben-Avraham, Z., Scheibner, E., © 1985. Reprinted by permission of AAPG.

in the western Pacific called the Izu Trench, located between the Mariana Trench and the Japan Trench. Plotted on this figure are the epicenters of recent earthquakes, with different symbols for each of the three types of focal depths. A line of cross section, A-A′, is indicated.

Data

Assume that you manage a seismic monitoring station set up in the Bonin Island group to the south and east of the Izu Trench (see Fig. 6.8). Over a week-long period, you record 24 significant earthquakes in the Izu Trench (along cross section A-A′) and calculate their focal depths and relative distances from your recording station. These data are compiled in Table 6.2.

a. Using the graph paper in Figure 6.9, plot the data points from Table 6.2. When plotting, use the appropriate symbol for each of the earthquake categories (see symbols on Fig. 6.9).

b. Group the graph points as shallow-, intermediate-, and deep-focus earthquakes. Then interpret the data by answering the following questions.

1. The graph plot shows the angle of descent for the down-going plate. What is the approximate angle for the sinking slab in this case?

TABLE 6.2 Data for Exercise 6–3

Event	*Depth (km)*	*Relative distance (km × 100)*
1	390	4.0
2	32	1.2
3	295	3.5
4	190	2.9
5	540	5.3
6	54	2.0
7	90	2.2
8	215	3.4
9	32	1.5
10	630	5.2
11	230	3.1
12	150	2.4
13	13	0.8
14	680	5.7
15	335	3.9
16	460	4.6
17	70	1.9
18	10	0.5
19	400	4.4
20	530	4.8
21	110	2.6
22	55	1.65
23	58	1.9
24	605	5.2

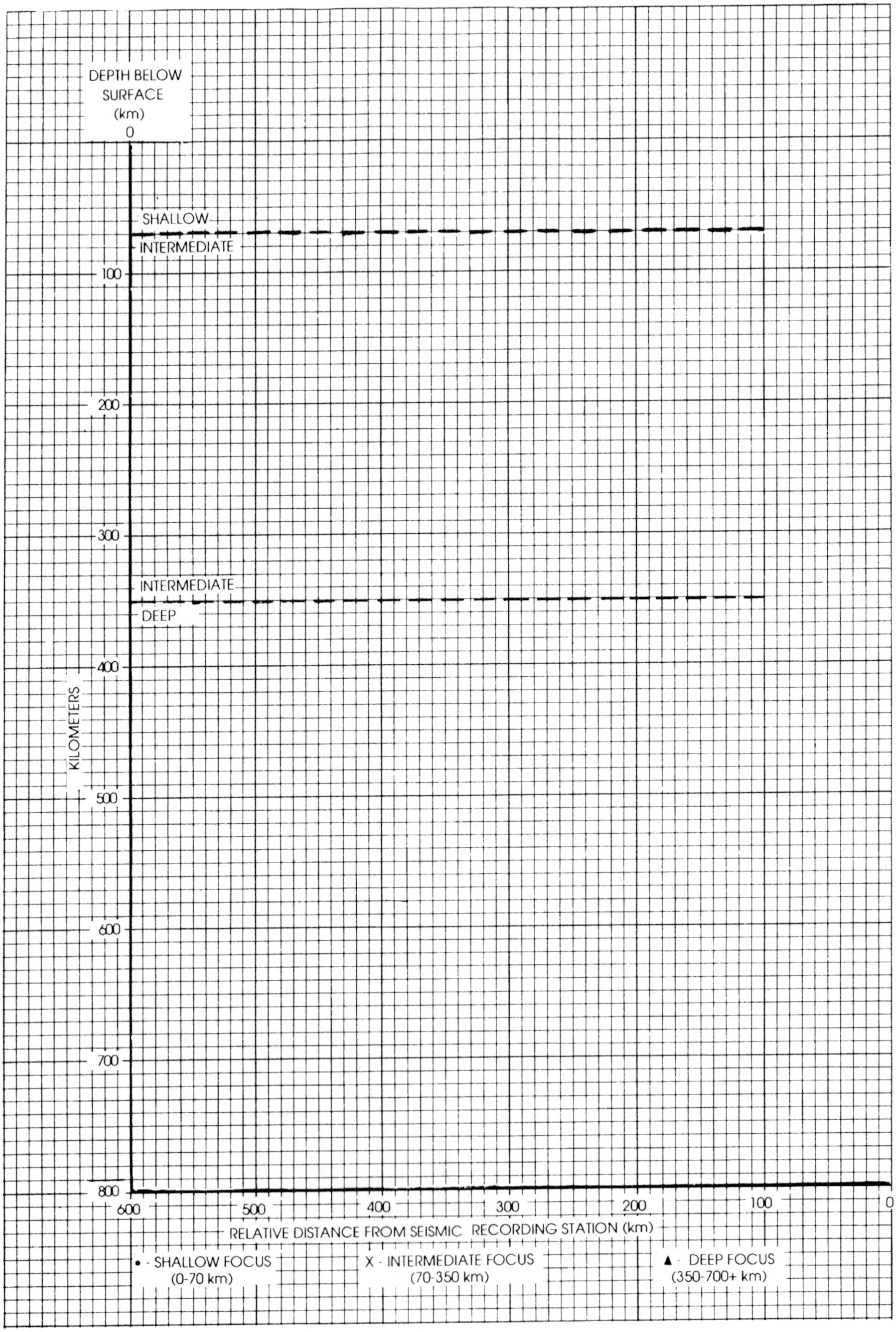

FIGURE 6.9 Graph paper for plotting earthquake focal depths, for Exercise 6–3.

2. Which plates are involved in this convergent boundary? Which is the upper plate? Which is the subducted plate? (Refer to the plate map in your textbook for this information.) Could the upper plate just as easily have been the subducted plate? Why or why not?

c. The relationship of the spacing between shallow-, intermediate-, and deep-focus earthquakes is similar to the rules for contour lines. The closer together they are, the steeper the slope or angle of plate descent, and vice versa. Using these relationships, analyze cross section B-B′ (the Mariana Trench) and cross section C-C′ (the Japan Trench) in Figure 6.8. Assume that the depth relationships across the three trenches are similar.

1. Compare the angle of descent of the plate at the Izu Trench with the angle at the Mariana and Japan trenches. Which is steepest? Which is shallowest?

2. Calculate the vertical exaggeration of your cross section. Remember, your scale is 1 inch = 100 kilometers.

Glossary*

Absolute time Geologic time measured in years, specifically time as determined by radioactive decay of elements.

Alluvium A general term for clay, silt, sand, gravel, or similar unconsolidated detrital material deposited during comparatively recent geologic time by a stream or other body of running water as a sorted or semisorted sediment in the bed of the stream or on its floodplain or delta or as a cone or fan at the base of a mountain slope.

***Amniotic egg* (amniote)** Pertaining to a vertebrate egg characterized by a large yolk and covered by shell that is lined with cellular membranes produced from embryonic tissue, which function to conserve water and for the exchange of gases.

Anticline A fold, generally convex upward, whose core contains the stratigraphically older rocks.

Arkose A feldspar-rich sandstone, typically coarse-grained and pink or reddish, that is composed of angular to subangular grains derived from the rapid disintegration of granite or granitic rocks and that often closely resembles granite. Arkose is commonly a current-deposited sandstone of continental origin, occurring as a thick, wedge-shaped mass of limited geographic extent.

Articulation (a) The action or manner of jointing or the state of being jointed, for example, the interlocking of two brachiopod valves by two ventral teeth fitting into sockets of the brachial valve (example: any brachiopod belonging to class Articulata); (b) any movable joint between the rigid parts of an invertebrate, as between the segments of an insect appendage, or of a vertebrate, as between the bones of limbs.

Artifact An object made or used by human beings.

Assemblage zone† A "biozone" characterized by the association of three or more taxa. It consists of a body of strata whose content of fossils, or of fossils of a certain kind, taken in its entirety constitutes a natural assemblage of association that distinguishes it in biostratigraphic character from adjacent strata.

Atmosphere The mixture of gases that surrounds the earth, being held there by gravity. It consists by volume of 78% nitrogen, 21% oxygen, 0.9% argon, 0.03% carbon dioxide, and minute quantities of helium, krypton, neon, and xenon.

Autotrophic Description of an organism that nourishes itself by utilizing inorganic material to synthesize living matter. Green plants and certain protozoans are autotrophic.

Batholith A large, generally discordant plutonic mass that has more than 40 square miles of surface exposure and no known floor. Its formation is believed by most investigators to involve magmatic processes.

Bedding The arrangement of a sedimentary rock in beds or layers of varying thickness and character; the general physical and structural character or pattern of the beds and their contacts within a rock mass, such as "cross-bedding" and "graded bedding"; a collective term denoting the existence of beds.

Bentonite A soft, plastic, porous, light-colored rock composed essentially of clay minerals of the montmorillonite group plus colloidal silica, produced by devitrification and accompanying chemical alteration of a glassy igneous material, usually a tuff or volcanic ash. The rock is greasy and soaplike to the touch and commonly has the ability to absorb large quantities of water accompanied by an increase in volume of about eight times.

Bioherm A moundlike, domelike, lenslike, or reeflike mass of rock built up by sedentary organisms (such as corals, algae, foraminifers, certain mollusks, and stromatoporoids) composed almost exclusively

---

*The glossary definitions have been derived wholly or slightly modified from Robert L. Bates and Julia A. Jackson (eds.), 1980, *Glossary of Geology*, 2nd ed., Falls Church, Va: American Geological Institute.

†Adapted from *American Commission on Stratigraphic Nomenclature, North American Stratigraphic Code*. American Association of Petroleum Geologists. ©1961, reprinted by permission.

of calcareous remains and enclosed or surrounded by rock of different lithology.

Biostrome A distinctly bedded, widely extensive or broadly lenticular, blanket-like mass of rock built by and composed mainly of the remains of sedentary organisms and not swelling into a moundlike or lenslike form.

Bioturbation The churning and stirring of sediment by organisms.

Breached anticline An anticline that has been deeply eroded in the center so that it is flanked by erosional scarps facing inward.

Byssus A tuft or bundle of long, tough hairlike strands or filaments, secreted by a gland in a groove of the foot of certain bivalve mollusks and issuing from between the valves, by which a temporary attachment of the bivalve can be made to rocks or other extraneous objects.

Casts *Paleontology*: Secondary rock or mineral material that fills a *natural mold*; specifically a replica or reproduction of the external details (size, shape, surface features) of a fossil shell, skeleton, or other organic structure, produced by the filling of a cavity formed by the decay or dissolution of some or all of the original hard parts of which the organism consisted. Cf: external mold, internal mold. *Sedimentology*: A sedimentary structure representing the infilling of an original mark or depression made on top of a soft bed and preserved as a solid form on the underside of the overlying and more durable stratum, e.g., a *flute cast* and a *load cast*.

Cementation The diagenetic process by which coarse clastic sediments become lithified or consolidated into hard, compact rocks, usually through deposition or precipitation of minerals in the spaces among the individual grains of the sediment. It may occur simultaneously with sedimentation or at a later time.

Clastic Pertaining to rock or sediment composed principally of broken fragments that are derived from preexisting rocks or minerals and that have been transported some distance from their places of origin; also said of the texture of such a rock.

Concordant *Intrusive rocks:* Description of a contact between an igneous intrusion and the country rock that parallels the foliation or bedding planes of the latter. *Stratigraphy*: Structurally conformable; said of strata displaying parallelism of bedding or structure.

Concurrent range zone* A zone defined by the overlapping ranges of specified taxa from one or more of which it takes its name.

Conglomerate A coarse-grained, clastic sedimentary rock composed of rounded (to subangular) fragments larger than 2 mm in diameter (granules, pebbles, cobbles, boulders) set in a fine-grained matrix of sand, silt, or any of the common natural cementing materials (such as calcium carbonate, iron oxide, silica, or hardened clay); the consolidated equivalent of gravel both in size range and in the essential roundness and sorting of its constituent particles. The rock or mineral fragments may be of varied composition, range widely in size, and are usually rounded and smoothed from transportation by water or from wave action.

Consumer An organism that is unable to manufacture its food from nonliving matter but that is dependent on the energy stored in other living things.

Correlation Demonstration of the equivalence of two or more geologic phenomena in different areas. Though there are different kinds of correlation, depending on the feature to be emphasized, lithologic correlation demonstrates correspondence of lithologic character and lithostratigraphic position. A correlation of two fossil-bearing beds demonstrates correspondence in their fossil content and in their biostratigraphic position and in chronostratigraphic position.

Country rock The rock enclosing or traversed by a mineral deposit; the rock intruded by and surrounding an igneous intrusion.

Decomposer An organism, usually microscopic, that breaks down organic matter and thus aids in recycling nutrients.

---

*Adapted from *American Commission on Stratigraphic Nomenclature, North American Stratigraphic Code*. American Association of Petroleum Geologists. ©1961, reprinted by permission.

Denudation The sum of the processes that result in the wearing away or the progressive lowering of the earth's surface by various natural agencies, which include weathering, erosion, mass wasting, and transportation; also the combined destructive effects of such processes. The term is wider in scope than erosion.

Desiccation A complete or nearly complete drying out or drying up or a deprivation of moisture or of water not chemically combined; for example, the loss of water from pore spaces of soils or sediments as a result of compaction or the formation of evaporites.

Diagenesis *Mineral:* Recombination or rearrangement of a mineral resulting in a new mineral; the geochemical, mineralogic, or crystallochemical processes or transformations affecting clay minerals before burial in the marine environment. *Sedimentation:* All the chemical, physical, and biologic changes undergone by a sediment after its initial deposition and during and after its lithification, exclusive of surficial alteration (weathering) and metamorphism.

Dike A tabular igneous intrusion that cuts across the bedding or foliation of the country rock.

Discordance *Intrusive rocks:* Description of a contact between an igneous intrusion and the country rock that is not parallel to the foliation or bedding planes of the latter.

Ecosystem A unit in ecology consisting of the environment with its living elements plus the nonliving factors that exist in and affect it.

Evaporite A nonclastic sedimentary rock composed primarily of minerals produced from saline solution as a result of extensive or total evaporation of the solvent.

External mold A mold or impression in the surrounding earth or rock, showing the surface form and markings of the outer hard parts of a fossil shell or other organic structure; also, the surrounding rock material whose surface received the external mold.

Extrusive Description of igneous rock that has been erupted onto the surface of the earth. Extrusive rocks include lava flows and pyroclastic material such as volcanic ash.

Facies The aspect, appearance, and characteristics of a rock unit, usually reflecting the conditions of its origin, especially as differentiating the unit from adjacent or associated units.

> *Lithofacies:* a mappable, really restricted part of a lithostratigraphic body, differing in lithology or fossil content from other beds deposited at the same time and in lithologic continuity.
>
> *Biofacies:* a local assemblage or association of living or fossil organisms, especially one characteristic of some type of marine conditions, for example, sandy-bottom facies.

Fauna The entire animal population, living or fossil, of a given area, environment, formation, or time span.

Fissility A general term for the property possessed by some rocks of splitting easily into thin sheets or layers along closely spaced, roughly planar, and approximately parallel surfaces, such as along bedding planes (as in a shale) or along cleavage planes (as in a schist) induced by fracture or flowage; its presence distinguishes shale from mudstone. The term is not applied to minerals but is analogous to cleavage in minerals, and it includes phenomena such as bedding fissility and fracture cleavage.

Flora The entire plant population, living or fossil, of a given area, environment, formation, or time span.

Folia In metamorphic rocks: thin, leaflike layers or laminae; specifically, cleavable folia of gneissic or schistose rocks.

Foliation *Structural geology:* A general term for a planar arrangement of textural or structural features in any type of rock, especially the planar structure that results from flattening of the constituent grains of a metamorphic rock.

Foraminifer Any protozoan belonging to the subclass Sarcodina, order Foraminifera, characterized by the presence of a test of one-to-many chambers composed of secreted calcite or agglutinated particles. Most foraminifers are marine, but freshwater forms are known. Range: Cambrian to present.

Glauconite A dull green earthy or granular mineral of the mica group. It has often been regarded as the iron-rich analog of illite. Glauconite occurs abundantly in greensand and seems to be forming in the marine environment at the present time. It is the most common sedimentary (diogenetic) iron silicate and is found in marine sedimentary rocks from the Cambrian to the present. Glauconite is an indicator of very slow sedimentation.

Graben An elongate, relatively depressed crustal unit or block that is bounded by faults on its long sides.

Graded bedding A type of bedding in which each layer displays a gradual and progressive change in particle size, usually from coarse at the base of the bed to fine at the top. It may form under conditions in which the velocity of the prevailing current declined in a gradual manner, as by deposition from a single short-lived turbidity current.

Guyot* A tablemount; a conical volcanic feature on the ocean floor that has had the top truncated to a relatively flat surface.

Habitat The particular environment or place where an organism or species tends to live; a more locally circumscribed portion of the total environment.

Index fossil A fossil that indentifies and dates the succession of strata in which it is found; especially a fossil taxon (generally a genus, rarely a species) that combines morphologic distinctiveness with relatively common occurrence or great abundance and that is characterized by a broad, even worldwide, geographic range and by a narrow or restricted stratigraphic range that may be demonstrated to approach isochroneity.

Index mineral A mineral developed under a particular set of temperature and pressure conditions, thus characterizing a particular degree of metamorphism.

Internal mold A mold or impression showing the form and markings of the inner surfaces of a fossil shell or other organic structure; it is made on the surface of the rock material filling the hollow interior of the shell or organism.

Intrusion *Igneous:* The process of emplacement of magma in preexisting rock; magmatic activity; also, the igneous rock mass so formed with the surrounding rock. *Sedimentary*: A sedimentary injection on a relatively large scale; for example, the forcing upward of clay, chalk, salt, gypsum, or other plastic sediment and its emplacement under abnormal pressure in the form of a diapiric plug.

Isotope One of two or more species of the same chemical element, that is, having the same number of protons in the nucleus but differing from one another by having a different number of neutrons. The isotopes of an element have slightly different physical and chemical properties, owing to their mass differences, by which they can be separated.

JOIDES Joint Oceanographic Institutions for Deep Earth Sampling. A group formed for scientific planning of a program to obtain deep sedimentary cores from the ocean bottoms.

Laccolith A concordant igneous intrusion with a known or assumed flat floor and a postulated dikelike feeder commonly thought to be beneath its thickest point. It is generally planoconvex in form and roughly circular in plan, less than 5 miles in diameter, and from a few feet to several hundred feet in thickness.

Laterite A highly weathered red subsoil or material rich in secondary oxides of iron, aluminum, or both, nearly devoid of bases and primary silicates, and commonly with quartz and kaolinite. It develops in a tropical or forested warm-to-temperate climate and is a residual product of weathering.

Lava A general term for a molten "extrusive"; also, a term for the rock that is solidified from it.

Lenticular Resembling in shape the cross section of a lens, especially of a double-convex lens. The term may be applied, for example, to a body of rock, a sedimentary structure, or a mineral habit or pertain to a stratigraphic lens or lentil (a rock-stratigraphic unit of limited geographic extent).

---

*From Harold V. Thurman, 1988, *Introductory Oceanography*. 5th ed., p. 515. New York: Macmillan Publishing Co.

Lithification The conversion of a newly deposited, unconsolidated segment into a coherent, solid rock, involving processes such as cementation, compaction, desiccation, and crystallization. It may occur concurrent with, soon after, or long after deposition.

Lithosphere The outer part of the earth that comprises its surface, as distinguished from the *biosphere*, which comprises the living things on the earth; the *hydrosphere*, which includes the waters and vapors of the earth; and the *atmosphere*, which is the air enveloping the earth.

Lithostratigraphic unit A body of rock that is unified by consisting dominantly of certain lithologic types or combination of types or by possessing other unifying lithologic features. It may consist of sedimentary, igneous, or metamorphic rocks or of two or more of these. It may or may not be consolidated. The critical requirement is a substantial degree of overall lithologic homogeneity.

Littoral Pertaining to the benthic ocean environment or depth zone between high water and low water; also pertaining to the organisms of that environment.

Magma Naturally occurring mobile rock material, generated within the earth and capable of intrusion and extrusion, from which igneous rocks are thought to have been derived through solidification and related processes. It may or may not contain suspended solids such as crystals and rock fragments and/or gas phases.

Maturity *Topography*: The second of the three principal stages of the cycle of erosion in the topographic development of a landscape or region, intermediate between youth and old age, lasting through the period of greatest diversity of form or maximum topographic differentiation, during which nearly all the gradation resulting from the operation of existing agents has been accomplished. *Sedimentation*: The extent to which a clastic sediment texturally and compositionally approaches the ultimate end product to which it is driven by the formative processes that operate upon it. *Streams*: The stage in the development of a stream at which it has reached its maximum vigor and efficiency, having attained a profile of equilibrium and velocity that is just sufficient to carry the sediment delivered to it by tributaries. It is characterized by: a load that is just about equal to the ability of a stream to carry it; lateral erosion predominating over down-cutting, with the formation of a broad, open flat-floored valley having a regular and moderate or gentle gradient and gently sloping, soil-covered walls with a few outcrops; absence of waterfalls, rapids, and lakes; a steady but deliberate current and muddy waters; numerous and extensive tributaries, some of whose headwaters may still be in the youthful stage; development of floodplains, alluvial fans, deltas, and meanders as the stream begins to deposit material; and a graded bed.

Metasomatic Pertaining to the process of metasomatism and to its results. The term is used especially in connection with the origin of ore deposits.

Metasomatism The presence of interstitial, chemically active pore liquids or gases contained within the rock body or introduced from external sources is essential for the replacement process which commonly occurs at constant volume with little disturbance of textural or structural features.

Monocline A local steepening in an otherwise uniform gentle dip.

Morphology *Paleontology:* A branch of biology or paleontology that deals with the form and structure of animals and plants or their fossil remains, the features included in the form and structure of an organism or any of its parts. *Geomorphology:* The shape of the earth's surface, the external structure, form, and arrangement of rocks in relation to the development of landforms.

Neritic Pertaining to the ocean environment or depth zone between low tide and 100 fathoms or between low tide level and approximately the edge of the continental shelf; also, pertaining to the organisms living in that environment.

Oceanic Pertaining to those areas of the ocean that are deeper than the littoral and neritic zones or to the oceanic environment in general.

Oolite A sedimentary rock, usually a limestone, made up chiefly of ooliths cemented together.

Oolith One of the small, round or ovate accretionary bodies in a sedimentary rock, resembling the roe of fish, and having diameters of 0.25–2 millimeters. It is usually formed of calcium carbonate but may be of dolomite, silica, or other minerals, in successive concentric layers, commonly around a nucleus such as a shell fragment, an algal pellet, or a quartz-sand grain, in shallow, wave-agitated water; it often shows an internal radiating fibrous structure, indicating outward growth or enlargement at the site of deposition. Ooliths are frequently formed by inorganic precipitation, although many noncalcareous ooliths are produced by replacement, in which case they are less regular and spherical, and the concentric or radial internal structure is less well developed than in accretionary oolites.

Oscillation ripple A symmetric ripple mark with a sharp, narrow, relatively straight crest between broadly rounded troughs, formed by the orbital or to-and-fro motion of water agitation by oscillation waves.

Outcrop That part of a geologic formation or structure that appears at the surface of the earth; also, bedrock that is covered only by surficial deposits such as alluvium.

Paleoenvironment An environment in the geologic past.

Paleogeography The study and description of the physical geography of the geologic past, such as the historical reconstruction of the pattern of the earth's surface or of a given area at a particular time or the study of the successive changes of surface relief.

Paleotectonic map A map intended to show geologic and tectonic features as they existed at some time in the geologic past, rather than the sum of all the tectonics of the region as portrayed on a tectonic map.

Palynology A branch of science concerned with the study of pollen of seed plants and spores of other embryophytic plants, whether living or fossil, including their dispersal and applications in stratigraphy and paleoecology.

Pentameral symmetry A five-rayed symmetry, as, for example, in starfish.

Phytoplankton The plant forms of plankton, for example, diatoms.

Pillow lavas A general term for those lavas displaying pillow structure and considered to have formed in a subaqueous environment; such lava is usually basaltic or andesitic.

Pillow structure A structure observed in certain extrusive igneous rocks that is characterized by discontinuous pillow-shaped masses ranging in size from a few centimeters to a meter or more. The pillows are close-fitting, the concavities of one matching the convexities of another. The spaces are few and are filled either with material of the same composition as the pillows, with clastic sediments, or with scoriaccous materials. Pillows have a fine-grained or glassy skin, are vesicular within, and in cross section exhibit a banded concentric structure. Pillow structure is generally believed to be the result of subaqueous/submarine deposition.

Plutonic rock A rock formed at considerable depth by crystallization of magma and/or by chemical alteration. It is characteristically medium to coarse-grained or granitoid in texture.

Producer *Ecology:* An organism (for example, most plants) that can form new organic matter from inorganic matter such as carbon dioxide, water, and soluble salts. *Petroleum*: A well that produces oil or gas.

Radioactive decay The spontaneous disintegration of the atoms of certain isotopes into new isotopes, which may be stable or undergo further decay until a stable isotope is finally created. Radioactive decay involves the emission of alpha particles, beta particles, and other energetic particles and usually is accompanied by emission of gamma rays and by atomic deexcitation phenomena. Radioactive decay always results in the generation of heat.

Radiometric dating Calculating an age in years for geologic materials by measuring the presence of a short-life radioactive element, for example, carbon-14, or by measuring the presence of a long-life radioactive element plus its decay product, for example, potassium-40/argon-40. The term applies to all methods of age determination based on nuclear decay of naturally occurring radioactive isotopes.

Range zone* A body of strata comprising the total horizontal and vertical range of occurrence of a specified taxon. (Each taxon has its own individual range, thus there are as many range zones as there are recognizable species, genera, and so on.)

Regolith A general term for the entire layer or mantle of fragmental and loose, incoherent, or unconsolidated rock material, of whatever origin (residual or transported) and of very varied character, that nearly everywhere forms the surface of the land and overlies or covers the more coherent bedrock. It includes rock debris (weathered in place) of all kinds, volcanic ash, glacial drift, alluvium, loess and eolian deposits, vegetal accumulations, and soils.

Relative time Geologic time deter-mined by the placing of events in a chronological order of occurrence, especially times as determined by organic evolution or superposition.

Remote sensing The collection of information about an object by a recording device that is not in physical contact with it. The term is usually restricted to mean methods that record reflected or radiated electromagnetic energy rather than methods that involve significant penetration into the earth. The technique employs such devices as the camera, infrared detectors, microwave frequency receivers, and radar systems.

Sill *Igneous rocks:* A tabular igneous intrusion that parallels the planar structure of the surrounding rock.

Stock *Intrusive rocks:* An igneous intrusion that is less than 40 square miles in surface exposure, is usually but not always discordant, and resembles a batholith except in size.

Stromatolite An organosedimentary structure produced by sediment trapping, binding, and/or precipitation as a result of the growth and metabolic activity of microorganisms, principally cyanophytes (blue-green algae). It has a variety of gross forms, from nearly horizontal to markedly columnar, domal, or subspherical.

Syncline A fold the core of which contains the stratigraphically younger rocks; it is generally concave upward.

Synoptic Pertaining to simultaneously existing meteorologic conditions that together give a description of the weather.

Tectonic Description of or pertaining to the forces involved in, or the resulting structures or features of, tectonics.

Tectonics A branch of geology dealing with the broad architecture of the outer part of the earth; that is, the regional assembling of structural or deformational features, a study of their mutual relations, origin, and historical evolution.

Terrestrial deposit A sedimentary deposit laid down on land above tidal reach, as opposed to a marine deposit, and including sediments resulting from the activity of glaciers, wind, rain wash, or streams; for example, continental deposit.

Thermal metamorphism A type of metamorphism resulting in chemical reconstitution controlled by a temperature increase and influenced to a lesser extent by confining pressure; there is no requirement of simultaneous deformation.

Varve A sedimentary bed or lamina or sequence of laminae deposited in a body of still water within 1 year's time; specifically, a thin pair of graded glaciolacustrine layers seasonally deposited, usually by meltwater streams, in a glacial lake or other body of still water in front of a glacier.

Ventifact A general term for any stone or pebble shaped, worn, faceted, cut, or polished by the abrasive or sandblast action of windblown sand, generally under desert conditions.

Vesicular Characteristically containing vesicles or many small cavities.

---

*Adapted from *American Commission on Stratigraphic Nomenclature, North American Stratigraphic Code*. American Association of Petroleum Geologists. ©1961, reprinted by permission.